International Review of **Cytology**

A Survey of **Cell Biology**

**VOLUME 171**

## SERIES EDITORS

| | |
|---|---|
| Geoffrey H. Bourne | 1949–1988 |
| James F. Danielli | 1949–1984 |
| Kwang W. Jeon | 1967– |
| Martin Friedlander | 1984–1992 |
| Jonathan Jarvik | 1993–1995 |

## EDITORIAL ADVISORY BOARD

Aimee Bakken
Eve Ida Barak
Howard A. Bern
Robert A. Bloodgood
Dean Bok
Stanley Cohen
Rene Couteaux
Marie A. DiBerardino
Charles J. Flickinger
Nicholas Gillham
Elizabeth D. Hay
P. Mark Hogarth
Anthony P. Mahowald
M. Melkonian
Keith E. Mostov
Audrey L. Muggleton-Harris

Andreas Oksche
Muriel J. Ord
Vladimir R. Pantić
Thomas D. Pollard
L. Evans Roth
Jozef St. Schell
Manfred Schliwa
Hiroh Shibaoka
Wilfred D. Stein
Ralph M. Steinman
M. Tazawa
Yoshio Watanabe
Donald P. Weeks
Robin Wright
Alexander L. Yudin

# International Review of Cytology

## A Survey of Cell Biology

Edited by

**Kwang W. Jeon**
Department of Zoology
University of Tennessee
Knoxville, Tennessee

**VOLUME 171**

**ACADEMIC PRESS**
San Diego  London  Boston  New York  Sydney  Tokyo  Toronto

*Front cover photograph*: Basement-membrane stromal relationships. (See Chapter 2 in Volume 173 for more details.)

This book is printed on acid-free paper.

Copyright © 1997 by ACADEMIC PRESS

All Rights Reserved.
No part of this publication may be reproduced or transmitted in any form or by any means, electronic or mechanical, including photocopy, recording, or any information storage and retrieval system, without permission in writing from the Publisher.
The appearance of the code at the bottom of the first page of a chapter in this book indicates the Publisher's consent that copies of the chapter may be made for personal or internal use of specific clients. This consent is given on the condition, however, that the copier pay the stated per copy fee through the Copyright Clearance Center, Inc. (222 Rosewood Drive, Danvers, Massachusetts 01923), for copying beyond that permitted by Sections 107 or 108 of the U.S. Copyright Law. This consent does not extend to other kinds of copying, such as copying for general distribution, for advertising or promotional purposes, for creating new collective works, or for resale. Copy fees for pre-1997 chapters are as shown on the title pages, if no fee code appears on the title page, the copy fee is the same as for current chapters.
0074-7696/97 $25.00

Academic Press
*a division of Harcourt Brace & Company*
525 B Street, Suite 1900, San Diego, California 92101-4495, USA
http://www.apnet.com

Academic Press Limited
24-28 Oval Road, London NW1 7DX, UK
http://www.hbuk.co.uk/ap/

International Standard Book Number: 0-12-364575-1

PRINTED IN THE UNITED STATES OF AMERICA
97  98  99  00  01  02  EB  9  8  7  6  5  4  3  2  1

# CONTENTS

Contributors .................................................................... ix

## Biology of Plant Cells in Microgravity and under Clinostating
### Elizabeth J. Kordyum

| | | |
|---|---|---|
| I. | Introduction ............................................................. | 1 |
| II. | Methodology of Experiments with Plants in Altered Gravity ....................... | 4 |
| III. | Structural–Functional Organization of Plant Cells in Altered Gravity ............... | 10 |
| IV. | Characteristics of Plant Cell Organelle Rearrangements in Altered Gravity ........... | 46 |
| V. | Concluding Remarks ....................................................... | 62 |
| | References ............................................................. | 63 |

## The Use of Antibodies to Study the Architecture and Developmental Regulation of Plant Cell Walls
### J. Paul Knox

| | | |
|---|---|---|
| I. | Introduction ............................................................. | 79 |
| II. | Antibodies as Molecular Probes for the Plant Cell Surface ....................... | 83 |
| III. | Generation and Use of Antibodies to the Major Cell Wall Polymers of Higher Plants ... | 85 |
| IV. | Uncovering Cell Surface Markers of Cell Development with Monoclonal Antibodies ... | 100 |
| V. | Monoclonal Antibodies to Algal Cell Walls .................................... | 105 |
| VI. | Cell Walls and Plant–Microbe Interactions .................................... | 107 |
| VII. | Conclusions and Prospects ................................................. | 109 |
| | References ............................................................. | 110 |

# Biophysical Aspects of P-Glycoprotein-Mediated Multidrug Resistance

Randy M. Wadkins and Paul D. Roepe

| | | |
|---|---|---|
| I. | Introduction | 122 |
| II. | Review of Drug Transport Studies | 128 |
| III. | Other Data Important in Analyzing Drug Transport | 137 |
| IV. | The Recent Controversy over ATP Transport | 155 |
| V. | Implications for Other Forms of "ABC-MDR" | 157 |
| VI. | Conclusions | 158 |
| | References | 159 |

# Normal and Pathological Tau Proteins as Factors for Microtubule Assembly

André Delacourte and Luc Buée

| | | |
|---|---|---|
| I. | Introduction | 167 |
| II. | Microtubules and Microtubule-Associated Proteins | 168 |
| III. | Role of Tau Proteins in Microtubule Assembly | 175 |
| IV. | Pathological Tau Proteins | 185 |
| V. | Lessons Given by Pathological Tau Proteins | 204 |
| VI. | Concluding Remarks | 208 |
| | References | 210 |

# Differentiation and Transdifferentiation of the Retinal Pigment Epithelium

Shulei Zhao, Lawrence J. Rizzolo, and Colin J. Barnstable

| | | |
|---|---|---|
| I. | Introduction | 225 |
| II. | Development of the RPE | 227 |
| III. | Instability of the RPE Differentiated State | 243 |
| IV. | Transdifferentiation of RPE | 248 |
| V. | Concluding Remarks | 256 |
| | References | 256 |

## The Role of Endothelins in the Paracrine Control of the Secretion and Growth of the Adrenal Cortex

Gastone G. Nussdorfer, Gian Paolo Rossi, and Anna S. Belloni

| | | |
|---|---|---|
| I. | Introduction | 267 |
| II. | Endothelin Biosynthesis | 269 |
| III. | Endothelin Receptor Subtypes and Their Localization | 270 |
| IV. | Effects of Endothelins on the Secretory Activity of the Adrenal Cortex | 275 |
| V. | Effects of Endothelins on the Growth of the Adrenal Cortex | 286 |
| VI. | Regulation of Endothelin Release | 291 |
| VII. | Involvement of Endothelins in Pathophysiology | 293 |
| VIII. | Concluding Remarks | 295 |
| | References | 297 |

Index ............................................................. 309

# CONTRIBUTORS

Numbers in parentheses indicate the pages on which the authors' contributions begin.

Colin J. Barnstable (225), *Vision Research Center, Department of Ophthalmology, Yale University School of Medicine, New Haven, Connecticut 06520*

Anna S. Belloni (267), *Department of Anatomy, University of Padua, I-35121 Padua, Italy*

Luc Buée (167), *Unité INSERM 422, 59045 Lille Cédex, France*

André Delacourte (167), *Unité INSERM 422, 59045 Lille Cédex, France*

J. Paul Knox (79), *Centre for Plant Biochemistry and Biotechnology, University of Leeds, Leeds LS2 9JT, United Kingdom*

Elizabeth J. Kordyum (1), *Institute of Botany, National Academy of Sciences of Ukraine, 252004 Kiev, Ukraine*

Gastone G. Nussdorfer (267), *Department of Anatomy, University of Padua, I-35121 Padua, Italy*

Lawrence J. Rizzolo (225), *Section of Anatomy, Department of Surgery, Yale University School of Medicine, New Haven, Connecticut 06520*

Paul D. Roepe (121), *Program in Molecular Pharmacology and Therapeutics, Raymond and Beverly Sackler Foundation Laboratory, Memorial Sloan-Kettering Cancer Center, New York, New York 10021*

Gian Paolo Rossi (267), *Department of Clinical and Experimental Medicine, University of Padua, I-35121 Padua, Italy*

Randy M. Wadkins (121), *Laboratory of Biophysical Chemistry, Cancer Therapy and Research Center, San Antonio, Texas 78245*

Shulei Zhao (225), *Department of Cell Biology, Baylor College of Medicine, Houston, Texas 77030*

# Biology of Plant Cells in Microgravity and under Clinostating

Elizabeth L. Kordyum
Institute of Botany, National Academy of Sciences of Ukraine, 252004 Kiev, Ukraine

Experimental data on plant cell reproduction, growth, and differentiation in spaceflight and under clinostating that partially reproduce the biological effects of microgravity are elucidated. The rearrangements of organelle structural and functional organization in unicellular plant organisms as well as in meristematic, differentiating, and differentiated cells of multicellular organisms in these conditions are considered. The focus is on the changes in the interrelations of prokaryotic and eukaryotic organisms under altered gravity. Ideas on the acceleration of differentiation and aging of cells in microgravity and clinostating and the organism's adaptive possibilities for carrying out its own functions are discussed.

**KEY WORDS:** Plant cell, Microgravity, Clinostating, Reproduction, Differentiation, Cytoplasmic membrane.

## I. Introduction

Cytoecology is playing a greater role in resolving questions on the resistance level and functional range of various living systems under altered ecological conditions. Investigations of cell reproduction, differentiation, and functioning—that is, the processes that determine the rate and direction of ontogenesis of organisms and their productivity—have opened the way for an elucidation of how these conditions affect living systems and the possibilities of adapting to them. The cabin of a space vehicle is a microcosm of an artificial biosphere pushed to its limits through vibration and acceleration on the active areas of trajectory and for motor functions, weightlessness (microgravity), disturbance of circadian rhythms, changes of magnetic and electrical fields, etc. Beginning in the 1960s, integrated investigations of organisms of different degrees of complexity, which were in a physiologically active state during spaceflight, made it possible to discover the diverse influences of space flight on the vital functions and growth of the living systems. A series of spaceflight and laboratory experiments that modeled, to a certain extent, the effects of separate orbital physical factors (vibration and acceleration during launching, microgravity, and altered magnetic field) established the dependence of the character of the changes in the living systems occurring under the influence of these factors on the nature and duration of the factor and the complexity and physiological state of the experimental objects.

On the basis of these experiments it was shown that weightlessness is the main active factor of spaceflights in the near cosmos. These flights are protected from the heavy charged particles of galactic cosmic radiation by the earth's magnetosphere (Dubinin and Vaulina, 1976; Gazenko *et al.*, 1974; Sytnik *et al.*, 1984). That is why the contribution of heavy ions to the biological effects of space flight is insignificant in the near cosmos but will be infinitely stronger in future long-term flights in the distant cosmos. Currently, the absence of protection from heavy ions is the main reason for limiting long-term flights with humans in the distant cosmos.

Therefore, an elucidation of the range and mechanisms of the biological effects of microgravity is one of the urgent fundamental tasks in space biology. At the same time, space biology provides an experimental basis for gravitational biology, which has developed only since the 1950s as a result of space programs (Smith, 1975, 1992). It is concerned with the role of gravity in the vital activity and evolution of living systems. It is known that the forms and functions of organisms are related to gravity and that gravity is a permanently active ecological factor. A comprehensive study of the role of gravity could not be made before the space age because of the constancy of gravity and our inability to alter it for significant periods on earth. The biological effects of gravity would be easier to understand if this factor were as variable as the other common environmental factors. A spaceship in orbital flight now makes possible the unique experiments that could not be carried out on earth for studies of the influence of gravity on living systems.

Plants are convenient objects for solving fundamental problems of space and gravitational biology. Plants are autotrophic organisms; all higher plants and the majority of lower plants are not motile and their spatial orientation is determined by gravi-, photo-, and chemotropisms. Consequently, the specific structural and chemical characteristics of plants determine their mode of nutrition and life. Unlike animals, including mammals, plants have an open type of ontogenesis, and the number of organs formed by one plant can be considerable. In addition, many plants are capable of unlimited vegetative reproduction.

A specific peculiarity of plant cell growth after division is growth by elongation, which results in the formation of a large central vacuole. Cell elongation and cell proliferation rarely coincide in time and are more often separated in space and in time (Ivanov, 1987). Plant cells have rigid walls composed largely of cellulose and are capable of changing their state in response to the regulatory system.

Differentiation of cells and tissues in higher plants is not as complete as it is in animals, and mature plant cells of different kinds have the capacity when cultured *in vitro* to revert to an unspecialized condition and then to divide to produce other kinds of cells, i.e., totipotency is a specific characteristic of plant cells. Simultaneously, as absorbents of carbon dioxide gas, regenerants of oxygen, and sources of various substances essential to the human organism, plants are an indispensable component of controlled ecological life-support systems for manned space vehicles. In addition, within the closed cabin of a space vehicle in extended flight (when astronauts are surrounded for weeks on end by the same walls, instrument panels, and the noise of functioning equipment), green leaves and flowers reminiscent of the earth's verdure play an acutely important role in the well-being of the astronauts. Therefore, the results of fundamental studies are the basis for (i) predicting reliable functioning of the autotrophic link in controlled ecological life-support systems and (ii) working out both new technologies for growing plants in space and cell biotechnologies aboard space vehicles. This is why the question raised about two or three decades ago of whether plants could successfully grow, develop, and propagate under conditions of spaceflight remains of great importance theoretically and practically. In particular, problems such as the degree of viability of seeds formed while in orbit and the possibility of obtaining several generations of higher plants during long-term spaceflight should be resolved.

Because the degree of specialization for gravity is of the greatest importance when elucidating the biological effects of microgravity, plants of various levels of organization were used in the space experiments. These included algae (lower plants), mosses, aquatic ferns, gymnosperms and angiosperms (higher plants), as well as cell, tissue, and protoplast cultures lacking organism correlations.

Cytological studies of plants growing and developing in space flights made it possible to establish that the processes of mitosis, cytokinesis, and tissue differentiation of vegetative and generative organs are largely normal. That is, these processes occur within the frameworks of cytodifferentiation and morphogenesis as they usually do on earth.

At the same time, under microgravity, essential reconstruction in the structural and functional organization of cell organelles occurs; these are indicative of changes in cell metabolism under the influence of altered gravity, i.e., a cell is gravisensitive. Consequently, an elucidation of the mechanisms involved in microgravity effects at the cellular and subcellular levels became the focus of increasing attention. In this chapter, the results of experiments carried out in the field of space cell phytobiology are reviewed and described. Because microgravity affects a variety of cell processes, there is a wide range of questions on the cell structure, reproduction, differentiation, biochemical content, metabolism, and adaptation capacities of different autotrophic organisms *in vivo* and *in vitro* under microgravity and clinostating.

## II. Methodology of Experiments with Plants in Altered Gravity

### A. Microgravity in Spaceflight

Weightlessness is a condition in which objects with demonstrable mass lack a detectable weight. The acceptance of the term "weightlessness" depends on one's concept of weight. From the physicist's point of view ($W = ma$, where $W$ is weightlessness, $m$ is mass, and $a$ is acceleration), a body can be weightless only in the absence of accelerative forces, which, by the law of universal gravitation, is theoretically impossible. Physicists conceive of weight as merely another force—abstractly and without material connotations because in their treatment the units of weight and mass are different.

Biologists apparently have not considered the matter in this light and hold more or less the popular concept of gravity. Scientifically, they use weight as an index of relative material magnitude—phenomenologically derived. Therefore, they are studying the biological effects of several conditions (parabolic flight, sounding rockets, and buoyant immersion) that are interrelated by dynamic weightlessness in space vehicles. To describe these varied conditions without demonstrable weight and their effects, it is obviously conductive to progress to adopt the generic term weightlessness. In this usage, weight and weightlessness are dealt with phenomenologically, without reference to their component physical factors. Such latitude in definition provides two criteria for weightlessness: (i) a condition in which no acceleration, whether gravity or other force, can be detected by an observer within the system in question; or (ii) a condition in which gravitational and other external forces acting on a body produce no stress, either internal or external, in the body (Smith, 1975).

Taking into account the first criterion for weightlessness, it is necessary to note the influence of power sources in the cabin of space vehicles as a result of the operation of equipment, switching on the motors for trajectory, activities of the crew, etc. The value of such influences can fluctuate according to theoretical calculations from $10^{-6}$ to $10^{-3}$ $g$ (Babsky et al., 1976; Brown and Chapman, 1984a). Therefore, the term "microgravity" is generally accepted. This is the condition in which the absolute sum of all mass-dependent accelerations does not exceed a certain small "noise" level, typically $10^{-5}$ to $10^{-4}$ $g$ (Albrecht-Buehler, 1992). The threshold for a plant gravitropic response examined with the use of both centrifuges and clinostats under the earth conditions and the on-board centrifuge in spaceflights appears to be $0.005g$ (Merkys, 1990; Shen-Miller et al., 1968; Sobick and Sievers, 1979).

At the beginning of space biology, the complexity of experimental equipment was determined by the goal of the experiments: determining whether spaceflight conditions affect a living organism. The change to investigating how spaceflight factors affect organisms in a physiologically active state and the uniqueness of the experiments have required a new approach to equipment that has to provide the conditions for the growth and development of organisms, i.e., to act as cultivators, and simultaneously to make it possible to carry out the necessary manipulations. The devices must also inoculate an organism into a nutrient medium in orbit at the appointed time, add different substances that affect biochemical and growth processes, and fix an organism at certain stages of its development. The growth and behavior of an organism has to be available for visual observation and observation with cameras and motion-picture and video equipment.

These specialized techniques and instruments were created during the past 25 years and are still being developed. For growing plants *in vivo* and *in vitro* devices of different types have been used—for example, inoculation–fixation systems (IFS-1 and IFS-2) and thermostats such as Biotherm, Svetoblock, Oasis, Malachite, Phyton, Aquarium, Biofixator, Biocontainer, Plant Growth Facility, Biobox, and so on (Brown and Chapman, 1984b; Cowles et al., 1984; Halstead and Dutcher, 1984; Mashinsky et al., 1984; Sytnik et al., 1983b; Vaulina et al., 1978; Volkmann et al., 1986). For example, one of the first biological space devices was the IFS-1, which was widely used for experiments with green algae, an aquatic fern, and a moss protonema growing in orbit from 4.5 to 96 days. This device (which consisted of one growth chamber and two ampoules) allowed the experiments to be carried out under aseptic conditions in darkness and by the light of a space vehicle cabin; it allowed both inoculation of a medium and fixation of an object by an astronaut during orbit (Sytnik et al., 1983b). The Svetoblock device was intended for growing small higher plants during long-term space experiments under aseptic conditions. In addition to a growth chamber, it is supplied with a lighting appliance and a temperature-sensing element. In this device, *Arabidopsis thaliana* plants with two cotyledon leaves sent into orbit blossomed for the first time during the 65-day space experiment (Kordyum and Chernyaeva, 1982; Kordyum *et al.*, 1983b, 1984c). The Plant Growth Facility consists of six growth chambers and it is intended for growing plants up to 30 days (Fig. 1).

In order to eliminate the influence of vibration and acceleration during launching and landing, it is desirable to deliver the experimental subjects in a dormant state (spores or seeds) or at low temperatures—4°C in the special refrigerators (cultures of algae or cells *in vitro*)—to transfer them in a physiologically active state during orbit, and to fix the

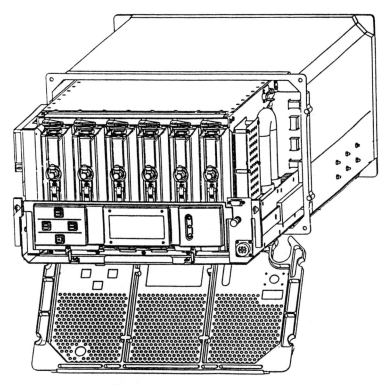

FIG. 1 The plant growth facility. It supports whole plant growth for up to 30 days by providing acceptable environmental conditions for normal growth. It can accommodate up to six plant growth chambers. It has a variable light photocycle to regulate the plant day/night cycle and computer-controlled lighting, temperature, carbon dioxide, humidity, and data archiving. Reproduced with permission from Cynthia M. Martin.

experimental material or its parts before landing. As a rule, space experiments with plants *in vivo* and *in vitro* are simultaneously accompanied by laboratory experiments in the same devices (the synchronous control). In some cases, when the temperature conditions in the biosatellite unexpectedly changed for a short time, a new laboratory control copied these conditions. Data obtained from spaceflight experiments studying the biological effects of microgravity are compared with the results of clinostat experiments carried out before, simultaneously, or after the spaceflight ones.

## B. Clinostating (Simulated Microgravity)

As mentioned previously, reduced gravity conditions for a long period are available only in an orbiting space vehicle. Therefore, on earth a clinostat is widely used. This is a mechanical "platform" that rotates at a uniform angular velocity around a horizontal axis and is equipped with clamps for attaching organisms. A complex organism or a cell that is mounted on a clinostat lives within a reference frame, a coordinate system, that rotates relative to earth and relative to $g$, the vector of gravitational acceleration (Kessler, 1992a).

For more than a century, a clinostat has been used to study plantgeotropic (now called "gravitropic") reactions, which are growth reactions underlying the graviperception system and which provide the spatial orientation of plant organs (Sack, 1991; Sievers and Hejnowicz, 1992; Sytnik *et al.*, 1984; Volkmann and Sievers, 1979). With the establishment of gravitational and space biology, a clinostat began to be used for investigating the influence of gravity on fundamental cell processes.

Because a clinostat rotates in the gravitational field, clinostating clearly does not eliminate, alter, or scalarize gravity. Rotation of a biological specimen on a horizontal clinostat at an appropriate rate only provides a continual reorientation of an object with regard to a gravity vector that prevents it from perceiving a gravitational stimulus or realizing the subsequent response. In order to be biologically effective, gravity as a vector must act for some minimum time—the presentation time. Nevertheless, in the literature for convenience it has now become popular to use the terms "simulated microgravity" (clinostating) or altered gravity (clinostating, centrifugation, or microgravity in spaceflight) as well as "real microgravity" (spaceflight).

Currently, as a research tool, two types of clinostat are used: a slow-rotating or slow clinostat (speed in the 1- to 5-rpm range) (Fig. 2) and a fast-rotating or a fast clinostat (speed in the 50- to 120-rpm range (Fig. 3). The physical principles of slow and fast clinostats as well as the validity of

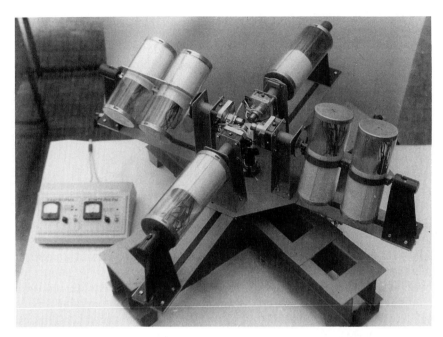

FIG. 2 The slow-rotating clinostat for experiments with lower and small higher plants or seedlings in simulated microgravity. The speed of rotation is regulated from 1 to 5 rpm.

their use have been repeatedly described and debated (Briegleb, 1983; Sobick and Sievers, 1979; see for details Albrecht-Buehler, 1992; Briegleb, 1992; Gruener and Hoeger, 1991; Kessler, 1992a). Here, we only note that use of the fast clinostat is limited to small living objects, which are strictly on the axis of rotation; in this case the centrifugal forces are immaterial. Organs and cells, which deflect the axis for some distance, will undergo centrifugal forces. A comparison of results from flight experiments and clinostating with unicellular and multicellular organisms and cell and tissue cultures showed that the clinostat is a valuable tool for evaluating an organism's sensitivity to altered gravity. The fast-rotating clinostat is considered to be an efficient and inexpensive tool for the simulation of microgravity on earth (Cogoli, 1992; Häder, 1994) and can effectively compensate for gravity (Hemmersbach-Krause and Briegleb, 1994; Sievers and Hejnowicz, 1992).

The slow-rotating clinostat enjoys a stable and well-known reputation as a reliable tool for reproducing a crucially important feature of microgravity, namely, the absence of a permanently orienting effect of a gravity vector (Kordyum, 1994; Merkys, 1973; Merkys and Laurinavichius, 1990; Hilaire et al., 1995). The absolute magnitude of the vector is incidentally preserved; consequently, the environmental properties are not changed. This naturally

FIG. 3 The central part of a fast-rotating clinostat microscope. All sorts of small, living specimen, suspended in a liquid or semiliquid translucent medium or included in special cuvettes for microscopy, can be observed under conditions of simulated microgravity. Different levels of light microscopy, real-time video observation, and image processing are available. Reproduced with permission from Wolfgang Briegleb.

limits the simulation of microgravity by clinostating to effects caused by the absence of a gravity vector. It does not allow full identification of the effects of microgravity, which has been repeatedly shown in space flights and clinostat experiments.

For instance, only a part of the structural changes at the cellular and subcellular levels that arise in real microgravity is reproduced under clinostating because a clinostat is unable to remove globally the scalar effects of gravity such as hydrostatic pressure and surface tension. That is why, currently clinostating is considered to reproduce only partially the biological effects of microgravity caused by the absence of gravity.

Despite these restrictions, clinostats are widely used to investigate the effects of altered gravity because they make it possible to carry out experiments in the necessary time parameters and to use a greater number of analytical methods that are equivalent to the tasks of experiments in comparison with spaceflight conditions.

The feasibility of performing complex experiments in orbit is frequently predicated on circumstances beyond the experimenter's control (Gruener,

1992). Currently, a new method of clinostating is proposed in which the sample is suspended in a fluid environment and the vessel is rotated at a speed at which the sample falls continuously through the fluid in which it is carried in a simulation of free fall (Claasen and Spooner, 1994; Moore, 1990).

A comparison of the indices of linear growth and gravitropic reaction of seedlings that depend on the magnitude of centrifugal force in microgravity shows that there is no direct correlation between growth and morphogenetic processes, with a gravisensory system determining the space orientation of plant organs (Merkys, 1990; Shvegzhdene, 1991). The realization of plant growth and development in microgravity and under clinostating makes it possible to analyze the nature of changes occurring under these conditions and to establish certain general patterns in their manifestation.

## III. Structural-Functional Organization of Plant Cells in Altered Gravity

Since the beginning of the 1960s, a large number of experiments with lower and higher plants have been performed on the earth's artificial satellites—the one-way spaceships Discoverer, Biosatellite II, Skylabs 3 and 4, Apollo, Voshod, Vostok, Zond, Sojuz; the biosatellites of the Cosmos series; the orbital stations Salyut and Mir; and the space shuttle. These space experiments were accompanied by synchronous ground controls. The orbital stations Salyut and Mir gave us an unique opportunity for carrying out fundamental investigations on the growth, development, and metabolism of living organisms, including plants, under conditions of long-term microgravity. A list of space experiments with different plants in a physiologically active state during orbital flight is given in Table I.

### A. Specialized Graviperceptive Cells

#### 1. Root Cap Statocytes

The absence of a gravitational vector, and as a result no realization of the gravitropic reaction of roots, was assumed to reveal a dysfunction in graviperceptive root cells, i.e., the statocytes in the central part of the root cap. Root statocytes are characterized by the structural polarity shown by the position of a nucleus in the proximal part of the cell and in the distal part of the endoplasmic reticulum (ER) complex (not all investigated species have a massive ER complex).

A nuclear membrane is in contact with the cytoplasmic membrane in the proximal part of a cell. Amyloplasts performing a statolithic function sediment in the distal part of the statocytes in the direction of the gravitational vector (Danilova, 1974; Moore and Evans, 1986; Sack, 1991; Sievers and Hensel, 1991; Sytnik et al., 1984; Volkmann and Sievers, 1979). The glycoprotein receptors on the surface of ER membranes are thought to include an interaction of sedimented amyloplasts with ER cisterns (Schneider and Sievers, 1981). This polar arrangement of organelles, which is genetically determined, is achieved and maintained by means of the cytoskeleton (Hensel, 1984, 1988; Lorenzi and Perbal, 1990a; Sievers et al., 1991).

In the first investigations using light optical microscopes it was shown that in microgravity, the amyloplasts–statoliths did not sediment in the distal part of the statocyte but rather spread over the entire volume of the statocyte cytoplasm of root caps in wheat seedlings after a 2-day experiment (Gray and Edwards, 1968, 1971; Lyon, 1968, 1971), in pea and lettuce seedlings after 3.5- to 5-day experiments (Merkys et al., 1976a), and in cucumber seedlings after a 4-day experiment (Sytnik et al., 1984).

Data obtained from electron microscopic analysis performed on different plants over the next few years are in perfect agreement and present a uniform and clear picture of the structural–functional organization of the root graviperceptive organ in microgravity. Histogenesis and cell differentiation in the embryonal root cap occur normally under these conditions; in the cap central column or columella, as in the control, the following zones are distinguished: the meristem, the differentiating statocytes, the mature statocytes (cells of the central statenchyma), and the peripheral secretory cells. Cell quantity in the different zones varies among species. In the peripheral zone of the large caps, for example, of maize, some workers (Moore et al., 1987a) distinguish the layer of cells transferring from the graviperceptive function to secretory ones. A continuous replacement of the cells of all the cap zones occurs constantly in the root growth process as a result of cap meristem proliferative activity and the removal of peripheral cells.

The root cap statocytes of cress and lens exhibited structural polarity 25 or 26 h after seed hydration under both microgravity and control conditions, i.e., a nucleus was situated in the proximal part of a cell and in distal ER cisterns (Perbal et al., 1986, 1987; Volkmann et al., 1986). The nucleus moved 0.87 $\mu$m away from the cytoplasmic membrane (Lorenzi and Perbal, 1990b). Amyloplasts–statoliths that do not sediment in the absence of the gravitational vector were in different parts or more concentrated in the center of the central statenchyma cells. In the lens statocytes, nearly four-fifths of the amyloplasts sedimented in the distal part of a cell in the controls; in microgravity, nearly three-fifths

TABLE I
Experiments with Lower and Higher Plants *in Vivo* and *in Vitro* Growing and Developing in Spaceflight

| Plant | Space vehicle | Duration of experiments | Parameters | Reference |
|---|---|---|---|---|
| **Algae** | | | | |
| *Anabaena azollae* Strasb. (in symbiosis with *Azolla pinnata* R. Brown | Orbital station Salyut-6 | 8 days | Morphology, ultrastructure | V. Kordyum *et al.* (1983), Popova (1986) |
| *Chlamydomonas reinhardii* Dang. | Shuttle D-1 mission Biosatellite Bion-9 | 7 days | Circadian rhythm, growth rate, ultrastructure | Gavrilova *et al.* (1991), Mergenhagen (1986), Mergenhagen and Mergenhagen (1987) |
| *Chlorella ellipsoidea* Gern. | Discoverer-17 | 14 days 2 days | Growth rate | Anderson *et al.* (1979) (cited in Mennigmann, 1989) |
| *Chlorella pyrenoidosa* Chil. | Orbital station Salyut-6 | 28 days | Growth rate, ultrastructure | V. Kordyum *et al.* (1980), Sytnik *et al.* (1979, 1983b) |
| *Chlorella vulgaris* Beijer. | Spaceship Soyuz-9 Biosatellite Cosmos-1887 Bion-10 Orbital station Salyut-6 Mir | 3 days 13 days 12 days 4.5, 5, 9, 10.5, and 18 days 7, 9, and 30 days; 4 and 12 months | Growth rate, morphology, mutability, pigment content, enzyme activity, ultrastructure, $Ca^{2+}$ content, fatty acid content, lipid peroxidation | Antonjan *et al.* (1992), E. Kordyum *et al.* 1979a, 1984b), Kordyum and Gavrish (1983), Kordyum *et al.* (1974), Popova and Kordyum (1991), Popova and Sytnik (1994, 1996), Popova *et al.* (1989, 1991, 1983), Sytnik *et al.* (1983b), Zhadko *et al.* (1994) |
| *Euglena gracilis* Klebs | Shuttle Columbia | 12 days | Gravitaxis orientation | Häder *et al.* (1995) |
| *Chara globularis* Thuill. | Parabolic flight TEXUS 18, 19, 21, 23, and 25 | 6 min | Statolith behavior and cytoskeleton in rhizoids | Buchen *et al.* (1993), Volkmann *et al.* (1991) |

| Species | Spacecraft | Duration | Parameters studied | References |
|---|---|---|---|---|
| **Mosses** | | | | |
| *Funaria hygrometrica* Hedw. | Orbital station Salyut-6 | 96 days | Growth rate, morphology, ultrastructure | Kordyum *et al.* (1981) |
| **Ferns** | | | | |
| *Azolla pinnata* R. Brown | Orbital station Salyut-6 | 8 days | Morphology, ultrastructure | V. Kordyum *et al.* (1983) |
| **Gymnosperms** | | | | |
| *Pinus elliotti* Engelm. | Shuttle SIS-3 | 8 days | Growth rate, morphology of seedlings, lignin and protein content, enzyme activity | Cowles *et al.* (1984, 1986, 1988) |
| *Pinus sylvestris* L | Biosatellite Cosmos-690 Cosmos-782 | 21 days 21 days | Orientation, morphology and growth rate of seedlings | Parfenov *et al.* (1975), Platonova *et al.* (1977, 1979) |
| **Angiosperms** | | | | |
| **Dicotyledons** | | | | |
| *Arabidopsis thaliana* (L.) Heynh. | Biosatellite Cosmos-1129 | 18 days | Morphology, growth rate of seedlings, mutability, formation of generative organs, rhizogenesis *in vitro*, ultrastructure of root apex cells (cap, meristem) and leaf mesophyll (the on-board 1g centrifuge was used in part of the experiments) | C. Brown *et al.* (1993); Gallegos *et al.* (1995a), Kordyum and Chernyaeva (1982), E. Kordyum *et al.* (1983b, 1984c), Kostina *et al.* (1986), Kuang *et al.* (1995, 1996), Merkys and Laurinavichius (1983, 1990), Merkys *et al.* (1983b, 1984, 1988), Musgrave *et al.* (1993), Parfenov and Abramova, (1981), Podluzky (1992), Sytnik *et al.* (1983a), Tarasenko (1985), Tarasenko *et al.* (1982) |
| | Cosmos-1167 | 7 days | | |
| | Bion-9 | 14 days | | |
| | Orbital station Salyut-6 | 4, 5, 7, and 65 days | | |
| | Salyut-7 | 69 days | | |
| | Mir | 28 days | | |
| | Shuttle STS-51 | 10 days | | |
| | STS-54 | 6 days | | |
| | STS-63 | 11 days | | |
| *Brassica napus* L. | Biosatellite Bion-9 | 14 days | Microcallus formation from the protoplasts, cell ultrastructure, enzyme activity, protein content, cell wall polysaccharide content, regeneration capacity | Iversen *et al.* (1992), Klimchuk *et al.* (1992), Rasmussen *et al.* (1992, 1994) |
| | Shuttle IML-1 | 8 days | | |

(*continues*)

TABLE I  (continued)

| Plant | Space vehicle | Duration of experiments | Parameters | Reference |
|---|---|---|---|---|
| Capsicum annuum L. | Biosatellite-II | 2 days | Epinasty | A. Brown et al. 1974, Johnson and Tibbits (1971) |
| Crepis capillaris Wallr. | Spaceship Soyuz-16 | 3 days | Growth rate and morphology of seedlings, root meristem cell structure, mitotic index, chromosome aberrations | Tarirbekov et al. (1979), Vaulina (1976), Vaulina and Kostina (1984), Vaulina et al. (1978) |
|  | Cosmos-782 | 21 days |  |  |
|  | Cosmos-936 | 18.5 days |  |  |
|  | Orbital station |  |  |  |
|  | Salyut-1 | 1.5–2 days |  |  |
|  | Salyut-5 | 1.5–2 days |  |  |
|  | Salyut-6 | 2 days |  |  |
| Cucumis sativus L. | Orbital station Salyut-6 | 4 days | Growth rate and morphology of seedlings, root cap structure (the on-board 1g centrifuge was used) | Merkys et al. (1983), Sytnik et al. (1984) |
| Daucus carota L. | Biosatellite Cosmos-782 | 21 days | Somatic embryogenesis, microcrocallus from protoplasts, cell ultrastructure, enzyme activity, protein content, cell wall polysaccharide content, regeneration capacity, growth, enzyme content, respiration activity, ultrastructure in tumor tissue induced by Agrobacterium tumefaciens | Baker et al. (1979), Butenko et al. (1979), Iversen et al. (1992), Klimchuk et al. (1992), Krikorian et al. (1981), Rasmussen et al. (1992, 1994), Rubin et al. (1979) |
|  | Bion-9 | 14 days |  |  |
|  | Shuttle IML-1 | 8 days |  |  |
| Haplopappus gracilis (Nutt.) A. Gray | Spaceship Soyuz-22 | 9 days | Mitotic index in the root meristem, chromosome aberrations, growth, ultrastructure and lipid peroxidation in the tissue culture | Klimchuk and Martyn (1994); Krikorian (1990), Sidorenko and Mashinsky (1978), Zhadko (1991), Zhadko et al. (1992) |
|  | Biosatellite Cosmos-1887 | 13 days |  |  |
|  | Bion-9 | 14 days |  |  |
|  | Bion-10 | 12 days |  |  |
|  | Shuttle STS-29 |  |  |  |

| | | | | |
|---|---|---|---|---|
| *Helianthus annuus* L. | Orbital station MIR | 9 and 28 days | Growth rate, morphology of seedlings, mitotic index in the root meristem, chromosome aberrations | Brown and Chapman (1984a,b), Krikorian and O'Connor (1984) |
| | Shuttle STS-2 STS-3 | 2.25 days 8 days | | |
| *Impatiens balsamina* L. | Biosatellite Cosmos-1987 | 13 days | Ultrastructure of the root (cap and meristem), the shoot, the hypocotyl, and cotyledons of seedlings | Kordyum *et al.* (1989, 1992), Nedukha *et al.* (1992), Philippenko *et al.* (1992) |
| *Lactuca sativa* L. | Spaceship Soyuz-12 | 2 and 3.5 days | Growth rate and morphology of seedlings, structure of the root apex (the on-board 1g centrifuge was used) | Merkys and Laurinavichius (1990), Merkys *et al.*, (1975, 1976a,b, 1983b, 1984, 1985), Shvegzhdene (1991) |
| | Soyuz-13 | 5 days | | |
| | Biosatellite Cosmos-690 | 20.5 days | | |
| | Orbital station Salyut-6 | 4 days | | |
| | Salyut-7 | 5 days | | |
| *Lens culinaris* L. | Shuttle D1-mission | 1 and 1.3 days | Growth rate and morphology of seedlings, ultrastructure of the root cap, DNA content (the on-board 1g centrifuge was used) | Darbelley *et al.* (1989), Driss-Ecole *et al.* (1994), Lorenzi and Perbal (1990a,b,), Perbal and Driss-Ecole (1994), Perbal *et al.* (1986, 1987) |
| | IML-1 | 8 days | | |
| *Lepidium sativum* L. | Orbital station Salyut-6 | 4 and 5 days | Growth rate, morphology of seedlings, structure of the root cap, ultrastructure of meristem and statocytes, cytoskeleton | Merkys *et al.* (1984), Volkmann *et al.* (1986, 1991) |
| | Shuttle D1-mission | 1 and 1.3 days | | |
| | Parabolic flight TEXUS 18, 19, 21, 23, and 25 | 6 min | | |
| *Lycopersicon esculentum* Mill. | Biosatellite Cosmos-1129 | 18.5 days | Orientation and morphology of seedlings, microcallus formation in the cell culture | Platonova *et al.* (1980), Tairbekov (1991) |

(*continues*)

TABLE I (continued)

| Plant | Space vehicle | Duration of experiments | Parameters | Reference |
|---|---|---|---|---|
| Melilotus alba L. | Cosmos-1167 | 7 days | Seedling growth, $CO_2$ and ethylene production | Gallegos et al. (1995b,c) |
| | Shuttle STS-57 | 8 days | | |
| | STS-60 | 8 days | | |
| Nicotiana tabacum L. Nicotiana rustica L. | Shuttle D2-mission | 10 days | Regeneration capacity, energy and carbohydrate metabolism in microcalli | Höffmann et al. (1994), Schönherr et al. (1994) |
| Pisum sativum L. | Spaceship Soyuz-12 | 2 days | Growth rate, orientation and morphology of seedlings, biochemical content of green biomass (content of pigments, amino acids, carbohydrates, mineral elements), ultrastructure of root apex cells (the cap, the meristem, and the elongation zone) and leaf mesophyll, lipid peroxidation, microviscosity of the cytoplasmic membrane, ultrastructure and lipid peroxidation in the tissue culture | Abilov et al. (1985), Aliev et al. (1986), Dubinin et al. (1977), Klimchuk and Martyn (1994), Kordyum and Sytnik (1983), E. Kordyum et al. (1983a), Laurinavichius et al. (1984), Merkys et al. (1975, 1976a,b), Sytnik et al. (1982, 1983a, 1984), Zhadko et al. (1994) |
| | Biosatellite Cosmos-690 | 20.5 days | | |
| | Bion-10 | 12 days | | |
| | Orbital station Salyut-4, | 24 days | | |
| | Salyut-6, | 7 and 18 days | | |
| | Salyut-7, | 29 and 33 days | | |
| | MIR | 8 days | | |
| Solanum tuberosum L. | Orbital station Mir | 8 days | Formation, anatomy and ultrastructure of minitubers formed in the organ culture | Kordyum et al. (1995), Sytnik et al. (1992) |
| Vigna radiata (L.) Wilczek | Shuttle STS-3 | 8 days | Growth rate and morphology of seedlings, lignin content, mitotic index in the root meristem, number and morphology of chromosomes, ultrastructure of the root apex (the cap and meristem) | Cowles et al. (1984, 1986, 1988), Krikorian and O'Connor (1984), Slocum et al. (1984) |

| | | | | |
|---|---|---|---|---|
| Monocotyledons | | | | |
| *Allium cepa* L. | Orbital station Salyut-4 Salyut-6 | 14, 30, and 32 days | Growth rate of plants; morphology of the inflorescence; anatomy of the root, leaves, and flower rudiments | Merkys *et al.* (1983 c), Kordyum *et al.* (1994) |
| *Avena sativa* L. | Shuttle STS-3 IML-1 | 8 days 8 days | Growth rate and morphology of seedlings, lignin content, mitotic index in the root meristem, number and morphology of chromosomes, ultrastructure of the root apex (the cap and meristem) | A. Brown *et al.* (1992), Cowles *et al.* 1984, 1986, 1988), Krikorian and O'Connor (1984), Slocum *et al.* (1984) |
| *Anoectochilus dawsonianus* Ldl. | Orbital station Salyut-6 | 60 days | All plants were dead at the end of the experiment | Cherevchenko *et al.* (1986) |
| *Dendrobium kingianum* Bidw. | Orbital station Salyut-6 | 110 and 171 days | Growth rate and morphology of plants | Nevruzova *et al.* (1987), Cherevchenko *et al.* (1986) |
| *Doritis pulcherrima* Ldl. | Orbital station Salyut-6 | 60 days | Growth rate and morphology of plants, fractional composition of proteins, activity of RuBPCase | Cherevchenko *et al.* (1986) |
| *Elodea* sp. | Skylab | | Cytoplasmic streaming | Sumerlin (1977) (cited in Halstead and Dutcher, 1984) |
| *Epidendrum radicans* Pav. et Ldl. | Orbital station Salyut-6 Salyut-7 | 110 days 171 days | Growth rate and morphology of plants, anatomy, ultrastructure of aerial root and leaf cells, fractional composition of proteins, activity of RuBPCase | Abilov *et al.* (1985), Nevruzova *et al.* (1987), Cherevchenko and Mayko (1983), Cherevchenko *et al.* (1986) |
| *Hemerocallis falva* L. | Shuttle STS-29 | 5 days | Mitotic index in the root meristem, chromosome aberrations | Krikorian (1990), Levine and Krikorian (1992) |
| *Hordeum vulgare* L. | Biosatellite Bion-10 | 12 days | Seedling growth, structure of statocytes | Laurinavichius *et al.* (1994) |
| *Muscari racemosum* (L.) DC. | Spaceship Soyuz-20 | 3 months | Microsporogenesis and male gametophyte development | Kordyum *et al.* (1979) |

(*continues*)

TABLE I (continued)

| Plant | Space vehicle | Duration of experiments | Parameters | Reference |
|---|---|---|---|---|
| *Oryza sativa* L. | Skylab | | Seedling growth rate | Summerlin, (1977) (cited in Halstead and Dutcher, 1984) |
| *Paphiopedilum hybrida* hort. | Orbital station Salyut-6 | 60 days | All plants were dead at the end of the experiment | Cherevchenko *et al.* (1986) |
| *Paphiopedilum insigne* Pfitz | Orbital station Salyut-6 | 60 days | All plants were dead at the end of the experiment | Cherevchenko *et al.* (1986) |
| *Spirodela polyrrhiza* Schleid. | Spaceship Soyuz-12 Soyuz-13 Biosatellite Cosmos-656 | 2 days 7 days 2 days | Germination of turions and offspring generation | Kutlakhmedov *et al.* (1978) |
| *Tradescantia paludosa* E. Anders et R. E. Woodson *Tradescantia* clone (02) | Spaceship satellite Vostok-3, 4, 5, and 6; Voskhod-1 Biosatellite-II | 2 days 2 days | Development of male and female gametophytes, mitosis, chromosome aberrations; somatic mutations | Delone *et al.* (1963, 1964, 1965, 1966, 1968), Marimuthu *et al.* (1972), Sparrow *et al.* (1968, 1971) |

| | | | | |
|---|---|---|---|---|
| *Triticum aestivum* L. | Biosatellite-II | 2 days | Growth rate and morphology of seedlings, structure of root cap cells, mitotic index in the meristem, enzyme activity, biochemical content of germinative caryposis endosperm and seedlings, microbial cenosis | A. Brown *et al.* (1992), Gray and Edwards (1968, 1971), Johnson (1971), Lyon (1968, 1971), Mashinsky *et al.* (1991), Saunders *et al.* (1971) |
| | Orbital station Mir | 19 days | | |
| | Shuttle IML-1 | 8 days | | |
| *Triticum compactum* Host. | Orbital station Salyut-6 | 7 days | Orientation and morphology of seedlings | Platonova *et al.* (1983) |
| *Triticum durum* L. | Orbital station Salyut-6 | 18 days | Nucleic acid content, ultrastructure of shoot apex and leaf cells | Nedukha *et al.* (1991), Sytnik and Musatenko (1980) |
| | Mir | 16 days | | |
| *Zea mays* L. | Biosatellite | 18.5 days | Growth rate and morphology of seedlings, mitotic index in root meristem, ultrastructure of root apex (the cap and meristem), dynamics of heat discharge, content of cell wall carbohydrates, root cap regeneration | Barmicheva *et al.* (1989), Darbelley, (1988), Darbelley *et al.* (1986, 1989), Moore *et al.* (1986, 1987a,b), Schulze *et al.* (1992), Tairbekov and Devyatko (1985), Tairbekov *et al.* (1979, 1986, 1988), Zabotina (1987) |
| | Cosmos-936 | 5 days | | |
| | Cosmos-1514 | 7 days | | |
| | Cosmos-1667 | | | |
| | Shuttle Columbia | 4.8 days | | |

of these organelles were in the proximal ones (Driss-Ecole and Perbal, 1989). Under the brief influence of microgravity during an average of 6 or 7 min in rocket flights, the position of amyloplasts–statoliths in the cress seedling root cap was only slightly influenced by the conditions during launch, e.g., vibration, acceleration, and rotation of the rocket. Within approximately 6 min of microgravity conditions, the center of the statolith complex moved approximately 3.6 $\mu$m in the direction opposite the originally acting gravity vector (Volkmann et al., 1991).

Researches assumed that on earth the position of statoliths in root statocytes depends on the balance of two forces, i.e., the gravitational force and the counteracting force mediated by microfilaments. An amyloplast envelope, as a rule, does not form noticeable outgrowths, and the organelles acquire a more rounded shape; their sizes diminish. A plastid stroma becomes more electron translucent, and starch grain volume also decreases. Such changes in the amyloplast structure and topography are described in pea (Kordyum and Sytnik, 1983; Sytnik et al., 1982, 1983a), Arabidopsis (Tarasenko, 1985; Tarasenko et al., 1982), cress (Volkmann et al., 1986), lens (Driss-Ecole and Perbal, 1989; Lorenzi and Perbal, 1990b; Perbal et al., 1987), oat and mung bean (Slocum et al., 1984), maize (Moore et al., 1986, 1987a,b), and balsamine (Kordyum et al., 1992). The cisterns of the ER complex are situated less parallel to the distal cell wall. Unlike the control, the existence of ER membrane clusters of spherical or ellipsoid shape, usually in the peripheral part of statocytes, is described in maize seedlings in microgravity (Moore et al., 1987b). In the root statocytes of 7-day-old pea seedlings in microgravity, the nucleus was situated at some distance from the proximal cell wall, and separate amyloplasts were localized under the nucleus or in the cell corners near the cytoplasmic membrane. A decrease of both lipid droplet volume and electron density of the ER cistern content was observed.

In the root statocytes of 18-day-old pea seedlings in microgravity, the relative volume of the plastidome per cell significantly decreased; part of the plastids became rounded or oval and contained one spherical starch grain localized in the center of an organelle on the sections. There were also plastids with or without several small starch grains (E. Kordyum et al., 1983a; Sytnik et al., 1982, 1984).

An increase of lipid droplet number and plasmalemmasome quantity has been revealed in the cells of different root cap zones in 7-day-old maize seedlings in microgravity. In the secretory zone, especially in the peripheral cells, a decrease of secretory activity and the appearance of electron-dense globuli at the cytoplasmic membrane were described (Barmicheva et al., 1989; Tairbekov et al., 1986). It was shown early on that such globuli appearing at the cytoplasmic membrane of the root cap cells (especially cells of the secretory zone) in 7-day-old pea seedlings growing in orbital flight in modified conditions contain $Ca^{+2}$ ions and thus present the membrane with $Ca^{+2}$-bound sites (Belyavskaya, 1984). The increased frequency of plasmalemmasomes seen in microgravity was also noted in the root cap cells of *Arabidopsis* and balsamine plants (Kordyum et al., 1992; Tarasenko, 1985).

Significant changes in cytoplasmic organelle volume, excluding the nucleus, have been revealed in the meristematic, differentiating, and differentiated cells of the root cap columella in 4.8-day-old maize seedlings in microgravity when compared with controls (Moore et al., 1987a). In the cap meristematic cells, the relative volumes of mitochondria, dictyosomes, and vacuoles decreased. At the same time, the relative volumes of lipid droplets and the hyaloplasm increased. In the statocytes the relative volume of plastids and lipid droplets significantly decreased. In the cells transferring from a graviperception function to mucilage secretion, and in the secretory cells, the relative volume of dictyosomes and quantity of mucilage diminished. The relative volumes of organelles in the cells were similar in microgravity and in controls. This agrees with Mauseth's (1985) opinion that the hyaloplasm state and volume are excellent indicators of metabolism intensity in the morphogenic process in vegetative organs. The author suggests a stimulating effect for microgravity on the root cap cell metabolism. When the seedlings were rotated on board a 1*g* centrifuge in spaceflight, the amyloplasts sedimented in the distal part of statocytes; the graviperceptive cell ultrastructure was similar to that in the ground control. After growth in microgravity, the roots proved to be less sensitive to gravistimulation than the roots growing under the stationary conditions or on the clinostats. Nevertheless, the statocytes preserved the capacity to perceive an acceleration on the centrifuge as a gravitational stimulus.

Destructive changes in the statocytes also occur when normal vertically grown cress roots are rotated continuously on a horizontal clinostat—the distal ER complex disintegrates, starch is hydrolyzed, and autolytic damage of the statocytes develops (Sievers et al., 1976). The volume of starch grains also decreased by 23% in amyloplasts of *Arabidopsis* root cap statocytes under clinostating compared with the control (Tarasenko, 1985). In simulated microgravity, progressive vacuolization was described in the root cap central statenchyma cells of cress (Hensel and Sievers, 1980) and *Arabidopsis* (Tarasenko, 1985); in the latter species the vacuole volume was 39% larger under clinostating than in the controls.

It is necessary to stress that 2- to 7-day-old seedlings are only accurate for investigating the structural–functional organization and differentiation of cells in the cap, such as the graviperceptive organ of the embryonal (main) root in seedlings of dicotyledon plants and embryonal roots in seedlings of monocotyledon plants. With extension of the seedling growth term in microgravity, cap cell vacuolization is significantly strengthened in spaceflight in comparison with the ground controls, i.e., primarily there is intensification of autolytic processes, leading to more rapid lysis of the cytoplasm and cessation of meristem activity in root ontogenesis. The patterns of histogenesis and cell differentiation established for root caps in microgravity lead to the conclusion that the graviperceptive apparatus of the intact embryonal roots has formed but does not function in the absence of a gravitational vector (the cap initials are genetically determined in a seed), preserving a capacity for graviperception for some time.

In an experiment with maize seedling decapitated roots, it was shown that the root does not regenerate a cap in microgravity (Moore et al., 1987b). The roots growing under stationary conditions regenerated caps that did not differ from those of the intact roots; amyloplasts in the statocytes sedimented in the distal part of a cell (Moore et al., 1987b). Decapitated roots lengthened more than 5 cm and their apices displayed a similarity to vacuolized callus-like cells. In the authors' opinion, the inability of decapitated roots to form caps in microgravity indicates that roots somehow sense the absence of gravity; therefore, root cap regeneration, similar to a gravitropic bend, can be an indicator of a root graviresponse.

In the spaceflight experiment on rhizogenesis in the *Arabidopsis* primary callus (Podluzky, 1992), the quantity of roots formed in microgravity during 8 and 15 days decreased and was only one-third that of the stationary ground control and half that of the clinostat material. Under clinostating, the root cap columella consisted of the same zones as in the stationary control, whereas the amyloplasts–statoliths spread over the entire volume of the cytoplasm. In microgravity, only the meristem and secretory cell zones found clear expression in the columella. The absence of mature statocytes provides evidence that the cap meristematic cells directly transferred to mucilage secretion, i.e., differentiated into secretory cells. The data obtained make it possible to assume that cells of the root apical meristem are gravisensitive, and a scalar value of the gravitational field is necessary to start the program of their differentiation into graviperceptive cells in roots developing *de novo* from a callus.

## B. Tip-Growing Cells

Unlike the multicellular graviperceptive organs, in tip-growing graviperceptive cells, the processes of graviperception and growth gravitropic reaction are the function of the same cell, i.e., they are connected spatially and functionally.

## 1. Chara Rhizoids

Rhizoids of the alga *Chara* are downward-growing cells with a clear polar zonation with respect to the localization of organelles. The apical and subapical zones contain thin bundles of microfilaments, and the basal zone contains thicker ones. The apical part contains statoliths and is gravitropically responsive. In a vertically oriented rhizoid, the statoliths—compartments filled with crystallites of barium sulfate—are maintained at a distance of 10–30 $\mu$m from the apical cell wall (Sievers *et al.*, 1991). It was concluded that the arrangement of microtubules is essential for polar cytoplasmic zonation and the functionally polar organization of the actin cytoskeleton but is not involved in the primary events of gravitropism in *Chara* rhizoids (Braun and Sievers, 1994). The actin microfilaments are involved not only in cytoplasmic streaming and the maintainance of functional cell polarity and tip growth but also in gravitropism through the positioning, transport, and sedimentation of statoliths (Braun and Sievers, 1994; Sievers *et al.*, 1991).

An experimental check of this assumption was made when during five rocket flight (TEXUS 18, 19, 21, 23, and 25) experiments, rhizoids of *Chara globularis* Thuill. were in microgravity approximately 6 min. The position of statoliths was only slighthly influenced by the conditions during launch, e.g., vibration, acceleration, and rotation of the rocket. Under microgravity, the shape of the statolith complex in the rhizoids changed from a transversely oriented lens into a longitudinally oriented spindle. The statoliths moved approximately 14 $\mu$m basipetally (Fig. 4)(Volkmann *et al.*, 1991). These authors suggested that the removal of gravity forces disturbs the initial balance between this force and the basipetally acting forces generated in a dynamic interaction of statoliths with microfilaments. This hypothesis was fully supported by flight experiments in which the rhizoids were treated immediately before the flight with cytochalasin D. By video microscopy, it was demonstrated that statoliths in all the rhizoids treated with cytochalasin D remained in their initial location, i.e., did not displace basipetally (Buchen *et al.*, 1993).

## 2. Moss Protonema Apical Cells

A graviperceptive apical cell of moss protonema displays a certain internal structural polarity. In such a cell, a young terminal region of active growth gradually transforms to a more mature basal region. In general, in a cell beginning from the apex, there are the following zones: the most distal apical zone, which does not contain plastids; the zone with numerous chloroplasts, which extends from the apical zone to a nucleus; and the basal, proximal zone, which is very vacuolized (Schnepf, 1986). A close association of

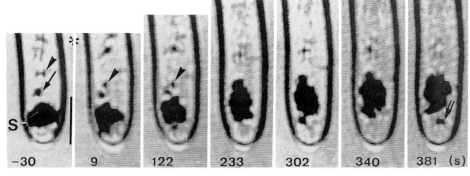

FIG. 4 A series of photographs of the apical part of a *Chara* rhizoid recorded during prelaunch (−97 s) and microgravity (9–381 s; TEXUS flight 21). For easier comparison, the cell tips were positioned on a horizontal line. Under 1g, the center of the lens-shaped statolith complex (S) is located approximately 17 μm above the vertex (−97 s). As a result of acceleration forces during launch, the statolith complex was displaced (9 s). During microgravity, the complex moved basipetally for approximately 14 μm (compare 9 with 381 s) and became spindle like. Small clusters of statoliths separated from the complex (−97–122 s; arrow and arrowheads, respectively). The more basal cluster moved acropetally (compare −97, 9, and 122 s; arrowheads). These clusters were incorporated into the statolith complex (233 s). At the end of the microgravity period, a small group of statoliths separated from the complex and moved acropetally (381 s; double arrow). Reproduced with permission from Andreas Sievers.

endoplasmic reticulum cisterns with the cytoplasmic membrane was described in *Funaria hygrometrica* cells; two membrane systems are situated very near one another, although their visible fusion was not observed (McCuley and Hepler, 1992). Amylochloroplasts in the apical cell of *Pottia intermedia* chloronema can have a different shape—rounded (in the distal part of a cell) and spindle like (in the proximal part of a cell). Rounded amylochloroplasts contain large starch grains, and spindle-like ones contain significantly less. Amylochloroplasts are believed to play the role of statoliths in the moss protonema (Chaban, 1991; Walker and Sack, 1990). In 15 min of *Pottia* protonema gravistimulation, the large plastids sedimented on the lower (physically) wall (Chaban, 1991).

Immunofluorescence has revealed that microtubules in the apical cell of the *Ceratodon purpureus* and *P. intermedia* protonema are oriented mainly longitudinally and pass through all zones beginning from the apex (Chaban *et al.*, 1995; Schwuchow *et al.*, 1990). In the proximal zone, microtubules are organized in comparatively thicker and less numerous bundles, which pass more strictly parallel to the cell long axis. Under clinostating, the zonal distribution of organelles, including amylochloroplasts,is preserved. The regular distribution of microtubules is somewhat broken; certain microtubules are oriented perpendicular to the cell long axis (Chaban *et al.*, 1995).

## C. Other Types of Cells

### 1. Algae

Despite the variety of morphological and ecological peculiarities found in algae, to date only species of unicellular green algae, namely *Chlorella* and *Chlamydomonas,* have been used in space experiments. During the past 20 years a series of space experiments were carried out with a *Chlorella* active culture—*Ch. vulgaris* (autotrophic strain LARG-1) and *Ch. pyrenoidosa* (heterotrophic strain g-11-1, a pigment mutant). Their duration varied from 4.5 days to 3 and 12 months. The culture grew on semiliquid and solid aseptic media both in light and in darkness and in a three-component water system (algae, fish, and bacteria) in light (Popova and Kordyum, 1991; Popova and Sytnik, 1994, 1996; Popova *et al.*, 1989; Sytnik *et al.*, 1983b).

The data obtained have shown an increase in biomass, reproduction, and viability of *Chlorella* on board space vehicles. At the same time, microgravity conditions affected the growth, structural, and biochemical indices, i.e., they had a diverse and important influence on the growth and vital functions of the physiologically active *Chlorella* culture. For instance, the cell quantity of *Ch. pyrenoidosa* (strain g-11-1) grown in a semiliquid nutrient medium for 28 days in spaceflight exceeded similar cells in the controls by 4.5 times; the biomass increased more than 3.5 times (Sytnik *et al.*, 1983b).

In addition to the stationary synchronous controls, other ground controls had dextran added to the medium and the medium was intermixed during the experiment. Both inoculation of the nutrient medium and fixation of *Chlorella* cells in orbit excluded the influence of vibration and acceleration during launching and landing and ensured complete correctness of the experiments (Sytnik *et al.*, 1983b).

A typical feature for *Chlorella* vegetative cells is the large cup-shaped chloroplast of 4 or 5 $\mu$m in size that occupies the main cell volume. Under autotrophic cultivation, the thylakoids are joined in bundles and form the granae in the separate chloroplast areas. The pyrenoid is perforated with one to three solitary thylakoids and surrounded by an amylogenic coating that consists of two lens-like starch grains. Under growth in darkness on an organic medium, the granae are absent and the bundles of thylakoids often bend. A capacity to synthesize chlorophyll in darkness is a typical feature of *Chlorella* cells. A few mitochondria 0.2–0.4 $\mu$m in size are situated in the peripheral cytoplasm layer between the chloroplast and the cytoplasmic membrane (Sytnik *et al.*, 1983b). The vacuoles are very small and their average relative volume per cell is 1.1%. The single dictyosome consists of four to six cisterns producing Golgi vesicules; its morphology changes with cell cycle stage. A round or oval nucleus contains one nucleolus in which a granular component dominates. A cytoplasmic membrane forms the invaginations in both the peripheral cytoplasm and a periplasmic space (Sytnik *et al.*, 1983b).

The changes in *Chlorella* cell ultrastructure can be seen in space experiments of different duration (Popova and Kordyum, 1991). In the short 4.5-day experiments with the strain LARG-1 cultivated on a semiliquid medium in darkness and light, there was a decrease in the relative volume of thylakoids and starch grains in chloroplasts and the cytoplasmic membrane had more complex folds (Kordyum *et al.*, 1979a; Sytnik *et al.*, 1983b). In addition to these changes, a disturbance of cytokinesis in single cells and a dilation of the intrathylakoid space as well as the appearance of electron-translucent areas in chloroplasts were described in space flight experiments for 10.5 and 18 days (Popova and Kordyum, 1991). Simultaneously with a decrease of a thylakoid relative volume, the tendency to a reduction of chlorophyll a and b content was revealed. Under more prolonged 28-day cultivation of the strain g-11-1 in darkness in microgravity, side by side with a decrease in the relative volumes of the plastid and starch grains a diminution of pyrenoid size was observed, along with the frequent absence of an amylogenic coating (Popova and Kordyum, 1991; Sytnik *et al.*, 1983b). An increase in condensed chromatin quantity was found along the nucleus periphery. There was an intensification of cell vacuolization, an increase in the relative volume of lipid droplets and mitochondria per cell as well as an augmentation of cristae quantity and their dilation, and an increase in mitochondrion matrix electron density (Sytnik *et al.*, 1983b).

The changes found in LARG-1 cells grown in a three-component aquatic system for 9 and 14 days in light in spaceflight were similar to those decribed in *Chlorella* cells cultivated on a semiliquid nutrient medium during 10–18 days in real microgravity. Unlike the ground controls, in the cells in the space experiments there was a decrease in relative volume of starch grains in the chloroplasts, a thinning or absence of amylogenic coating around pyrenoids, a more pronounced vacuolization (vacuole relative volume reached 10–16% of a cell volume), an increase of condensed chromatin volume in the nuclei of vegetative cells and autospores, dilation of perinuclear space, and an increase in variability of the mitochondrion size and structure. Single cases of cytokinesis disturbances (formation of binucleate or unequal autospores) did not affect the density of spaceflight populations. Interestingly, there was an increase in the number of *Chlorella* cells infected with bacteria (*Pseudomonas* sp.) in microgravity up to 3% compared with that of the ground controls, which was approximately 1%. Some algal cells in microgravity were entirely filled with bacteria. This phenomenon can indicate a decrease of alga cell immunity and/or a strengthening of bacterial pathogenity in microgravity (Popova and Kordyum, 1991; Popova et al., 1989).

During cultivation of the strain LARG-1 on a solid medium in darkness for 8 and 30 days, and during 4 and 12 months in orbit, the changes in cell structural–functional organization in many respects were similar to those described in the previous experiments, although the changes could be seen earlier, by the eighth day (Popova and Sytnik, 1994, 1996; Popova et al., 1991). In the experimental population, some mitochondria reached 1.3–1.5 $\mu$m in length (in the ground controls, 0.1–0.3 $\mu$m); cristae often were regularly arranged. The total volume of mitochondria was 5.3% per cell; in the control it was 2.5% per cell. The specific features of chloroplast ultrastructure in spaceflight experiments of 30 and more days were numerous unequal dilations of an intrathylakoid space and the formation of plastid envelope outgrowths that frequently were attached to the cytoplasmic membrane. The accumulation of granular material of an average electron density is described in an expanded periplasmic space and in the cytoplasm membrane invaginations. In ground controls, local dilations of an intrathylakoid space were rarely observed only on the solid medium (Popova et al., 1991); their relative volume per chloroplast was 0.03% on average. In microgravity, the amount of local dilation of the intrathylakoid space increased dramatically, and the relative volume amounted to 3.2%. The size of such dilations varied on average from 50 to 200 nm in the same chloroplast (Popova et al., 1991, 1992a). The relative starch volume decreased under cultivation during 30 days in spaceflight; it was 31.5% in space chloroplasts and 62.1% in the ground controls. This decrease of starch volume was correlated with an increase in a specific amylase activity that was twice that of ground controls. Simultaneously, mono- and disaccharide content also doubled (Popova et al., 1993, 1995).

The cell ultrastructure changes during the 4- and 12-month spaceflights were qualitatively similar on the whole to those in the 30-day experiment (Popova et al., 1992b); however, they increased in quantity. Mainly there were large condensed mitochondria with a dense matrix and parallel or concentrically arranged cristae as well as mitochondria with a matrix of a low electron density and swollen cristae. Unlike the ground controls, in the 12-month experiment there was a well-developed granular endoplasmic reticulum with cisterns that were filled with a substance of an average electron density (Popova and Sytnik, 1994; Popova et al., 1992b). *Chlorella* cultures grown for 4 and 12 months in the orbit had a dark green coloration. The cultures preserved their viability; their growth continued after they were transferred to fresh nutrient medium. Differences in the ultrastructure and sizes of mitochondria similar to those in spaceflight were also detected under the prolonged effect of clinostating up to 25–35 days (Fig. 5), whereas short-term clinostating has essentially no effect.

The localization of ATPase activated by $Mg^{2+}$ ions in *Chlorella* cells (strain LARG-1) was studied under normal conditions and under clinostating using the electron-cytochemical method. A localization of the cytochemical reaction product in the form of electron-dense granules of lead phosphate marked the membranes of various cell compartments. This was similar under clinostating and control conditions. A stronger reaction was noted on the membranes of the mitochondrion cristae, especially in the large

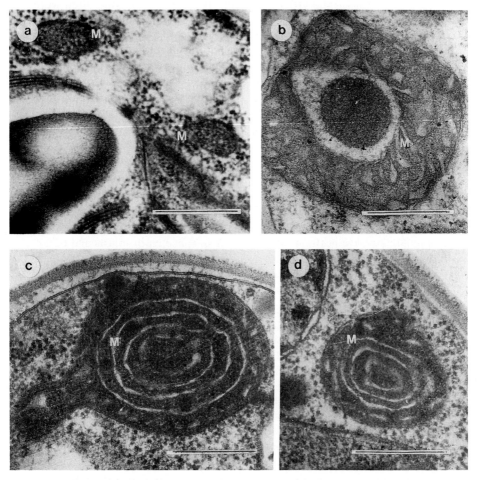

FIG. 5 Mitochondria in *Chlorella vulgaris* (strain LARG-1) cells under stationary growth conditions on a solid medium in darkness. (a) control, (b,c) clinostating, and (d) spaceflight. Bars = 0.5 μm.

mitochondria appearing in *Chlorella* cells under long-term clinostating (Popova, 1994).

Changes in the plastid starch grain volume as well as in the integral membrane system of the chloroplasts exhibited the same tendency under clinostating—a decrease in the starch volume, thylakoid swelling, and local unequal dilations of the intrathylakoid space (Popova and Kordyum, 1991). An enhancement of amylase activity, especially at the end of the growth logarithmic stage, has also been shown for *Chlorella* cells under clinostating (Popova and Shnyukova, 1989).

The results obtained provide evidence that some cell ultrastructure rearrangements were correlated with the duration of the culture grown in microgravity and the content of the nutrient medium. It has been clearly shown that microgravity affects cell metabolism and ion homeostasis more strongly when *Chlorella* is grown on a solid medium. It is suggested that this phenomenon is connected with a strengthening of the indirect effect of microgravity because *Chlorella* is an aquatic organism and its interaction with environmental changes in microgravity is stronger under modified growth conditions (Popova *et al.*, 1995). Therefore, in microgravity cultivation of *Chlorella* in liquid and semiliquid media is preferable.

Conservation of the pronounced endogenous rhythm of cell photoaccumulation in microgravity was established in *Chlamydomonas reinhardtii* (a wild-type strain with a circadian period of 25 or 26 h and a mutant strain with a shorter period of 18–20 h). Mobile cells of this unicellular green alga have the capacity to move rapidly with the aid of flagella in the lighted area of a liquid cultural medium. The maximum photoaccumulation in spaceflight was nearly twice that of ground controls. These data show that in microgravity during a dark period the cells did not distribute over the entire volume of the culture vessel as in ground controls but were not far from a light cone, perhaps because of a loss of orientation in darkness in the absence of the gravitational vector. In addition, more cells remained in the lighted area for a protracted period; as a result, it is suggested that photosynthesis intensity and cell survival increase (Mergenhagen, 1986; Mergenhagen and Mergenhagen, 1987). Based on the index of colony formation on an agar medium, cell viability after a 7-day space flight was considerably higher than that in ground controls (Mergenhagen, 1986). Mutant forms were not detected. A greater amplitude in photoaccumulation reaction and an increase in proliferation and viability of the experimental cells is evidence, in the authors' opinion, that *Chlamydomonas* has better cultural conditions in spaceflight than on earth (Mergenhagen and Mergenhagen, 1987).

At the same time, no differences have been found in the viability of CALU N495+ cells after a 14-day spaceflight compared with controls (Gavrilova *et al.*, 1991). An algal culture grew on the surface of an agar nutrient medium. The cell sizes in spaceflight exceeded those in the controls at all stages of a vital cycle; most of the cells preserved the flagella and the number of deformed cells decreased. Cell ultrastructure in spaceflight and in the controls was similar; only an increased variability of cell organelle sizes was noted in microgravity.

A random orientation of the unicellular green alga *Euglena gracilis* was observed in simulated microgravity and in spaceflight. No adaptation of the cells to microgravity could be observed over the duration of the spaceflight (Häder *et al.*, 1995).

The ultrastructure of the blue-green, nitrogen-fixing alga *Anabaena azollae*, which is a prokaryotic symbiont of the water fern *Azolla pinnata*, has been described in detail after an 8-day spaceflight (V. Kordyum *et al.*, 1983). This alga grows in the closed cavities of chlorophyllous leaves. The material was fixed in orbit for electron-microscopic study. In the color of the filaments and cell shape and ultrastructure, the experimental and control populations were similar. Swollen thylakoids were noted in only a relatively small number of cells and there were some single cases of wavy or curved cell partitions formed during cytokinesis. The electron density of such partitions was considerably lower than that of controls (V. Kordyum *et al.*, 1983; Popova, 1986).

## 2. Mosses

In a 96-day spaceflight experiment with *Funaria hygrometrica* moss protonema, it was shown that the moss spores germinate in microgravity, and the protonema, consisting of a photosynthetic chloronema and caulonema, develops (Kordyum *et al.*, 1981). Changes in the length of protonema filaments as well as in the cell size and shape were revealed. As a rule, in spaceflight, the protonema filaments were shorter than those in controls, cell size decreased, and part of the cells became pyriform or dumbbell shaped. The differences in the subapical cell ultrastructure in microgravity were mainly in dilation of the perinuclear space, a decrease of both chloroplast size and starch grain relative volume in the chloroplasts to the point of complete disappearance, an increase in electron density of both the stroma and the outer membrane of the chloroplast envelope as well as thylakoid swelling, a consolidation of the mitochondrion envelope outer membrane, ER vesiculation and considerable (nearly three times) thinning of the cell walls (Kordyum *et al.*, 1981).The changes described in the cell organelle structure were similar to those in cell aging. The same rearrangements in protonema cell ultrastructure have been observed after long-term clinostating (20–30 days) (Nedukha, 1984, 1986). Thylakoid swelling under clinostating is assumed to be the result of alterations of the ion concentrations in the chloroplast stroma caused by changes in the chloroplast envelope and thylakoid membrane permeability for ions. The increase in number of peroxisomes observed in moss cells under clinostating is evidence for activation of photorespiration under these conditions (Tolbert and Essner, 1981).

An increase of cellulosopectolytic enzyme activity (endo-1,4-$\beta$-glucanase, exo-1,4-$\beta$-glucanase, polygalacturonase, and pectin esterase) was revealed in protonema cells grown under long-term clinostating (Nedukha, 1992) and is evidence for an intensification of cellulose and pectin hydrolysis that results in the appearance of free galacturonic acids and the breaking off of pectin metoxylic groups with a release of $Ca^{2+}$ ions. Electron-cytochemical study of pectinase activity revealed a greater size and frequency of occurrence of the needle-like precipitate throughout the apoplast under clinostating for 30 days compared with controls (Nedukha, 1989). Pectinase activity in a moss protonema was localized only within the walls between the cellulose fibrils. An increase of endo- and exogluconase activity was accompanied by a decrease in total cellulose content, including its crystalline form (Nedukha, 1992).

Use of the pyroantimonate method showed that localization of free and weakly bound calcium in the protonema differed under clinostating in comparison with controls. The cytochemical reaction was more intense in the intracellular depots and in cell walls. In addition, the reaction product appeared in the periplasmic space and in the hyaloplasm, where it was never found in controls (Nedukha, 1989). An inhibition of $Ca^{2+}$-ATPase activity at the cytoplasmic membrane of the protonema cells has been also demonstrated under clinostating (Nedukha, 1989).

## 3. Ferns

To date, only a single space experiment has been performed with the water fern *A. pinnata*. Based on the morphological characteristics and cell ulrastructure, the spaceflight plants did not differ from the controls (V. Kordyum *et al.*, 1983). Bacteria in the cultural medium and on the surface of fern leaves showed a considerable increase over controls, although the number of species was reduced. The data obtained provide evidence that microgravity influences macroorganisms such as *A. pinnata* that control the development of associated microorganisms, symbionts, and pathogens under normal conditions. It is suggested that it affects metabolic rearrangements in the fern and thus the interactions of the partners in the symbiotic system (V. Kordyum *et al.*, 1983).

## 4. Gymnosperms

The main focus of space experiments with *Pinus sylvestris* seedlings has been on plant orientation in microgravity. Seeds were placed in the substrate with an embryo at right, at left, upward or downward (Parfenov et al., 1975; Platonova et al., 1977, 1979). A disturbance of seedling orientation was observed after the 21-day space flight in the absence of illumination, i.e., the seedlings lost the capacity to display both positive and negative gravitropism. Only seedlings that were grown from seeds with the embryo placed downward in the substrate oriented more or less normally. Cotyledon leaf cells were rounder in microgravity than in controls.

Microgravity influenced the growth of *Pinus elliotti* seedlings (Cowles et al., 1984, 1986, 1988). Under normal orientation (shoots grew upward toward light), the average length of shoots and roots decreased 10–22 and 35–40% respectively compared with controls; respiration intensity also changed in microgravity. Lignin content was reduced by 4 or 5%; however, this was not statistically significant. A reduction in the activity of lignin biosynthesis enzymes was statistically significant. In the upper part of a hypocotyl grown completely in microgravity, lignin content decreased 24% compared with controls. Enzyme activity was also reduced, which can indicate, in the authors' opinion, that microgravity critically affects lignin synthesis.

## 5. Angiosperms

***a. Seed Germination and Seedling Growth*** In space flight, the seed germination rate under optimal growth conditions is, as a rule, 80–90%; it can even reach 100%. Temperature and humidity, but not microgravity, are the limiting factors for germination. Seedling growth occurred normally. Data on seedling growth rate (based on the measurements of length of the roots, shoots, and coleoptiles) are not always in agreement. In addition to data about the similarity of seedling growth rate in spaceflight and in controls, there is information on its acceleration or its delay in microgravity (Claasen and Spooner, 1994; Halstead and Dutcher, 1987; Kordyum et al., 1994).

Morphogenesis occurs normally in microgravity; no changes are seen in the arrangement of plant seedling organs. However, there are differences in the number of lateral roots formed in microgravity compared with controls. For example, in experiments with cress, the percentage of seedlings in which lateral roots had formed by the end of spaceflight was 89.8%; in the on-board centrifuge it was 71.6%, and in the ground controls it was 74.9% and 70.2% (Merkys et al., 1984).

***b. Root Meristem*** Data on the mitotic activity of seedling root meristem cells in microgravity are not entirely in agreement. For the majority of the species investigated, there are data on a decrease in mitotic index to different degrees, in particular in wheat (Gray and Edwards, 1971), pea, lettuce, and barley (Dubinin et al., 1977; Merkys and Laurinavichius, 1990; Merkys *et al.,* 1976a, 1983b); in sunflower, oat, and mung bean (Krikorian and O'Connor, 1984); in *Haplopappus* and daylily (Krikorian, 1990). It is interesting to note that in maize seedlings grown 7 days on board the biosatellite Cosmos-1667, mitotic activity decreased 15–30% (Barmicheva *et al.,* 1989). In the same experiment there are opposite data, showing an increase in mitotic index compared with controls (Darbelley, 1988). An increase in mitotic activity is also noted in the lens seedling root meristem (Driss-Ecole and Perbal, 1989). Moreover, it was shown that a short rotation in the on-board 1g centrifuge for 3 h affected mitotic activity, which confirms, in the authors' opinion, the gravisensitivity of mitotic division. Statistically significant differences in the mitotic activity of crepis root meristem cells between space flight and control samples have not been found (Kostina *et al.,* 1986; Tairbekov *et al.,* 1979; Vaulina and Kostina, 1984).

A reduction in mitotic index is supposed to be related to a decrease in proliferative pool and/or a change in dividing and interphase cells, perhaps at the expense of acceleration of mitosis (Barmicheva *et al.,* 1989). An increase of mitotic activity could be caused either by a lengthening of the duration of mitosis or by a shortened cell cycle (Darbelley, 1988).

The high sensitivity of mitotic activity to the action of gravity, which was clearly demonstrated by Driss-Ecole and Perbal (1989), may explain certain contradictions in the data obtained by different authors because the short-lived but considerable overloads that occur during spacecraft maneuvers and other events are not taken into consideration when a material is fixed on board a spacecraft.

According to data obtained by Vaulina and Kostina (1984), the frequency of cells with chromosome aberrations in the crepis seedling root meristem was higher in nearly all space flight experiments. In controls it was 0.15, 0.09, 0.25, and 0.06; in spaceflight seedlings it was 1.04, 0.42, 1.02, and 1.7. Moreover, the number of rearrangements of the chromatid types increased. In one of the experiments, 0.09% of cells had a large number of damaged chromosomes (Vaulina and Kostina, 1984). It should be noted that in this experiment the seeds germinated in orbit after 234 days in spaceflight. Krikorian and O'Connor (1984) report the presence of aneuploid cells ($2n = 33$ instead of $2n = 34$) in the root meristem of two sunflower seedlings in microgravity. The formation of chromosomal bridges and chromosome fragmentation are described in sunflower, oat, *Haplopappus,* and daylily (Krikorian and O'Connor, 1984; Krikorian, 1990).

As noted previously, a considerable number of cytogenetic investigations have been performed with seedlings grown from seeds that were in space from 2 to 827 days (Vaulina, 1976; Halstead and Dutcher, 1987; Nevzgodina *et al.,* 1990). Therefore, I do not consider these questions in detail. In addition, it should be noted that (i) currently, according to available data, microgravity (although it is not the least) is not primary among the factors that evoke disturbances in the genetic apparatus; and (ii) chromosome aberrations and disturbances of mitotic division in experiments of limited duration in the near cosmos are an insignificant percentage of the whole and therefore do not affect cell and tissue differentiation and morphogenesis of plant vegetative and generative organs.

There are only isolated communications on rounded cells in the root meristem in crepis and maize (Halstead and Dutcher, 1987; Tairbekov et al., 1979) as well as multinuclear cells in crepis (Vaulina and Kostina, 1984). The ultrastructure of root meristem cells in angiosperms species investigated turned out to be similar in their main features in microgravity and in ground controls. The formation of a typical mitotic spindle and kinetochore microtubules is described in root meristem dividing cells in 8-day oat seedlings in microgravity. Cytokinesis occurred, as usual, by means of cell plate formation with participation of the Golgi apparatus elements and the endoplasmic reticulum (Slocum et al., 1984).

Certain differences in the cell organelle ultrastructure in microgravity have been observed in 7-day seedlings and 18-day pea plants (Kordyum and Sytnik, 1983; Sytnik et al., 1984). In 7-day seedlings, the mitochondrion matrix density and cristae number increased in periblema meristem cells at the level of the two- and three-layer cap. A specific modification of the peripheral cistern of a dictyosome distal secretory pole that was not found in ground controls was also revealed in these cells (Kordyum and Sytnik, 1983). Vacuolization of the root meristem cells in 7-day seedlings in spaceflight was similar to that in controls, but it increased considerably in 18-day plants in microgravity. Mitochondria swollen to a different degree were observed in root meristem cells in 8-day oat and mung bean seedlings (Slocum et al., 1984); large mitochondria of unusual shape in which matrix density and crista number varied considerably have been described in root meristematic cells in *Impatiens balsamina* seedlings grown for 13 days in microgravity in darkness; DNA fibrils were seen in the electron-translucent areas of the mitochondria (Kordyum et al., 1992, 1994).

It is established that there is a considerable decrease in the length of the root meristem zone in microgravity (Barmicheva et al., 1989; Darbelley, 1988; Kordyum et al., 1989; Merkys and Laurinavichius, 1990; Shvegzhdene, 1991; Sytnik et al., 1984). Cell elongation in these conditions begins closer to the root apex than in ground controls. For example, according to Darbelley (1988), the root meristematic cells in maize seedlings began to elongate 0.5 mm distant from the limits of a root dormant center in microgravity; in ground controls this was not less than 1 mm. A reduction in meristem activity in orbital flight decreases apical dominance at earlier stages of root ontogenesis than in ground controls.

***c. Root Elongation and Differentiation Zones*** In microgravity as in ground controls, in the growing embryonal root, besides the meristem, the growth zones of elongation and differentiation are clearly expressed. Histogenesis occurs without deviations from normal. According to Slocum *et al.* (1984), the ultrastructure of cells in the root cortex and the central cylinder in oat and mung bean seedlings in microgravity was similar to that of controls. Unlike the controls, in microgravity swollen mitochondria were observed in the cortex cells.

Elongation zone length in lettuce seedling roots was reduced in orbital flight in comparison with both the ground controls and the on-board 1*g* centrifuge in flight (Merkys *et al.*, 1981, 1983b). The presence of cells growing by elongation nearer to the root dormant center in microgravity is in conformity with the acceleration of cell differentiation processes. At the morphological level, this process is clearly pronounced in the formation of root hairs, i.e., rhizoderm differentiation, at a considerably shorter distance from the dormant center (Merkys *et al.*, 1983b; Shvegzhdene, 1991). Under the more protracted influence of microgravity, meristem activity stopped earlier than in the ground control, and root apices of the same age in orbital flight contained only the differentiated vacuolized cells (Kordyum *et al.*, 1989; Sytnik *et al.*, 1984; Tarasenko *et al.*, 1982). An earlier decrease or a naturally complete removal of apical dominance leads to the abundant formation of lateral roots described in many investigated species (Halstead and Dutcher, 1987; Merkys *et al.*, 1984).

Under clinostating, in the root elongation and differentiation zones of 7-day pea seedlings there were more cells with 4C nuclei than in controls. Toward Day 14 of clinostating, this difference disappeared at the expense of an increase in cells with 4C nuclei predominantly in the root differentiation zone in the controls (Zaslavsky and Danevich, 1983; Zaslavsky and Fomicheva, 1986). Thus, under clinostating, the nuclear–cytoplasmic relationships characteristic for root differentiated cells exhibit somatic polyploidization in a considerable part of the root cell population earlier than in controls.

The anatomical organization of the differentiation zone of the roots in 28-day pea plants and 14-day onion plants as well as the aerial roots of epiphytic orchids grown in spaceflight for 110 and 171 days was similar to that in the control plants. Differences are found only in the quantitative characteristics of the indices investigated—in the diameters of the roots, central cylinders, and vessels in the directions of both augmentation and diminution compared with controls (Aliev *et al.*, 1986; Cherevchenko *et al.*, 1986; Nevruzova *et al.*, 1987). A certain inhibition of linear and radial growth of the aerial roots is noted; diminution of aerial root diameter occurred at the expense of parenchyma reduction.

***d. Shoot Apical Meristem*** The structural organization of the shoot growth point in 19-day etiolated balsamine seedlings did not differ from the controls after a 13-day spaceflight. A two-layer tunica and corpus cells were clearly distinguished in the growth point. Although the ultrastructure of tunica cells in spaceflight and in the control was similar in the main features, at the same time an increased vacuolization, greater ER development, and blade-shaped nuclei were observed (Nedukha et al., 1992).

*e. Shoot Primary Structure* The ultrastructure of the hypocotyl epidermal cells of 19-day balsamine seedlings grown for 13 days in spaceflight and in ground controls was similar on the whole. Only an insignificant thinning of an outer wall of the epidermal cells up to 1.4 ± 0.2 $\mu$m versus 1.7 ± 0.4 $\mu$m in the control was noted, as well as changes in cuticule structure in which a pronounced flakiness was found; denser layers with a thickness of about 60 nm alternated with more electron-transparent ones (Nedukha et al., 1992). No significant differences have been found in the anatomical organization of pea plant shoots grown 29 days in spaceflight. Only changes in degree of sclerification have been found in microgravity; sclerenchyma cell walls became considerably thinner in spaceflight than in controls (Nevruzova et al., 1987). In orbital flight, the linear and radial growth of *Epidendrum radicans* shoots was retarded; both cell size and parenchyma tissue volume were slightly diminished (Cherevchenko et al., 1986).

*f. Leaf Photosynthetic Cells* In the outer walls of epidermal cells on the cotyledon adaxial surface of 19-day etiolated balsamine seedlings grown 13 days in spaceflight, there was no cross-multilayer structure, which is characteristic for such cells in controls, and clear zones without cellulose fibrils were found (Nedukha et al., 1992). Thinning of the cuticular layer in comparison with controls and loosening of the epidermal cell wall were also observed in wheat leaves after a 16-day spaceflight (Nedukha et al., 1991). On the basis of data showing that similar changes occur in plant cell walls with an increase in cuticular transpiration (Miroslavov, 1974), it is suggested that microgravity intensifies this process.

Leaf disposition, or phyllotaxis, and leaf tissue differentiation occur within normal limits in microgravity. The changes observed involve the quantitative characteristics of leaf morphoanatomical structure (Cherevchenko et al., 1986; Kordyum et al., 1989; Nedukha et al., 1991; Nevruzova et al., 1987). For example, the size of both the leaves and the epidermal cells diminished in *E. radicans* plants after a 171-day spaceflight (Cherevchenko et al., 1986). Some diminution of the leaf plate size and thickness in pea plants were found after a 29-day spaceflight (Nevruzova et al., 1987). The correlation of the epidermis and palisade parenchyma in leaves of 14-day *Allium cepa* plants was not changed in microgravity (Kordyum et al., 1994).

Tracheid secondary walls of the vascular bundles in wheat leaves after a 16-day spaceflight were electron transparent and lost a granular–fibrillar component (Nedukha et al., 1991).

The ultrastructural organization of leaf mesophyll cells in spaceflight is typical of the main features of photosynthetic cells. At the same time, a diminution of mesophyll cell size in microgravity has been shown; for example, in wheat on average it was $146 \pm 17$ $\mu$m in spaceflight and $263 \pm 22$ $\mu$m in controls. The population density of organelles was also changed in the palisade parenchyma cells; chloroplast density, which was $7.58 \pm 0.32$ organelles per cell in the controls, decreased to $5.13 \pm 0.28$ in microgravity. Mitochondrion density increased to $4.35 \pm 0.32$ per cell in comparison with the control density of $2.31 \pm 0.36$ per cell. Peroxisome density changed little and was $1.11 \pm 0.23$ per cell in controls and $1.57 \pm 0.23$ per cell in microgravity (Nedukha et al., 1991).

Changes in chloroplast size and structure in leaf mesophyll cells in microgravity have been found in all angiosperm species investigated in this respect. Moreover, the degree of these changes could rise with prolonged space experiments. Chloroplast size in 16-day wheat plants diminished in comparison with controls and on an average was 3.2 $\mu$m on the long axes of an organelle and 1.2 $\mu$m on the short ones; in controls it was 4.3 and 2.6 $\mu$m, respectively. The number of granae on the section also decreased and was $17 \pm 2$ (in the controls it was $33 \pm 2$). The diameter of plastoglobuli was on an average 0.75 nm in spaceflight and 50 nm in control plants (Nedukha et al., 1991). In the palisade parenchyma cells of 28-day *Arabidopsis thaliana* leaves from the middle tiers, the chloroplast structure was characteristic for a mature leaf in the control. In spaceflight plants, unlike controls, there was a widening and vesiculaton of intergranal thylakoids, lightening of the stroma, a decrease in ribosome quantity, an increase in plastoglobuli volume, and a diminution in cell plastidome volume. Membrane elements of various shapes and droplets of osmiophilic substances in the cytoplasm were noted (Kordyum et al., 1989). Along with chloroplasts preserving a native membrane system in palisade parenchyma cells of the 29-day pea plant leaves in microgravity, chloroplasts with friably packed or damaged granae, whose thylakoids appeard as vesicles with an electron-transparent content, were also observed (Aliev et al., 1985; Abilov, 1986).

Plastids with a curved shape and smaller sizes, unlike the controls, were observed in palisade parenchyma cells on the first two leaves of *E. radicans* plants after a 110-day spaceflight along with chloroplasts of the usual form. Starch grain volume decreased or starch grains were absent on the sections. Ribosome density decreased in the chloroplast stroma; DNA-containing areas of the stroma was not clearly revealed. Disturbances of the organization of the internal membrane system and stroma vesiculation as well as different stages of organelle lysis were also observed (Abilov et al., 1985).

Proteins of the stroma and thylakoid membrane fractions of isolated chloroplasts from leaves of 14-day *Vicia faba* plants were investigated after 15 and 24 h of clinostating. Proteins of the stroma fraction (18 and 57 kDa) showed qualitative modifications. Thylakoid membrane-bound proteins (15 and about 20 and 30 kDa) showed a tendency toward a decreased immunoresponse to ubiquitin antibodies. Protein samples of the stroma fraction showed an increased ubiquitination of the large subunit of the RuBPCase. It is assumed that protein turnover changes in soluble and membrane-bound proteins of the chloroplasts under clinostating (Wolf *et al.*, 1993).

*g. Biochemical Content and Enzyme Activity* Some deviations in protein and carbohydrate metabolism as well as some changes in a mineral content have been established in a biomass of pea plants grown for 24–33 days in microgravity (Merkys *et al.*, 1976b; Laurinavichius *et al.*, 1984). There were changes in the quantity of amino acids in proteins; lysine and tyrosine content decreased 10–15% and histidine and phenylalanine content increased approximately the same percentage.

The dry mass of 18-day wheat plant shoots in microgravity did not differ insignificantly from that in the controls. RNA content in the roots and in the leaves was similar in spaceflight and in control plants, but it was two times smaller in the growth cone in microgravity in comparison with controls (Sytnik and Musatenko, 1980).

The quantity of ethanol-soluble sugars in pea plants increased in microgravity while starch content decreased (Laurinavichius *et al.*, 1984). These data are considered to indicate enhancement of hydrolytic processes. Unlike the controls, there was a dramatic disturbance of the balance of the most important mineral elements, namely phosphorus content in the flight samples increased 2.5 times and potassium content 1.5 times, and calcium, magnesium, manganese, zinc, and iron content decreased considerably (Laurinavichius *et al.*, 1984). In 19-day wheat plants there was a fivefold decrease in calcium content in space flight plants compared with controls (Mashinsky *et al.*, 1991). These researchers see this as an indication of a disturbance of ion selection uptake, which is under the metabolic control of the root cells.

Chlorophyl a and b content decreased considerably in microgravity; chlorophyll a was reduced 63.4% compared with controls and chlorophyll b was reduced 70.3%. This can be considered as evidence of a delay of pigment biosynthesis or an acceleration of its destruction (Laurinavichius *et al.*, 1984). Analogous data on a 1.5-fold decrease of chlorophylls and carotenoids in 19-day wheat plants in microgravity are given by Rumyanzeva *et al.* (1990). A rise in carotenoid content can be explained, according to these authors, by its function as a heat filter (these pigments absorb surplus quantum energy), which could increase owing to the leaves overheating in spaceflight. There are contradictory data on chlorophyll a and b increases in 29-day pea plants and, on this basis, stimulation of pigment biosynthesis in orbital flight is assumed (Abilov et al., 1985; Aliev et al., 1985).

More intensive water accumulation has been found in 7-day etiolated maize seedlings in microgravity. The dry mass of the remaining caryopses with the remanents of reserve nutrient substance was lower than in the controls (Tairbekov et al., 1988), which indicates more intensive utilization of these substances in seedling growth processes in microgravity. These authors suggest that there is a tendency for metabolic processes to be enhanced in the absence of gravity. An increase in the number of soluble and structural proteins in E. radicans after 60- and 171-day spaceflights has been observed (Cherevchenko et al., 1986).

A significant decrease (54%) of cellulose content, which is the main component of a plant cell wall, as well as some increase in hemicellulose content was found in 24-day pea plants (Laurinavichus et al., 1984). This shows, in the authors' opinion, the serious shifts in cell wall polysaccharide biosynthesis that take place in microgravity and that can be viewed as a result of the disturbance of cell metabolism.

Lignin content in 7-day mung bean seedlings diminished on average 18% in microgravity in comparison with controls; at the same time it did not essentially change (the differences were statistically unreliable) in oat seedlings (Cowles et al., 1984). No significant differences in lignin content were revealed in 7-day etiolated maize seedlings grown in spaceflight and in ground controls (Tairbekov et al., 1988). In microgravity, pectin content in maize seedlings decreased somewhat compared with controls, whereas hemicellulose content increased. A decrease in uronic acid content and an increase in glucose was found in the pectin fraction. In the hemicellulose fraction, there was an increase in glucose and galactose content and a decrease in xylose and arabinose content. The existence of a connection between the tendency to a monosaccharide change, the structural rearrangements of polysaccharide molecules, and their quantitative content is assumed (Tairbekov et al., 1988; Zabotina, 1987).

An increase in peroxidase activity level was found in 2-day wheat seedlings in space (Conrad, 1968, 1971). The activity of RuBPCase, being a key enzyme of photosynthesis, did not essentially change in E. radicans and Doritis pulcherrima in spaceflight, which can indicate the normal occurence of initial photosynthesis stages in plant leaves.

***h. Intracellular Calcium Balance*** Unlike the controls, the appearance of calcium pyroantimonate deposits in the hyaloplasm of *Chlorella* cells (Popova, 1992), the protonema cells of the moss *F. hygrometrica* (Nedukha, 1989), and in root cap statocytes of pea (Belyavskaya, 1992) was a characteristic feature under clinostating. These data obtained at the structural level have been confirmed by the determination of $Ca^{2+}$ ion concentration in cress root hairs under clinostating using a sensitive calcium indicator, indo-1, and a calcium channel blocker and $Ca^{2+}$-ATPase inhibitor. An increase in $Ca^{2+}$ ion concentration over that of controls was detected in cells under clinostating; the steady-state ion gradient was clearly demonstrated (Kordyum and Danevich, 1995). The calcium ion increase under clinostating is supposed to be due to an inhibition of $Ca^{2+}$-ATPase activity. Earlier findings claimed a reduction in $Ca^{2+}$-ATPase activity in the cytoplasmic membrane of pea root statocytes under clinostating using electron cytochemistry and biochemistry (Kordyum et al., 1984a). Detection of an increase in the $Ca^{2+}$ concentration under clinostating shows that this factor belongs to the environmental stimuli correlated directly with the functioning of a $Ca^{2+}$ second messenger system in cells.

*i. Respiration–Heat Discharge* Determination of oxygen content and carbon dioxide in the growth chambers with sown seed and later with the seedlings of mung bean and oat before and after an 8-day spaceflight showed, as was to be expected, an increase in $CO_2$ level and a decrease in $O_2$ level in the experiment and in the control; however, the values of these parameters varied according to species. The most significant $CO_2$ accumulation and $O_2$ decrease (exceeding that of the controls) was found in the chamber with mung bean seedlings. In the chamber with oat seedlings, the $CO_2$ level was lower than that in controls and the $O_2$ level was higher. The ethylene concentration in the chamber with mung bean seedlings was 1–1.5 ppm (Cowles et al., 1984). These data provide evidence that seedlings with various respiration levels respond differently to experimental conditions; in microgravity respiration intensity increased in mung bean seedlings; a higher respiration level, compared with oats, is typical for this plant; and in oat seedlings it somewhat decreased.

No statistically significant differences in dynamics and quantity of heat discharged by maize seedlings (the intensity of energy metabolism in microgravity has been elucidated) during 5 days in spaceflight and in the ground control have been found (Tairbekov and Devyatko, 1985).

*j. Microviscosity and Lipid Peroxidation* A high sensitivity of microviscosity indices of the cytoplasmic membrane to the influence of altered gravity has been shown using the cytoplasmic membrane fraction and the liposomes obtained from the membrane lipids. Microviscosity increased to 5.12–5.62 P after 24 h of clinostating; it was 0.46–0.69 P in the controls. After 48 h of clinostating, microviscosity decreased rapidly to its initial level. These data are in accord with changes in the fatty acid content when after 48 h of clinostating the index of unsaturation (the ratio of unsaturated fatty acids to saturated ones) increased (Polulyakh, 1988; Polulyakh and Volovick, 1989; Polulyakh et al., 1989).

Determination of cytoplasmic membrane microviscosity at specific intervals during up to 8 days of clinostating allowed us to demonstrate the oscillatory character of this indicator of the cytoplasmic membrane state. Moreover, the oscillations showed a tendency to fade. Toward the eighth day of clinostating and space flight, the oscillations were close to those in the controls (Kordyum et al., 1994).

Lipid peroxidation intensity was determined in the cells and in the cytoplasmic membrane fraction of pea and wheat seedlings after 2-, 9-, and 14-day spaceflights and under clinostating in the same intervals in addition to the other time periods beginning at 15 min. In the laboratory experiments, seeds and seedlings of different ages were placed on the clinostat; in spaceflight experiments, the seeds were germinated in orbit. The dynamics of alterations in lipid peroxidation in microgravity and under clinostating was similar on the whole, with an initial decrease followed by a stable increase (Zhadko, 1991; Zhadko et al., 1992, 1994).

*k. Vegetative Propagation* The inhibition of the early stages of awakening of the *Spirodela polyrrhiza* turions in a dormant state in spaceflight has been established (Kutlakhmedov et al., 1978). The meristematic zone of the turion, in which intensive cell division and formation of the daughter rudiments occur, is an organ of *Spirodela* vegetative propagation. Upon awakening of the turions, the meristematic cells enter mitosis; with this event, morphogenesis begins, i.e., the formation of rudiments and then their separation. The initial stage of turion awakening as a transitional process from a dormant state to an active growth stage is highly sensitive to external factors including microgravity; under the influence of microgravity, an appreciable delay in offspring formation takes place.

*l. Cells of Generative Organs* Only one species has flowered and fruited in spaceflight conditions (i.e., carried out a nearly complete ontogenesis in microgravity) when plants were delivered for orbit in the phase of beginning bud formation (Kuang et al., 1995, 1996; Musgrave et al., 1993; Parfenov and Abramova, 1981), two cotyledon leaves (Kordyum and Chernyaeva, 1982; Kordyum et al., 1983b, 1984c), and grown from seeds in spaceflight (Merkys and Laurinavichius, 1983). This is *Arabidopsis thaliana;* plants of this species are annual and their life cycle continues on an average of 40–45 days. E. Kordyum et al. (1983b) showed that *A. thaliana* plants are not obligate self-fertilizators, as was thought earlier, but have a gynomonoecic sexual form; the development of hermaphoditic (bisexual) and potentially bisexual female flowers (with sterile stamens) on the same plant owing to cross-pollination is possible in nature.

In spaceflight, the morphogenesis of generative organs occurred normally; nevertheless, certain processes leading to sterility of the gyneceum and the androceum were revealed at different stages of their development (Kordyum and Chernyaeva, 1982; Kuang et al., 1995; Musgrave et al., 1993). Degeneration of the gyneceum elements was observed at the stages of microspore tetrads, one-, two-, and four-nuclear embryo sacs as well as in the formed ones (Kordyum et al., 1984c); moreover, the first indications of arrested development were observed, as a rule, in the ovule somatic cells. On the other hand, there was a tendency to normal division and differentiation of the sporophyte cells and to disturbances in gametophyte (a pollen grain) development. In spaceflight, the initial stages of bud tubercle differentiation in *Allium cepa,* as in *A. thaliana,* did not differ from the controls (Kordyum et al., 1994).

An acceleration of male gametophyte development in microgravity is described in *Muscari racemosum* (Kordyum et al., 1979b). Bulbs with racemose rudiments at the stage of archesporial cells were delivered for orbit and grew during a 3-mounth flight. After the flight, there were three-cell pollen grains in the anthers; in stationary growth conditions on earth, two-cell pollen grains are characteristic for this species as for the family Liliaceae as a whole.

Anomalous mitoses (nearly 3%), which have not been observed in controls, have been described in *Tradescantia paludosa* microspores. Changes in mitotic spindle direction; the formation of three- and four-pole spindles; and trailing and nondispersing chromosomes as well as chromatid rearrangements, recombinations, and fragments, including spherical ones, have been described (Delone et al., 1963, 1964, 1966, 1968). An increase in the occurence of embryo sac and pollen abortion have been observed in irradiated flight samples of Tradescantia clone compared with controls (Marimuthu et al., 1972; Sparrow et al., 1968; 1971).

***m. Organ Culture*** Stem segments of potato miniplants were cultivated for 8 days in spaceflight. During this period, minitubers formed in microgravity and in the ground control and contained well-developed amylogenic storage tissue. The comparative analysis of an ultrastructural organization of the storage parenchyma cells showed a decrease of starch grain size in the amyloplasts in microgravity to $3.1 \pm 0.16$ $\mu$m in diameter; in the ground control it was $4.2 \pm 0.22$ $\mu$m in diameter. Local enlargements in lamellae as well as an increase in the mitochondrion matrix density and the presence of well-developed cristae were the typical differences of these organelles in spaceflight (Kordyum et al., 1992, 1995).

***n. Cell and Tissue Cultures*** In one of the first experiments with plant cells *in vitro,* a *Haplopappus gracilis* tissue culture was grown for 9 days in spaceflight in the spacecraft cabin at 18–22°C. Because the optimal temperature for the given clone is 28°C, the data showing some retardation in the culture growth and the absence of dividing cells were difficult to interpret (Sidorenko and Mashinsky, 1978).

Nearly simultaneously it was shown that somatic embryogenesis in a *Daucus carota* cell culture occurred normally in microgravity. However, unlike the controls, the embryoids in spaceflight had longer roots. Microgravity conditions are assumed to induce earlier formation of the organs and their more rapid development (Butenko et al., 1979; Krikorian and Steward, 1978; Krikorian et al., 1981). A greater increase in biomass occurred in a *Pimpinella anisum* cell culture during a 7-day spaceflight; somatic embryogenesis in microgravity and in the ground control was similar in this experiment (Theimer et al., 1986). At the same time, the number of meristematic centers, proliferative activity, and cell sizes in the *Arabidopsis* tissue culture were lower after a 63-day spaceflight than in the ground control (Merkys et al., 1988).

In the experiments with the *D. carota* crown gall tissue culture, there was an increase of cell quantity in the meristematic growth centers (Baker and Elliot, et al., 1979) as well as a change in the isoenzyme spectrum (Kleinshuster and Magon, 1979). Clinostating had a stimulating effect on the growth rate of this object (Wells and Baker, 1969; Kleinshuster et al., 1975) as well on changes in isoenzyme spectrum in simulated microgravity (Hanchey et al., 1975). The ultrastructural organization of the tumor cells after spaceflight and in the ground control was similar (Rubin et al., 1979).

No essential differences in the main features of ultrastructural organization, ploidy level, and differentiation process were revealed in the cells of the *Haplopappus* tissue culture grown in spaceflight for 9–28 days (Klimchuk and Martyn, 1994; Kordyum et al., 1994; Zhadko, 1991). At the same time, the available changes in a mitochondrion ultrastructure indicate an increase in their functional load. An analysis of culture growth parameters in spaceflight and under clinostating showed a reduced growth rate in spaceflight in comparison with the ground control and clinostating; under slow clinostating (2 rev/min), the increases in biomass and dry mass content at the different growth phases were similar to those of controls; under fast clinostating (50 rev/min) the increase in biomass was 20% higher than in the controls; dry mass content did not differ much from the controls. It was established that a greater increase in biomass under fast clinostating is accompanied by higher water content.

Essential changes in lipid peroxidation intensity in a *Halopappus* tissue culture have been revealed in real and simulated microgravity (Zhadko et al., 1992). Malonic dialdehyde content increased 88% after a 9-day spaceflight and 64% after a 14-day spaceflight, but it decreased 30% after a

28-day spaceflight in comparison with controls. In mitochondria, malonic dialdehyde content was 43% higher than controls.

The content of lipid peroxidation final products—lipofluorescent compounds—increased at the same time 30, 36, and 41% after 9-, 14-, and 28-day spaceflights, respectively. Under clinostating, there were also essential oscillative changes in both malonic dialdehyde content and spontaneous chemiluminescence intensity as well as antioxidant content (Kordyum *et al.*, 1994).

*o. Protoplast Culture* The higher plant protoplasts are convenient objects for studying the effects of microgravity on isolated single plant cells, including the processes of cell wall biogenesis, cell division, microcallus formation, and plant regeneration. Currently, there are results from several spaceflight and clinostat experiments with the protoplasts of *Brassica napus, D. carota, Solanum tuberosum, Nicotiana tabacum,* and *N. rustica* (Höffman *et al.*, 1994; Iversen *et al.*, 1992; Klimchuk *et al.*, 1992; Nedukha *et al.*, 1994; Rasmussen *et al.*, 1992, 1994). In the space experiments with *B. napus* and *D. carota* protoplasts, there was a delay in new cell wall synthesis in comparison with ground controls and on the 1$g$ centrifuge on board the shuttle. Unlike the controls, no plants were regenerated from protoplasts exposed to microgravity. After 4 days in orbit, cells were swollen in appearance and formed small microcalli with only few cells (an average of 2–4 cells) per microcallus compared with samples developed under 1$g$ conditions (an average of 8–12 cells per microcallus); in real microgravity, cells had larger vacuoles and a slightly reduced starch content compared with controls. An increased concentration of soluble proteins per cell and a reduced specific activity of peroxidase in the cytoplasm has been found (Rasmussen *et al.*, 1994). These authors assumed that the retardation in the regeneration processes is caused by an initial effect on the cytoskeleton system, by stress caused by exposure to microgravity conditions, or, most likely, by a combination of these effects.

In another experiment with the same objects during a 14-day spaceflight, the growth of rapeseed cells in flight was only 56% compared with ground controls; the growth of carrot cells in orbit was 82% of ground controls. Peroxidase activity and the amount of protein was lower in the flight samples than in the ground controls. The number of different isoenzymes was also decreased in the flight samples. A 54% decrease in the production of cellulose was found in rapeseed, and a 71% decrease was found in carrot. Hemicellulose production was also decreased in the flight samples compared with ground controls.

Ultrastructural analysis of the microcalli from protoplasts cultured in orbit demonstrated that hydrolysis and disappearance of reserve starch occurred in the flight cell plastids. The mitochondria were more varied in appearance in the flight samples than in the ground controls. Fluorescence analysis showed a decrease in the membrane-bound calcium content in cell cultures in spaceflight compared with ground controls. A well-documented effect is a decrease in cell wall thickness during spaceflight in 14-day-old samples of rapeseed, which can be correlated with the decreased content of cellulose and hemicellulose. All the results obtained so far indicate that, in general, developmental processes in protoplast cultures are retarded in microgravity, (Rasmussen *et al.*, 1992). No differences have been found in the regeneration capacity of protoplasts of *N. tabacum* and *N. rustica* in spaceflight compared with controls (Hoffmann *et al.*, 1994).

An inhibition of cell wall regeneration by protoplasts under the influence of altered gravity was also shown in clinostat experiments with potato protoplasts isolated from mesophyll cells. Ultrastructural analysis of protoplasts cultivated on the clinostat and under stationary conditions has shown that cell wall regeneration occurs in the same way, but under clinostating this process begins 1 day later. In addition, under clinostating, only 10% of the protoplasts regenerated in the cell wall; in the control it was 42% (Nedukha *et al.*, 1994).

In conclusion, it is necessary to again note that genetic determination of both morphogenesis and cell differentiation is preserved and carried out in microgravity. In the short experiments (up to 8–10 days) investigating the specialized graviperceptional cells in multicellular plant forms, complete information was obtained on cell differentiation and the structural–functional organization of the root cap, which is the root graviperceptional apparatus, in microgravity. The assumption of the gravisensitivity of root apical meristem cells was confirmed experimentally.

A wide spectrum of ultrastructural rearrangements in cells of different types in multicellular and free-living unicellular forms as well as their biochemical content has been demonstrated. These rearrangements, which reflect changes in the level and direction of plant metabolism in microgravity, provide evidence for the diverse and significant influence of this factor on the growth of plants and their vital functions. The alterations observed in microviscosity and lipid peroxidation furnish the background for the generally voiced hypothesis that there are possible changes in cell membrane permeability, primarily in the cytoplasmic membrane in microgravity. In light of these ideas, it has been possible to explain the facts of cell metabolism and ultrastructure rearrangements when morphogenesis and cell differentiation are normal, as well as the complicated ways microgravity affects multicellular organisms through intercellular interactions in the tissue system.

An increase in the quantity of bacteria found in the *Anabaena azolla* symbiotic system and an increase of the pathogenicity of *Pseudomonas* bacteria of a genera Pseudomonas under common cultivation with *Chlorella* and a fungus that infects the buds of *Arabidopsis thaliana* plants, together with a hypothetical decrease in *Chlorella* and *Arabidopsis* cell immunity, can indicate the changes in the interaction of eukaryotic and prokaryotic organisms in microgravity.

## IV. Characteristics of Plant Cell Organelle Rearrangements in Altered Gravity

A. Regularities of Organelle Structural–Functional Rearrangements

As has been shown in the previous section, the breadth and content of plant responses in microgravity and under clinostating vary according to the growth phase, physiological state, and taxonomic position of the object as well as the duration of action of these factors. It is necessary to stress that the depth of cell rearrangements increases with the length of exposure in altered gravity conditions.

At the same time, the responses have, to some degree, a similar character, reflecting the changes in cell organelle functional load. The rearrangements of structural–functional organization of organelles in meristematic and differentiated cells of multicellular forms and free-living unicellular ones that occur under the influence of microgravity and clinostating are summarized in Table II.

The cell's energy system is highly sensitive to environment fluctuations. The structural changes that occur in the mitochondria under different environmental influences are characteristic; they include an increase or decrease of the matrix electron density as well as organelle swelling and changes in the quantity, sizes, and regulation of cristae compared with controls (Bakeeva and Chenzov, 1989; Mashansky and Rabinovich, 1987). Mitochondrion populations in cells of lower and higher plants reveal a considerable polymorphism in altered gravity (Kordyum *et al.*, 1989; Rubin *et al.*,1979; Sytnik *et al.*, 1983a). The appearance of single large mitochondria with a dense matrix and concentrically arranged, often extended cristae can be seen as evidence of the disruption of the processes of oxidation phosphorylation. At the same time, an increase in matrix density as well as an increase in the volume of both regularly disposed cristae and the mitochondrion population in a cell is believed to compensate for the higher energy expenditures under the influence of altered gravity and to indicate some activation of respiration and an elevation of the energy metabolism level in a cell. An analogous dependence has been shown for animal cells; under decreasing activity of oxidizing-restoring enzymes, the mitochondrion

TABLE II
Main Typical Rearrangements in the Plant Cell Ultrastructure in Altered Gravity

| Organelle | Feature altered |
| --- | --- |
| Mitochondria | Increase in the volume per cell and in the ratio of the cristae volume to the matrix and its density, appearance of the mitochondrial reticulum and single large organelles of a diverse shape with numerous narrow or slightly dilated rings or parallely disposed cristae; organelle swelling |
| Leukoplasts | Appearance of organelles of elongated and branched shape in meristematic cells |
| Amyloplasts | Decrease in organelle size and starch grain volume, unequal local enlargements of lamellae give the impression of a network |
| Chloroplasts | Decrease in organelle size and starch grain volume, changes in grana size, curved organelles, single or numerous unequal local enlargements of thylakoids (thylakoid vesiculation), disorientation of intergranal thylakoids, appearance of electron-translucent areas in the stroma, decrease in ribosome quantity, increase in plastoglobuli size and number, envelope outgrowth, organelle swelling |
| Dictyosomes | Increase in variability, unusual modification of the peripheral cistern of a secretory pole |
| Endoplasmic reticulum | Increase or decrease in the volume, topography change, increase in activity, enhanced vesiculation of the agranular type |
| Vacuoles | Essential increase in the volume and in the variety of intravacuolar membranous elements |
| Cell wall | Thinning and loosening |
| Other cell elements | |
| Cytoplasmic membrane | Formation of folds of a complicated shape and invaginations filled with fine granular substance |
| Lipid droplet | Volume increase or decrease |

population in a cell increases at the expense of organelle swelling or new organelle formation (Kurnosenko, 1975).

Data on the rearrangements in cell energy metabolism in altered gravity obtained at the structural level are confirmed by the elevated activity of succinate dehydrogenase as an indicator of aerobic respiration and the expansion of multiple molecular forms of this enzyme at the expense of the appearance of new ones, compared with controls. These changes are stimulated, most likely, by the necessity for a greater energy supply (Nedukha et al., 1985; Popova et al., 1987).

In plant cells, the most noticeable changes in mitochondrion structure under the influence of altered gravity, as has been mentioned, occur in dark growing conditions. Taking into account the regulatory character of cooperation between mitochondria and chloroplasts (two energy-related organelles in a plant cell), it can be assumed that the rate of oxidation phosporylation and ATP activity of the mitochondria is elevated in the absence of chloroplast photochemical activity. One of the possible causes of an increase in mitochondrial functional activity as well as the change in the capacity for swelling and contraction is the activity of enzymes in the mitochondria. Pyroantimonate and clinostating (Belyavskaya, 1992; Nedukha, 1989; Popova, 1992) have been used to show that increased oxidation rate can be accompanied by an increase in cation (primarily $Ca^{2+}$) accumulation in the mitochondria (Zaitseva et al., 1991) under the influence of altered gravity.

The ultrastructure of such plant cell-specific organelles as plastids— leuko-, amylo-, and chloroplasts—clearly reflects their functional load. The degree of development of the chloroplast membrane system development can be, to a certain extent, an indicator of the organelle's photosynthetic activity. Amyloplast formation is typical for highly specialized plant cells. In microgravity and under clinostating, plastids in the cell population, such as mitochondria, reveal a certain heterogeneity in relation to the degree of structural rearrangement. On the basis of the structural changes described in chloroplasts in altered gravity (see Section III) and the data on alterations in chloroplyll content, a decrease in photosynthetic activity can be assumed.

However, this hypothesis is not valid if one takes into account the increase in thylakoid number in the granae and chloroplast heterogeneity in relation to the degree of structural rearrangements as well as the elevation of photosynthesis intensity in microgravity (Ward and King, 1979). Therefore, a general conclusion that there are changes in the functional state of photosynthetic organelles in altered gravity is more correct. For the determination of photosynthetic activity as the result of many factors (Björkmann, 1981), its direct measurement in orbital flight and under clinostating is necessary.

In connection with the question of photosynthetic intensity in microgravity, recall the diminution in size and loosening of the matrix of pyrenoids in *Chlorella* cell chloroplasts in microgravity. A pyrenoid is a pool of RuBP-Case, which is synthesized autonomously in a chloroplast and can be in a quiescient or active state or simultaneously in partially active and quiescient states (Vladimirova et al., 1982). The data on the existence of a correlation between the peculiarities of the pyrenoid structure and photosynthetic activity in algae can also be taken as indirect evidence of changes in photosynthetic activity in microgravity.

A decrease of starch grain volume in plastids of different types, including their complete disappearance under the influence of microgravity and clinostating, is in accord with an amylase activity that was elevated nearly 3.6 times evoking starch hydrolysis, the extension of its soluble forms (Popova et al., 1995; Popova and Shnyukova, 1989), and a 37% decrease in ATPglucosopyrophosphorylase activity, which is an enzyme of starch biosynthesis (C. Brown et al., 1992).

In light of biochemical data on qualitative changes in the poly- and monosaccharide composition of a cell wall, in particular a decrease in pectin content, primarily uronic acids, and such components of hemicelluloses as xylose and arabinose, which lead to a compression of the wall (Tairbekov et al., 1988), the phenomenon of cell wall thinning and curving (the wall of parenchyma cells) in microgravity and clinostating can be explained. Altered gravity conditions also slow the process of wall regeneration by protoplasts compared with controls (Nedukha et al., 1994).

Formation of more complex cytoplasmic membrane folds and a rise in plasmalemmasome quantity, which increases the cytoplasmic membrane surface and is quite often accompanied by unequal dilations of the periplasmatic space, probably increases the plant's functional load. This process is expressed especially clearly in free-living unicellular plant forms in microgravity and under clinostating when the relationship of an object with its environment is changed. For multicellular forms in this case, it is a matter of the changes of an internal, apoplastic, or free space.

It is known that considerable development of the cytoplasmic membrane surface occurs when the cell is actively functioning (Kursanov and Paramonova, 1976) or after the influence of unfavorable factors to restore normal vital functions (Steinbiss, 1978). The formation of the cytoplasmic membrane folds and invaginations is assumed to be connected with transport changes, i.e., with phenomena of endo- and exocytosis. Probably, the sites of membrane folds have a higher activity, which can be seen by a small increase in the distance between two electron-dense layers of the cytoplasmic membrane and an increased contrast at these sites.

Progressive cell vacuolization compared with controls, especially in the long-term spaceflight and clinostat experiments, has been described for all plant organisms investigated. During differentiation, autolytic processes of both local (reversible) and total (irreversible) character in the cytoplasm, depending on cell type, are clear. In the plant cell, catabolic processes are known to be carried out by the cell multifunctional compartment—the vacuolar apparatus. In cells of one type (for example, conducting elements of the xylem–tracheids and vessels), genetically programmed autolytic processes lead to complete atrophy of protoplasts. In other cell types, certain nuclear–cytoplasmic relationships change during the activity of more highly specialized types (leaf mesophyll cells). Autolysis is manifested as intensive vacuolization and is carried out in different ways—by the expansion of ER cisternae, derivatives of dictyosomes, and formation of cytosegresomes, which are cytoplasm areas surrounded by cisternae of agranular ER followed by transformation into vacuoles. Rearrangements of the vacuolar apparatus are normal in the processes of cell growth, differentiation, and aging and are accompanied by changes in the level and direction of cell metabolism. An intensification of autolytic processes in the cytoplasm under clinostating is assumed to be connected with an increase in ethylene production that reduces the phospholipid exchange level and leads to changes in membrane permeability and hydrolytic enzyme release (Hensel and Iversen, 1980). It should be noted that progressive vacuolization in microgravity and under clinostating has been convincingly demonstrated in animal cells (Neubert, 1979; Petrukhin *et al.*, 1976; Rokhlenko and Savik, 1981).

Data on a decrease of dictyosome volume in meristematic and differentiated cells of a root cap (Moore *et al.*, 1987a), a specific modification of the peripheral cistern of the dictyosome distal secretory pole in root meristematic cells (Kordyum and Sytnik, 1983), as well as an increase of variability in the Golgi apparatus in cells of the root elongation zone under clinostating (Kordyum, 1989) undoubtedly are evidence of changes in dictyosome functional activity. Essential differences in the structure of the Golgi apparatus in microgravity have been revealed in animal cells; for example, the transcompartment in newt oocytes is reduced at the previtellogenesis stage in microgravity, which indicates an inhibition of transport processes (Belousov *et al.*, 1991).

The concentration of dictyosomes and Golgi vesicles in the cytoplasm increases during high secretory activity. The chemical composition of Golgi vesicles is modified in various types of cells and in the vital processes of the same cell. The diversity of the synthesis and assembly of secretion products, which is connected with different kinds of dictyosome activity, and restriction of the actual material by only the structural level make it impossible to determine precisely the functional meaning of rearrangements described at the structural level. The exception is the secretory cells of a root cap, where a decrease in dictyosome volume in a cell and the absence of hypertrophy are clearly correlated with a visible reduction of slime quantity in microgravity (Moore *et al.*, 1987a; Tairbekov *et al.*, 1988) and under clinostating (Kordyum, 1989).

Taking into consideration the role of the Golgi apparatus in the synthesis and transfer of polysaccharides, including those in the cell wall, it is highly probable that there is a connection between the available data and changes in the polysaccharide composition in cell walls in altered gravity. It is possible that structural rearrangements of the dictyosomes affect other functions of the Golgi apparatus, in particular the assembly, growth, and specialization of the membranes, which can result in disturbances of the dynamic equilibrium between metabolic phases in the cells as well as the intermembrane contacts; synapse formation is sensitive to a gravity vector (Gruener and Hoeger, 1991).

In orbital flight the statocytes show an increase in endoplasmic reticulum volume and some changes in cistern topography (Volkmann *et al.*, 1986).

The ER membrane clusters are spherical or ellipsoid (Moore et al., 1987b) and there is a change in the density of the cistern content (Sytnik et al., 1984). Under clinostating there is an increase in agranular ER cistern vesiculation (Kordyum, 1989). In altered gravity, the statocytes do not fulfill their graviperception function; thus, the structure and distribution of ER elements as a part of the cell gravisensory system are naturally changed.

A decrease in granular ER volume, cistern swelling, and dilation, an increase in agranular ER volume, and changes in cistern topography were observed in animal cells of different types in spaceflight (Baranska et al., 1991; Baranski et al., 1979; Rokhlenko and Savik, 1981). The ER is known to play a central role in the biosynthesis and transport of enzymes and constitution of proteins, lipids, and carbohydrates. It makes organelle interactions possible and fulfills a reproduction function in biogenesis of vacuoles and microbodies. The degree of granular ER development in a cell and the ribosome concentration in the hyaloplasm, according to generally accepted notions, determines the functional activity of the protein-synthesizing apparatus. Therefore, the volume, structure, and topography of ER elements can be indicators of a cell's metabolic state and deserve close attention in investigations of the biological effects of microgravity.

Microbodies (peroxisomes), which utilize oxygen in a cell through oxidized reactions (an adaptive function of microbodies is also considered; Belitzer, 1978), do not reveal any significant differences in their structure and topography compared with controls. An increase in a microbody population was observed in the microcallus cells of *D. carota* in microgravity (Klimchuk et al., 1992).

Reserve lipids—an energy store in a cell—are represented by neutral triglycerides and are revealed as droplets of different sizes; their quantity is connected with the cell functional state and age. The data on both increases in lipid droplet volume and decreases (the root statocytes) in most plant and animal cells investigated provide evidence for a change in cell lipid metabolism under altered gravity (Brown et al., 1992; Kordyum et al., 1994; Moore et al., 1987b).

Based on data showing an increased condensed chromatin volume in the nuclei of some plant and animal cells (Baranska et al., 1991; Popova and Kordyum, 1991; Sytnik et al., 1984) the volumes of diffused and condensed chromatin and their ratio can be useful as structural indicators of the functional state of interphase nuclei in microgravity and under clinostating. These data can indicate a decrease in the transcription level in a cell that is in accord, for example, with lowered RNA content in the cells of the shoot apices of 18-day wheat plants in microgravity (Sytnik and Musatenko, 1980). Nevertheless, such an assumption is not universally true because in many spaceflight and clinostat experiments the heterochromatin volume did not increase; the data on biochemical determination of RNA content are completely opposite (Delone et al., 1991).

The rearrangements of the cell organelle ultrastructure in microgravity and under clinostating, viewed by us as indicating the general condition of the cell's functional state, to a significant degree are in accord with the results of electron cytochemical and biochemical analyses (currently they are fragmentary). They provide evidence on the essential changes in energetic, carbohydrate, and lipid metabolism in cells as well as membrane ion transport under these conditions. Of course, to understand the fine mechanisms of metabolic changes and their regulation in cells under the influence of altered gravity, a deep analysis of the rearrangements of the organelle structure and functional load, closely tied to the types of cells and tissues investigated and their growth, differentiation, and vital functions, can be the most informative in the future.

Because the ultrastructural patterns of cell organelles reflect the degree of their functional load, and the organelle's reactions under microgravity are mostly similar in different plant organisms, to analyze the sequence of cell ultrastructure changes in growth and differentiation under altered gravity we chose to use the cells of the root growth zones (Kordyum, 1989). In pea root apices as in other dicotyledons, there are meristematic initials of the cortex and root central cylinder and initials of the columella and peripheral zone of a root cap. The latter are located on the distal part of the root center at rest, the former are on the proximal part. Therefore, the differentiation of meristematic cells proceeds in two directions: to the basal part of the root proper—the elongation zone and differentiation zone, and to the apical part—the zones of differentiating cells, central statenchyma, and secretory cells of a root cap. These two directions of root cell differentiation are manifest in the ultrastructural organization and topography of cell organelles of different root tissues in accordance with their main functions.

An analysis of structural and functional changes in organelles of meristematic, elongating, and differentiated cells under clinostat conditions demonstrated certain prominent trends: (i) There is heterogeneity in the organelles in cell populations with respect to the degree of rearrangements (reflecting the distinctions in organelle biogenesis stages). (ii) At the same time there is the coincidence of a spatial succession in development. It is manifested, for example, in changes in leukoplast structure in the root meristem; part of the plastids take on an ameboid (pleiomorphic) shape (the third stage of plastid development; Whatley, 1983), which is characteristic for leukoplasts in the elongation and differentiation zones. (iii) There is increased reactivity under changes in functional load during cell growth and differentiation; in particular, this is expressed in the increased variability of the Golgi apparatus represented by dictyosomes of three development stages in the elongation zone and in the absence of hypertrophy of dictyosomes in secretory cells of a root cap. (iv) There is enhanced activity when a cell loses its specific functions (replacement of functions). For example, it is seen in the statocytes as activation of the endoplasmic reticulum with progressive vacuolization and an increasing number of dilated granular cisternae with protein content.

These established patterns of structural and functional changes in organelles during cell growth, differentiation, and vital activity under the influence of altered gravity have a common character that is independent of taxonomy or tissue type. At the same time, these changes can provide information on the dis-, hyper-, or hypofunction of various cells, tissues, and organs (Kordyum, 1989).

## B. Cell Differentiation and Acceleration of Aging

In a series of spaceflight and clinostating experiments carried out with growing and developing plants of different degrees of complexity, general data (Claasen and Spooner, 1994; Halstead and Dutcher, 1987; Kordyum *et al.,* 1994; Merkys and Laurinavichius, 1990) have been obtained that show the acceleration of development and aging at cellular and subcellular levels. These include the following:

- The formation in spaceflight of three-cell pollen grains in *Muscari racemosum.* Two-cell pollen grains are characteristic of this species on earth
- An increase in biomass of the unicellular green algae and in the plant cell culture
- An increase in lipid droplet volume
- An increase in cell vacuolization
- Premature disturbance of a chloroplast structure and accumulation of osmiophilic droplets in the cytoplasm of leaf mesophyll cells
- The appearance of leukoplasts of an elongated and branched form in root meristem cells that is characteristic for cells in the root elongation zone under stationary growth conditions
- The growth of root meristem cells by elongation and rhizoderm differentiation that begins spatially nearer to the root dormant center
- An increase in the volume of condensed chromatin in nuclei
- Enhanced activity of hydrolytic enzymes—amylase and the cellulose–pectolytic complex of the cell wall
- An increase in respiration rate (in several experiments); an increase in the mitochondrion volume per cell and enhancement of succinate dehydrogenese activity as well as an expansion of its multiple molecular forms

- Enhanced lipid peroxidation intensity; a tendency toward a continuous increase of lipofluorescent compound under the prolonged influence of altered gravity.

It is known that aging is a part of the integral, genetically determinated processes of growth and development, taking place in time and space, during which an organism's structural and functional characteristics change.

It is generally believed that these processes are based on changes in cell competence. Taking into account the fact that one of the most important criteria of physiological cell competence is its compartmentalization as determined by the adequate stability of barrier membrane functions (Van Steveninck, 1976), two tendencies in the aging processes are noted: (1) (i) irreversible alterations in membrane permeability and loss of compartmentalization, i.e., true aging (for example, the senescence and fall of leaves), and (ii) the further elaboration of physiological competence by means of stimulation of protein synthesis, including new genetic information and acquisition of new physiological and biochemical functions, i.e., adaptive aging or adaptation. *In situ,* it occurs as a part of natural cell differentiation (for example, acquisition by meristematic cells of features characteristic of various mature cells of a root and a stem).

Analysis of changes reported in the structural and functional organization of plant cells induced by microgravity and clinostating shows that both adaptive and real aging processes are stimulated. Cell differentiation in ontogenesis of multicellular organisms, namely plants, may be defined as a process during which genetically determined meristematic cells acquire a certain structural and functional organization that results in the formation of different types of tissues. Alterations in the structure of cell organelles reflect the degree of their functional load and hence they are much alike in the process of differentiation. However, a certain sequence of ultrastructural rearrangements and their combinations leads to a specific structural cell organization that is characteristic of differentiated cells in different tissues (E. Kordyum *et al.,* 1980). A root is a good model for demonstrating the spatial sequence of growth zones—meristem, elongation, and differentiation. Therefore, acceleration of plant cell differentiation has been shown with this model. Acceleration of adaptive aging on the structural level manifests itself in the regularity of the coincidence of spatial and time sequences in development; that is, cells that are less differentiated (growing) or meristematic gain features characteristic of the more differentiated cells. Acceleration of the process of differentiation leads to accelerated development of the entire organ and reduction of its ontogenesis, i.e., earlier maturity and true senescence. The root's main meristem ceases to function earlier, compared with controls, and therefore root growth due to cell division and its ontogenesis is shortened.

The loss of physiological competence (true senescence) manifests itself in a rapid decrease in the rate of synthesis of nucleic acids and proteins; an increase in lytic enzyme activity, respiration rate, and metabolic transport; and loss of weight. Natural senescence is characterized by an increase in membrane permeability; that is, decreasing compartmentalization and lysis of the middle plate of the cell wall. Free ribosomes and ribosomes attached to ER membranes break down in the cell aging process. In particular there is the loss of chlorophyll and destruction of the chloroplast structure, including a decrease in the volume of the thylakoid system, thylakoid swelling, accumulation of plastoglobuli, the presence of membrane elements of various forms, an increase in the volume of elements of the agranular ER with subsequent transformation of the latter into vacuolar structures, gradual loss of nucleoli and chromatin, and transparency of nucleoplasm (Gamaley and Kulikov, 1978).

Features of early aging observed in microgravity are manifested clearly in the ultrastructure of leaf mesophyll cells, in which expansion and vesiculation of intergrana thylakoids, transparency of chloroplast stroma, and an abrupt increase in the number of plastoglobuli occur in chloroplasts not in contact with mitochondria and peroxisomes. Lipolysis of organelles, the presence of membrane elements of various forms, and numerous osmiophilic drops of various sizes are seen (Gamaley and Kulikov, 1978). The induction of lipid peroxidation, especially an increase in such products as lipofluorescent compounds, is evidence of accelerated aging in microgravity. When lipid peroxidation proceeds at a low stationary level, it is one of the ways unsaturated fatty acids are metabolized; an elevation of lipid peroxidation level leads to the development of free radical pathology. One of the characteristic features of this pathology is premature aging of an organism, which results from the formation of toxic cell products—free radicals, unstable hydroperoxides, and active carbonic compounds (Vladimirov, 1987). The accelerated aging observed in *Drosophila melanogaster* and *Tribolium confusum* in spaceflight may be explained by the theory of free radical damage of mitochondria (Miquel, 1986; Miquel *et al.*, 1979). It is assumed that microgravity influences the development and aging of the flies by an increasing oxygen consumption, i.e., respiration (because since in spaceflight and during clinostating adult flies change their behavior), which changes the rate of development, including its final stage—aging. Numerous data obtained in the majority of the experiments conducted (but not in all of them) also show an increase in the functional load of plant cell mitochondria in microgravity and during clinostating. Contrary to the assumption that cell energy expenditure should be decreased because there is no gravity vector, i.e., in an organism the sense of mass itself in the dormant stage is absent as well as the necessity to overcome it in motion (Planel *et al.*, 1982; Tairbekov *et al.*, 1980), most of the data presented show the opposite, at least at certain stages of ontogenesis of the organism. It is likely that the energy increase maintains intracellular homeostasis, i.e., cellular functioning, and thus the adaptation to changing gravity through enhanced autophagous catabolic processes leads to acceleration of differentiation and, finally, cell aging.

It should be noted that although there are still many questions concerning the mechanisms by which microgravity and clinostating affect organisms of various levels of complexity, the acceleration of (i) growth and differentiation of meristematic cells and (ii) the accelerated aging of differentiated mature cells are well demonstrated. Increased metabolism in microgravity leads to an acceleration of aging and, therefore, to a shortening of the meristematic activity and ontogenesis of the organism.

Acceleration of growth (propagation rate) and an increase in population density in spaceflight have been demonstrated in the majority of experiments with unicellular eukaryotic and prokaryotic organisms in addition to *Chlorella*. Thus, the acceleration of agamic propagation was shown in the infusorian *Paramecium tetraurelia* (Planel *et al.*, 1990; Richoilley *et al.*, 1986; Tixador *et al.*, 1981). The use of an on-board centrifuge with 1*g* made it possible to prove conclusively that an increase in cell proliferative activity occurs under the influence of microgravity. Shortening of the cell cycle and, as a result, the acceleration of agamic propagation, has been shown in another infusorian species, *Tetrahymena pyriformis* (Irlina *et al.*, 1989). An increased growth rate and biomass has been described in bacteria, namely *Salmonella typhimurium* (Mattoni *et al.*, 1971), *Proteus vulgaris* (Sytnik *et al.*, 1983b), and *Bacillus subtilis* (Mennigmann and Lange, 1986). Results obtained in space experiments with unicellular algae, protozoa, and bacteria confirm the patterns established early in *P. vulgaris* (Sytnik *et al.*, 1983b)—under optimal conditions the growth of microorganisms is more intensive in spaceflight than in ground controls.

## C. Cell Adaptation Strategy in Microgravity

Microgravity is an abnormal environmental condition that played no role in the evolution of living systems. Nevertheless, the chronic effect of microgravity as an unfamiliar factor does not prevent the development of adaptive reactions at the cellular level under certain temporal ranges.

It is known that cells of a multicellular organism not only take part in reactions of the organism but also carry out processes that maintain their integrity. An analysis of different characteristics and reactions of living systems suggests that only at the cellular level and at the level of the organism is there an independent biological importance, in contrast to all other intraorganism levels (Shmalgauzen, 1940).

In light of these principles, the problem of identification of biochemical, physiological, and structural patterns that can have adaptive significance at the cellular and subcellular level in microgravity has attracted attention. Alterations of enzyme activity—both increases (this tendency prevails with enzymes of catabolic processes) and decreases—have been demonstrated. An unstable balance between anabolism and catabolism at the initial stages of the microgravity effect has been reported, with catabolic processes prevailing in later stages (Yuganov et al., 1974).

Changes in metabolism during cell division, growth, and differentiation under conditions of altered gravity result in structural alterations of organelles, whose general pattern is to enhance the organelle's functional load and activity when a cell loses its specific functions (Kordyum, 1989). An increase of endopolyploid cells in the root at earlier stages of its growth under clinostating than in controls (Zaslavsky and Danevich, 1983) can have an adaptive significance.

The data on biochemical content and enzyme activity in plants under the influence of altered gravity clearly indicate changes in carbohydrate and lipid metabolism and energy metabolism as a whole (see Section III). Alterations in carbohydrates are often considered universal in the process of adaptation of organisms to various environmental factors (Udovenko, 1979). A clear interaction between the rate of utilization of starch and sugar and the level of respiration in cells has been shown. Considering the role of calcium as a secondary messenger of transfer for endogeneous and exogeneous signals to intracellular systems (Hepler and Wayne, 1985; Moore and Evans, 1986) and changes in its concentration and form in microgravity and during clinostating (Kordyum et al., 1994), it is likely that the increased activity of a number of hydrolytic and synthetic enzymes is connected with alterations in calcium ion concentration in the cytosole and the cell wall under these conditions. It has been found that altered gravity belongs to environmental stimuli that are directly correlated with the functioning of the $Ca^{2+}$ second messenger system in cells (Kordyum and Danevich, 1995).

The enhancement in synthesis of amorphous callose glucan during clinostating testifies to enzymatic adaptation, reflected in the induction in 1,3-$\beta$-glucan synthase synthesis (Nedukha et al., 1988). It is known that callose participates in cell responses to external effects and an acceleration of its synthesis may serve as a criterion of cell adaptability to various effects. It has been noted that phosphatidic acid content increases with changes in gravity and this in turn indicates the activation of degradation processes and phospholipid biosynthesis during clinostating. Increased microviscosity of the plasmalemma at an early stage of clinostating followed by its decrease to the initial level has been found using the membrane fraction enriched by plasmalemma and liposomes obtained from lipids of the pea root plasmalemma. Increased total content of unsaturated fatty acids in the plasmalemma and hence the unsaturation index (Polulyakh, 1988; Polulyakh *et al.*, 1989) may be considered one of the mechanisms maintaining fluidity of the lipid bilayer of the plasmalemma within certain limits during clinostating (Shinitzky, 1984).

Our data clearly indicate that microgravity and clinostating induce rearrangements of the structural and functional membrane organization, on which a series of sequential changes of plant cell metabolism is based (Kordyum, 1993; Kordyum *et al.*, 1994). Changes in oxidative homeostasis reflected in lipid peroxidation intensity and antioxidant system activity also occur in these conditions. Indices of lipid peroxidation decrease in the first stages of clinostating but in further stages exhibit a stable tendency to a slight but continuous increase during prolonged experiments. At the same time, indices of microviscosity have an oscillatory character, with a progressive decrease in the amplitude of oscillations. The dynamics of these events demonstrated that the adaptation occurs on the principle of self-regulating systems. However, the return to and the maintenance within normal limits of the most important indices of cell homeostasis are carried out against a background of an accelerated aging process.

Experimental data have shown that increased metabolism in animals and plants is an unspecific cell response to various stress effects, leading to a shortening of the life span (Elstner and Osswald, 1994; Gamaley and Kulikov, 1978; Udovenko, 1979). At the same time, an increase in energy expenditures and enhanced hydrolytic and often (but not always) synthetic processes are related to physiological adaptation. In microgravity, the maintenance of a fluid plasmalemma lipid bilayer within a certain range, which is demonstrated by an increase in the fatty acid index and the activation of antioxidant systems (microviscosity and oxidative homeostasis), is of strategic importance in primary adaptation of the cell to microgravity. Prolonged or secondary adaptation to microgravity is provided by the intensification of the metabolic processes involved in cell activity.

| *Primary adaptation* | *Secondary adaptation* |
|---|---|
| Maintenace of fluidity of the cytoplasmic membrane lipid bilayer | Intensification of cell metabolism |
| Increase in the unsaturated fatty acid content | Increase in cell organelle functional loads |
| Activation of antioxidant systems | Changes in enzyme activity |

## D. Cytoplasmic Membrane as a Possible Primary Site of Microgravity Effects

During the past 15–20 years it has been shown in space and clinostat experiments that essential rearrangements in cell structural organization and the vital functions of bacteria, protozoa, plants, fungi, animals, and man occur *in vivo* and *in vitro* under the influence of altered gravity (Claasen and Spooner, 1994; Cogoli *et al.,* 1984, 1990; Gmünder and Cogoli, 1988; Kessler, 1992b; Planel *et al.,* 1990; Rijken *et al.,* 1992). Naturally, the question of the possible mechanisms of the biological effects of microgravity is not separate from the question of how the cell perceives gravity. It should be noted that this question touches on various cells not specialized for perception of the gravity vector. The structure of graviperceptive organs is very diverse, but gravireceptors are well known: These are statoliths of different types that change their position in the direction of the gravity vector and thus initiate the next steps of the gravitational response. The intracellular statoliths clearly correspond to indices of the primary receptor (Briegleb, 1984; Briegleb and Block, 1986). The structural and functional organization of graviperceptive organs and cells is determined genetically.

In most cell types not specialized for perception of gravity, the primary receptors are not clearly defined. Therefore, the question of the mechanisms of gravity and, consequently, the action of microgravity at the cellular level, is open. Currently, there are certain assumptions concerning possible cell gravity-reactive systems as well as cell interactions with the changed environment of microgravity. Among them, the concept of positional homeostasis suggested by Nace (1983) explains the fixed stable position and optimal cell orientation in the gravitational field as a state of mechanical stress of the cytoskeletal elements and that maintains cell membrane integrity; this state is supported by energy expenditures. The cytoskeleton is considered by Tairbekov (1991) an integral unspecialized cell gravireceptor.

In the opinion of Schatz *et al.* (1992, 1994), gravity can affect membrane processes through interaction with the interfacial layer and this structure can amplify and transform the weak mechanical response to a measurable degree. Gravitational and microgravitational effects at the cellular level are considered by Cogoli *et al.* (1990) as the combined results of both the direct influence of gravity (owing to mass in cell organelles) and the indirect effects (as the consequence of decreasing or absent convection in microgravity), as well as a relaxation of intracellular contact and cell adhesion to the substrate. Kovalenko (1974) suggests a possible interaction of gravitational fields with electromagnetic oscillations and an effect on intracellular processes that occurs at the level of molecules, atoms, and electrons.

To explain the direct influence of such relatively weak forces as gravity at the cellular and molecular levels, bifurcation is suggested as a state in which an unlinear thermodynamic system in disequilibrium is compelled to choose between two possible paths. As a result, such a system becomes exquisitely sensitive to external influences; even a slight change is favorable to one of the states. Assuming that this is the case, a number of researches have discussed the possibility of changes in intracellular processes based on the interaction of molecular systems of unbalanced unlinear types as a result of the direct influence of microgravity (Kondepudi and Storm, 1992; Mesland, 1992; Todd, 1992). Mesland views gravity as interacting with living systems at the levels of gene sensors and directly at the level of intracellular dynamic processes. Because this factor is permanent, an organism's genome contains the information determining the organism's existence in the gravitational field as "encoded information."

In regard to the possible mechanisms by which microgravity operates at the cellular and subcellular levels, we think that the structure and functions of the cell are genetically programmed to take into account ecological factors, including the action of gravity. In other words, cell responses to the permanent influence of gravity (sensation of mass, sedimentation, and hydrostatic pressure) are carried out by gene expression in the cell itself with its complex molecular mechanisms of metabolism An informational role of gravity is also possible because a decrease in gravity to nearly zero does not prevent cell growth and differentiation.

In the literature there are some views of the cell as a mechanical construction and a chemical reactor; such a cell constantly resists gravitational forces or the sedimentation of intracellular particles under the influence of gravitational vectors. Based on these views, the concepts of the possible mechanisms of action of microgravity are very simplified in light of current knowledge of cell biology.

Now we have new information on the interaction of cytoskeleton elements with membranes, their participation in cytoplasmic streaming and cell organelle motion, and the intracellular transport of substances as well as the dynamics of these processes (Bray, 1992; Kreis and Vale, 1993; Wyatt and Carpita, 1993). This information precludes the random distribution of organelles and molecules in a cell or their sedimentation under the influence of a gravitational vector. Thus, for example, in microtubules and microfilaments, there is temporary immobilization of multienzyme complexes. The successive transmutations of substrates are attended by their transport along the channel formed by enzymes (Turkina *et al.,* 1995).

The concept of "positional homeostasis" is based on an assumption of the cell strength increasing in the gravitational field as a result of two opposing processes: the random distribution of intracellular inclusions and their tendency for sedimentation in a cell. That is why this concept, which has focused the attention of investigators on the role of the cytoskeleton in cell graviresponses, does not reflect the real spatial organization of a cell in a whole organism.

Finally, there is affirmation that diffusion as a free and gravitationally independent transfer of substances through biological membranes is necessary for intracellular processes to occur (Tairbekov and Parfenov, 1983; Tairbekov, 1991). However, this affirmation is not in agreement with the absence of diffusion in a pure state in living systems because the diffusion of electrolytes through a membrane is determined by the membrane charge potential. In physical chemistry, the transfer of substances through a biological membrane is a specific process that is not the usual idea of diffusion because even in the simplest case this process has two steps: the first is the transfer of a substance from a solution on the membrane surface; the second takes place within the membrane. In addition, the transfer of a substance through a membrane depends on membrane structure (Nakagaki, 1991).

In asymmetric membranes, such as the cytoplasmic membrane, transfer processes and potential values acquire considerably more complex character because of the presence of membrane potential gradients, charge distribution, etc. The presence of ion channels and their selectiveness as well as differences in the mechanisms of ion motion in channels suggest energy expenditures to carry out ion transport. Although there are many unclear aspects of water transport through membranes, experiments with artificial membranes and theoretical calculations indicate that it is difficult to attribute water transport to free diffusion (Boyer, 1985).

An analysis of the arrangement of the structural and functional organization of different types of cells under microgravity that is based primarily on changes in cell metabolism allows us to state with conviction that microgravity induces rearrangements in the cytoplasmic membrane's physical–chemical organization. These changes underlie changes in its permeability (primarily ion transport) and the further chain of sequential changes in cell metabolism. This assumption is based on the following: First, hydrostatic pressure and surface tension may interfere with biological processes by changing membrane fluidity (Shinitzky, 1984), which is one the main properties of membranes. Recently, it has been found that there are changes in viscous membrane homeostasis in altered gravity; this conclusion is based on alterations in the indices of cytoplasmic membrane microviscosity (Kordyum *et al.*, 1994). Second, an enhanced lipid peroxidation intensity in altered gravity has been demonstrated (Zhadko *et al.*, 1994). Third, the formation of more complex folds of the cytoplasmic membrane has been observed. Fourth, $Ca^{2+}$-bound sites were found that were connected with the cytoplasmic membrane and changes in $Ca^{2+}$-ATPase activity.

Nevertheless, we are still far from knowing what induces changes and the precise succession of events that occur in the cytoplasmic membrane under the influence of microgravity. Some speculations in this respect assume that under conditions of reduction or the absence of hydrostatic pressure, a change in the surface tension of the cytoplasmic membrane can play an inductory role. The effect of such an inductor increases owing to its heterogeneity over the length of the cytoplasmic membrane. In the gravitational field, surface tension and gravitational force are summed in the direction of the same side and are subtracted in the direction of the opposite sides. In the absence of gravity, only surface tension is present. In clinostat conditions, the resulting action of these two changes is felt at every point of the membrane.

In addition, the events termed "piezoeffects" can also complicate the tension state of the membrane surface. Briefly, they describe a mechanical tension at the level of single molecules that arises and disappears in the cytoplasmic membrane based on cytoplasmic streaming involving cytoskeleton elements. Some data point to the cytoplasmic membrane as a primary site of microgravity effects, i.e., the state and permeability of the cytoplasmic membrane are gravidependent. Experimental proof of this assumption is given by Hanke (1994) who used a planar lipid bilayer doped with alamethicin and found that its conductance was dependent on the angle of the bilayer with the gravitational vector. This author concludes that changes in the membrane current are not the result of an effect on pore conductance or the membrane-aqueous solution interface but rather result from an interaction of gravity with the pore-forming mechanisms. In addition, it has shown that the ion channel alamethicin is mechanically sensitive and can be used as a model for elucidating the mechanisms underlying the mechanical sensitivity in tension-sensitive ion channels (Opsahl and Webb, 1994). These results open new approaches to elucidating the primary effects of altered gravity at the cellular level.

## V. Concluding Remarks

Currently, there is considerable information on the phenomenology of the effects of altered gravity at the cellular level and their interpretation based on the discovery of cell gravisensitivity, including plant cells. The concept that microgravity has an essential effect on cell metabolism is among the most important in this area. Modifications of metabolism are reflected in cell ultrastructure rearrangements and occur (in experiments of certain duration) within the framework of the ontogenesis program, which is genetically determined, i.e., within the physiological response limits. Hence, proliferating and metabolizing cells are the most sensitive to the influence of altered gravity.

The available experimental data and theoretical ideas on cell space biology are the basis for further investigations of the structure, reproduction, differentiation, and functioning of plant cells in microgravity. The main focus in these investigations should be on events (i) occurring at the membrane level (physical–chemical properties of cell membranes, especially the cytoplasmic membrane and the tonoplast as well as ion transport) and (ii) providing the transduction of primary microgravity effects in the integrated intracellular processes, the modifications of which are demonstrated using electron microscopic, physiological, biochemical, molecular–biological, and other methods. That is why the most significant questions are those that involve the functioning of second messenger systems in plant cells (including Ca, inositol phospholipids, protein kinase C, and cyclic mononucleotides), gene expression and its regulation at different levels, as well as the state of the phytohormone complex and phytohormonal regulation. As before, investigations of cell organelle topography and organization as well as the cytoskeleton should pay attention to the structural basis of metabolic processes and how cell functions are carried out. Undoubtedly, the photosynthetic apparatus and its functioning is one of the most important.

It should be noted that relatively short space experiments (2–14 days) are effective for studying the effects of microgravity on membrane and intracellular transport, the regulation of gene expression and enzyme activity, cell reproduction and differentiation, and the structural–functional organization of graviperceptive cells. The results of such fundamental studies provide the basis for developing space cell biotechnologies, including plant cell and tissue cultures. However, to estimate a cell's adaptive capacity as well as that of the whole organism and to elucidate the possibility of obtaining successive generations in orbit, prolonged space experiments of up to 2 or more months are necessary. The results of such studies will be the basis for developing space planting technologies in controlled ecological life-support system (CELSS). It should be stressed that a wider use of the ideas and methods of modern cell biology in identifying the tasks and organization of space experiments, in analyzing the experimental material, and in interpreting the results is necessary for understanding how gravity works at the cellular, subcellular, and molecular levels.

## References

Abilov, Z. K.(1986). Adaptative, physiological and morphological changes in chloroplasts of plants different periods of time cultivated at "Salyut-7" station. *26th COSPAR Plenary Meet.*, Toulouse, France, p. 301.

Abilov, Z. K., Alekperov, U. K., and Mahinsky, A. L.(1985). The morphological and functional state of a photosynthetic apparatus in cells of plants different periods of time cultivated in space flight conditions. *In* "Mikroorganismi v Iskusstvennikh Ekosistemakh" (I. I. Gitelson, Ed.), pp. 29–32. Nauka, Novosibirsk, Russia.

Albrecht-Buehler, G. (1992). The simulation of microgravity conditions on the ground. *ASGSB Bull.* **5**(2), 3–10.

Aliev, A. A., Abilov, Z. K., Mashinsky, A. L., Ganieva R. A., and Ragimova, G. K. (1985). Ultrastructural and some physiological peculiarities of the photosynthetic apparatus in pea cultivated for 29 days on board the orbital station "Salyut-7." *Izv. Acad. Nauk AzerbSSR Ser. Biol. Nauk* **6**, 18–23.

Aliev, A. A., Mashinsky, A. L., Alekperov, U. K., Ganieva, R. A., and Ragimova, G. K. (1986). Studies of the ultrastructure in plants cultivated in space flight conditions. *Izv. Acad. Nauk AzerbSSR,* Ser. Biol. Nauk **4,** 3–7.

Antonjan, A. A., Meleshko, G. I., Naidina, V. P., and Sychev V. N. (1992). Characteristics of the fatty acid composition of lipids in algae in weightlessness. *In* "Resultati Issledovanii na Biosputnikakh" (O. G. Gazenko, Ed.), pp. 391–394. Nauka, Moscow.

Babsky, V. G., Kopachev, N. D., and Mishkis, A. D. (1976). "Gidromekhanika Nevesomosti." Nauka, Moscow.

Bakeeva, L. E. and Chenzov, Yu. S. (1989). Mitochondrial reticulum: Structure and some functional properties. *In* "Obschie Problemi Biologii" (P. B. Petrov, Ed.), pp. 3–102. VINITI, Moscow.

Baker, R., Baker, B. L., and Elliot, L. (1979). Experiments with carrot crown gall tissue: Development and anatomy of tumor. *In* "Biologicheskie Issledovaniya na Biosputnikakh 'Kosmos' " (E. A. Ilyin and G. P. Parfenov, Eds.), pp. 137–142. Nauka, Moscow.

Baranska, W., Skopinski, P., Kujawa, M., and Prodan, N. G. (1991). Morphometric evaluation of adrenal cells of rats exposed for 7 and 13 days on the biosputniks 1667 and 1887. *In* "Biosatellites 'Cosmos' " (A. I. Grigorjev, Ed.), p. 136. Institute of Biomedical Problems, Moscow.

Baranski, S., Baranska, V., and Savina, E. A. (1979). Morphometric studies of an adrenal cortex at the ultrastructural level. *In* "Vliyanie Faktorov Kosmicheskogo Poljota na Organismi Zhivotnikh" (O. G. Gazenko, Ed.), pp. 67–72. Nauka, Moscow.

Barmicheva, E. M., Grif, V. G., and Tairbekov, M. G. (1989). Growth and structure of cells in the maize root apex under space flight conditions. *Cytology* **31,** 1324–1328.

Belitzer, N. V. (1978). "Lisosomnaya Sistema i Mikrotelza v Kletkakh Rastenii i Zhivotnikh." Sc. D. thesis, Institute of Cytology of the Akad. Nauk SSSR, Leningrad.

Belousov, L. V., Luchinskaya, N. N., and Ostroumova, T. V. (1991). Reaction of newt oocytes at space flight conditions. *In* "Biosatellites 'Cosmos' " (A. I. Grigorjev, Ed.), pp. 21–22. Institute of Biomedical Problems, Moscow.

Belyavskaya, N. A. (1984). Ultrastruktura i zitohimiya statozitov kornei gorokha v norme, pri klinostatirovanii i v usloviyakh gipogravitazii. Ph.D. thesis, Moscow State University, Moscow.

Belyavskaya, N. A. (1992). The function of calcium in plant graviperception. *Adv. Space Res.* **12**(1), 83–91.

Björkman, O. (1981). Responses to different quantum flux densities. *Encycl. Plant Physiol. New. Ser.* **12A,** 57–107.

Boyer, J. S. (1985). Water transport. *Annu. Rev. Plant Physiol.* **36,** 473–516.

Braun, M., and Sievers, A. (1994). Role of the microtubule cytoskeleton in gravisensing *Chara* rhizoids. *Eur. J. Cell Biol.* **63,** 289–298.

Bray, D. (1992). "Cell Movements." Garland, New York.

Briegleb, W. (1983). The clinostat—A tool for analysing the influence of acceleration on solid–liquid system. *Proc. Workshop Space Biol.* Cologne, Germany, pp. 97–101.

Briegleb, W. (1984). Acceleration reactions of cells and tissues—Their genetic–phylogenetic implications. *Adv. Space Res.* **4**(12), 5–7.

Briegleb, W. (1992). Some qualitative and quantitative aspects of the fast-rotating clinostat as a research tool. *ASGSB Bull.* **5**(2), 23–32.

Briegleb, W., and Block, I. (1986). Classification of gravity effects on "free" cells. *Adv. Space Res.* **6**(12), 15–19.

Brown, A. H., and Chapman, D. K. (1984a). Circumnutation observed without a significant gravitational force in space flight. *Science* **225,** 230–232.

Brown, A. H., and Chapman, D. K. (1984b). Experiments on plants grown in space: A test to verify the biocompatibility of a method for plant culture in a microgravity environment. *Ann. Bot.* **54**(Suppl. 3), 19–31.

Brown, A. H., Chapman, D. K., and Lin, S. W. W. (1974). A comparison of leaf epinasty induced by weightlessness or by clinostat rotation. *BioScience* **24**, 518–520.
Brown, A. H., Chapman, D. K., Heathcote, D., and Johnson, A. (1992). Plants tropistic responses in zero gravity. *World Space Congr.*, Washington, DC, p. 530.
Brown, C. S., Piastuch, W. C., and Knott, W. M. (1992). Soybean cotyledon starch metabolism is sensitive to altered gravity conditions. *World Space Congr.*, Washington, DC, p. 526.
Brown, C. S., Obenland, D. M., and Musgrave, M. E. (1993). Space flight effects on the growth, carbohydrate concentration and chlorophyll content in *Arabidopsis*. *ASGSB Bull.* **7**(1), 83.
Buchen, B., Braun, M., Hejnowicz, Z., and Sievers, A. (1993). Statoliths pull on microfilaments. Experiments under microgravity. *Protoplasma* **172**, 38–42.
Butenko, R. G., Dmitrieva, N. N., Ongko, V., and Basyrova, L. B. (1979). Influence of weightlessness on somatic embryogenesis. *In* "Biologicheskie Issledovaniya na Biosputnikakh 'Kosmos'" (E. A. Ilyin and G. P. Parfenov, Eds.), pp. 118–125. Nauka, Moscow.
Chaban, K. I. (1991). Gravisensitivity of the *Pottia intermedia* protonema. *Dopov. Acad. Nauk UkrRSR Ser. Biol.* **12**, 122–125.
Chaban, K. I., Smertenko, A. P., Ripezky, R. T., and Blum, Ja. B. (1995). Studies of the role of microtubules in gravitropism of moss protonema cells. *Cytol. Genet.* **29**, 3–11.
Cherevchenko, T. M., and Mayko, T. K. (1983). Some results of studies of the effects of weightlessness on the growth of epiphytic orchids. *Visnyk Acad. Nauk UkrSSR* **1**, 31–35.
Cherevchenko, T. M., Mayko, T. K., Bogatyr, V. B., and Kosakovskaya, I. V. (1986). Prospects of utilization of tropic orchids for the space studies. *In* "Kosmicheskaya Biologiya i Biotechnologiya" (K. M. Sytnik, Ed.), pp. 41–54. Naukova Dumka, Kiev.
Claasen, D. E., and Spooner, B. S. (1994). Impact of altered gravity on aspects of cell biology. *Int. Rev. Cytol.* **156**, 301–373.
Cogoli, A., Tschopp, A., and Fuchs-Bislin, P. (1984). Cell sensitivity to gravity. *Science* **225**, 228–230.
Cogoli, A., Bechler, B., and Lorenzi, G. (1990). Response of cells to microgravity. *In* "Fundamentals of Space Biology" (M. Asashima and G. M. Malacinski, Eds.), pp. 97–111. Springer-Verlag, Berlin.
Cogoli, M. (1992). The fast rotating clinostat: A history of its use in gravitational biology and a comparison of ground-based and flight experiment results. *ASGSB Bull.* **5**(2), 59–68.
Conrad, H. M. (1968). Biochemical changes in the developing wheat seedling in weightless state. *BioScience* **18**, 645–649.
Conrad, H. M. (1971). A study of the effect of weightlessness on the biochemical response of a monocotyledonous seedling. *In* "The Experiments of Biosatellite II" (J. F. Saunders, Ed.), pp. 189–212. NASA, Washington, DC.
Cowles, J. R., Scheld, H. W., Lemay, R., and Peterson, C. (1984). Growth and lignification in seedlings exposed to eight days of microgravity. *Ann. Bot.* **54**(Suppl.3), 33–48.
Cowles, J. R., Lemay, R., Omran, R., and Jahns, G. (1986). Cell wall related synthesis in plant seedlings grown in the microgravity environment of the space shuttle. *Plant Physiol.* **80**(Suppl.4), 9.
Cowles, J. R., Lemay, R., and Jahns, G. (1988). Microgravity effects on plant growth and lignification. *Astrophys. Lett.* **27**, 223–228.
Danilova, M. F. (1974). "Structurnii Osnovi Poglocheniya Veschestv Kornem." Nauka, Leningrad.
Darbelley, N., (1988). Effects de la stimulation gravitropique et de la microgravitŭ sur la prolifŭration et la differŭnciation cellulaires dans les racines primaires. *Bull. Soc. Bot. Fr.* **135**, 229–250.
Darbelley, N., Driss-Ecole, D., and Perbal, G. (1986). Differénciation et prolifération cellulaires dans les racines du maïs cultivé en microgravité (Biocosmos, 1985). *Adv. Space Res.* **6**, 157–160.

Darbelley, N., Driss-Ecole, D., and Perbal, G. (1989). Elongation and mitotic activity of cortical cells in lentil roots grown in microgravity. *Plant Physiol. Biochem.* **27,** 341–347.

Delone, N. L., Popovich, P. R., Antipov, V. V., and Visotsky, V. G. (1963). Effect of space flight conditions on microspores of *Tradescantia paludosa* on board the spaceships-satellites "Vostok-3" and "Vostok-4." *Kosm. Issled.* **1,** 312–325.

Delone, N. L., Bikovsky, V. F., Antipov, V. V., Visotsky, V. G., and Rudneva, N. A. (1964). Initiation of the disturbances in mitosis in microspores of *Tradescantia paludosa* under the influence of different duration of space flight on-board the spaceship-satellite "Vostok-6." *Dokl. Acad. Nauk SSSR Ser. Biol.* **159,** 439–441.

Delone, N. L., Rudneva, N. A., and Antipov, V. V. (1965). Effect of space flight conditions on board the spaceships-satellites "Vostok-5" and "Vostok-6" on the chromosomes of primary roots in seeds of some higher plants. *Kosm. Issled.* **3,** 480–487.

Delone, N. L., Egorov, B. B., and Antipov, V. V. (1966). Effect of space flight factors on board the spaceship-satellite Voskhod on the microspores of *Tradescantia paludosa. Kosm. Issled.* **4,** 156–161.

Delone, N. L., Trusova, A. S., Morozova, E. M., Antipov, V. V., and Parfenov, G. P. (1968). Effect of space flight on board the satellite "Cosmos-110" on the microspores of *Tradescantia paludosa. Kosm. Issled.* **6,** 299–303.

Delone, N. L., Antipov, V. V., and Voronkov, Yu. I. (1991). Influence of weightlessness on gene expression. *In* "Biosatellites 'Cosmos' " (A. I. Grigorjev, Ed.), pp. 34–35. Institute of Biomedical Problems, Moscow.

Driss-Ecole, D., and Perbal, G. (1989). Importance of the 1g controls in interpreting the results of an experiment on plant gravitropism. Preprint of the 40th Congress of IAF, Malaga.

Driss-Ecole, D., Schoevaert, D., Noin, M., and Perbal, G. (1994). Densitometric analysis of nuclear DNA content in lentil roots grown in space. *Biol. Cell.* **81,** 59–64.

Dubinin, N. P., and Vaulina, E. N. (1976). Evolution and gravity. *Problemi Kosmicheskoi Biol.* **33,** 7–17.

Dubinin, N. P., Glembotsky, Yu. L., and Vaulina, E. N. (1977). Biological experiments on the orbital station Salyut-4. *Life Sci. Space Res.* **15,** 267–272.

Elstner, E. F., and Osswald, W. (1994). Mechanisms of oxygen activation during plant stress. *Proc. R. Soc. Edinburgh.* **102B,** 131–154.

Gallegos, G. L., Hilaire, E. M., Peterson, B. V., and Guikema J. A. (1995a). Effects of microgravity and clinorotation on stress ethylene production in two starchless mutants of *Arabidopsis thaliana. J. Gravit. Physiol.* **2,** 153–154.

Gallegos, G. L., Odom, W. L., and Guikema, J. A. (1995b). Effect of microgravity on stress ethylene and carbon dioxide production in sweet clover (*Melilotus alba* L.). *J. Gravit. Physiol.* **2,** 155–156.

Gallegos, G. L., Peterson, B. V., Brown, C. S., and Guikema, J. A. (1995c). Effects of stress ethylene inhibitors on sweet clover (*Melilotus alba* L.) seedling growth in microgravity. *J. Gravit. Physiol.* **2,** 151–152.

Gamaley, Yu. V., and Kulikov, G. V. (1978). "Rasvitie Khlorenkhimi Lista." Nauka, Leningrad.

Gavrilova, O. V., Gabova, A. V., Goryainova, L. N., and Philatova, E. V. (1991). Chlamydomonas is on board the biosatellite "Cosmos 2044." *In* "Biosatellites 'Cosmos' " (A. I. Grigorjev, Ed.), pp. 30–31. Institute of Biomedical Problems, Moscow.

Gazenko, O. G., Ilyin, E. A., and Parfenov, G. P. (1974). Space biology (some results and prospects). *Izv. Acad. Nauk SSSR* **4,** 461–475.

Gmünder, F. K., and Cogoli, A.(1988). Cultivation of single cells in space. *Appl. Microgravity Technol.* **1**(3), 115–122.

Gray, S. W., and Edwards, B. F. (1968). The effect of weightlessness on wheat seedling morphogenesis and histochemistry. *BioScience* **18,** 638–655.

Gray, S. W., and Edwards, B. F. (1971). The effect of weightlessness on the growth and orientation of roots and shoots of monocotyledonous seedlings. In "The Experiments of Biosatellite II" (J. F. Saunders, Ed.), pp. 123–165. NASA, Washington, DC.
Gruener, R. (1992). Introduction. *ASGSB Bull.* **5**(2), 1–2.
Gruener, R., and Hoeger, G. (1991). Vector averaged gravity alters myocyte and neuron properties in cell culture. *Aviat. Space Environ. Med.* **62**, 1159–1165.
Häder, D.-P. (1994). Gravitaxis in the flagellate *Euglena gracilis*—Results from Nizemi clinostat and sounding rocket flights. *J. Gravit. Physiol.* **1**, 82–84.
Häder, D.-P., Rosum, A., Schafer, J., and Hemmersbach, R. (1995). Gravitaxis in the flagellate *Euglena gracilis* is controlled by an active gravireceptor. *J. Plant Physiol.* **146**, 474–480.
Halstead, T. W., and Dutcher, F. R. (1984). Status and prospects. *Ann. Bot.* **54**(Suppl.3), 3–18.
Halstead, T. W., and Dutcher, F. R. (1987). Plants in space. *Annu. Rev. Plant Physiol.* **38**, 317–345.
Hanchey, P., Baker, B. L., and Baker, R. (1975). Isosime patterns in gravity-compensated crown gall tissue. *Phytopathology* **65**, 1136–1138.
Hanke, W. (1994). Planar lipid bilayers as model systems to study the interaction of gravity with biological membranes. *30th COSPAR Scientific Assembly,* Hamburg, Germany, p. 283.
Hemmersbach-Krause, R., and Briegleb, W. (1994). Behavior of free-swimming cells under various accelerations. *J. Gravit. Physiol.* **1**, 85–87.
Hensel, W. (1984). A role of microtubules in the polarity of statocytes from roots of *Lepidium sativum* L. *Planta* **162**, 404–414.
Hensel, W. (1988). Demonstration by heavy-meromyosin of actin microfilaments in extracted cress (*Lepidium sativum* L.) root statocytes. *Planta* **173**, 142–143.
Hensel, W., and Iversen, T.-H. (1980). Ethylene production during clinostat rotation and effect on root geotropism. *Z.Pflanzenphysiol.* **97**, 343–352.
Hensel, W., and Sievers, A. (1980). Effects of prolonged omnilateral gravistimulation on the ultrastructure of statocytes and on the graviresponse of roots. *Planta* **150**, 338–346.
Hepler, P. K., and Wayne, R. O. (1985). Calcium and plant development. *Annu. Rev. Plant Physiol.* **36**, 397–439.
Hilaire, E., Guikema, J.A., and Brown, C.S. (1995). Clinorotation affects soybean seedling morphology. *J. Gravit. Physiol.* **2**, 149–150.
Höffman, E., Schönherr, K., and Hampp, R. (1994). Regeneration of plant cell protoplasts under microgravity: A D-2 spacelab experiment. *30th COSPAR Scientific Assembly,* Hamburg, Germany, p. 277.
Irlina, I. S., Gabova, A. V., Rajkov, I. B., and Tairbekov, M. G. (1989). Influence of space flight conditions on reproduction speed, morphology of cells, RNA and protein content in infusoria *Tetrahymena pyriformis. Cytology* **31**, 829–838.
Ivanov, V. B. (1987). Cell proliferation in plants. *Itogi Nauki Tekhniki* **5**, 3–216.
Iversen, T.-H., Rasmussen, O., Gmunder, F., Baggerud, C., Kordyum, E. L., Lozovaya, V. V., and Tairbekov, M. G. (1992). The effect of microgravity on the development of plant protoplasts flown on Biokosmos 9. *Adv. Space Res.* **12**(1), 123–131.
Johnson, S. P. (1971). Biochemical changes in the endosperm of wheat seedlings in the weightless state. In "The Experiments of Biosatellite II" (J. F. Saunders, Ed.), pp. 21–221. NASA, Washington, DC.
Johnson, S. P., and Tibbitts, T. W. (1971). The liminal angle of a plagiogeotropic organ under weightlessness. In "The Experiments of Biosatellite II" (J. F. Saunders, Ed.), pp. 223–248. NASA, Washington, DC.
Kessler, J. O. (1992a). The internal dynamics of slowly rotating biological systems. *ASGSB Bull.* **5**(2), 11–22.
Kessler, J. O. (1992b). Theory and experimental results on gravitational effects on monocellular algae. *Adv. Space Res.* **12**(1), 33–42.

Kleinshuster, S. D., and Magon, K. (1979). Experiments with carrot crown gall tissue: Activity of glutaminesynthase. In "Biologicheskie Issledovaniya na Biosputnikakh 'Kosmos'" (E. A. Ilyin and G. P. Parfenov, Eds.), pp. 147–148. Nauka, Moscow.

Kleinshuster, S. J., Baker, B. L., and Baker, R. (1975). Responses of crown gall tissue to gravity compensation. *Phytopathology* **65,** 931–935.

Klimchuk, D. A., and Martyn, G. M. (1994). Structural–functional characteristics of plant cells *in vitro* in microgravity and in clinostation. In "Kosmicheskaya Biologiya i Aerokosmicheskaya Medizina" (A. I. Grigorjev, Ed.), p. 94. Slovo, Moscow.

Klimchuk, D. A., Kordyum, E. L., Danevich, L. A., Tairbekov, M. G., Iversen, T.-H., Baggerud, C., and Rasmussen, O. (1992). Structural and functional organization of regenerated plant protoplasts exposed to microgravity on Biokosmos 9. *Adv. Space Res.* **12**(1), 133–140.

Kondepudi, D. K., and Storm, P. B. (1992). Gravity detection through bifurcation. *Adv. Space Res.* **12**(1), 7–14.

Kordyum, E. L. (1989). Plant cell in the process of the adaptation to simulated microgravity. *Adv. Space Res.* **9**(11), 33–36.

Kordyum, E. L. (1993). Effects of microgravity and clinostating on plants. *G. Bot. Ital.* **27,** 379–385.

Kordyum, E. L. (1994). Effects of altered gravity on plant cell processes: Results of recent space and clinostatic experiments. *Adv. Space Res.* **14**(8), 77–85.

Kordyum, E. L., and Chernyaeva, I. I. (1982). Peculiarities in formation of generative organs of Arabidopsis thaliana (L.) Heynh. under space flight conditions. *Dokl. Acad. Nauk UkrSSR Ser. Biol.* **8,** 67–70.

Kordyum, E. L., and Danevich, L. A. (1995). Calcium balance changes in tip growing plant cells under clinorotation. *J. Gravit. Physiol.* **2,** 147–148.

Kordyum, E. L., and Sytnik, K. M. (1983). Biological effects of weightlessness at cellular and subcellular levels. *Physiologist* **26**(6), 141–142.

Kordyum, E. L., Mashinsky, A. L., Popova, A. F., and Sytnik, K. M. (1979a). Ultrastructure of *Chlorella vulgaris* (strain LARG-1) cells grown for 5-days under space flight conditions. *Dokl. Acad. Nauk UkrSSR Ser. Biol.* **6,** 476–479.

Kordyum, E. L., Popova, A. F., and Mashinsky, A. L. (1979b). Influence of orbital flight conditions on formation of genitals in *Muscari racemosum* and *Anethum graveolens*. *Life Sci. Space Res.* **17,** 301–304.

Kordyum, E. L., Nedukha, E. M., and Sidorenko, P. G. (1980). "Structurno–Funkzionalnaya Kharakteristika Rastitelnoi Kletki v Prozessakh Differenzirovki i Dedifferenzirovki." Naukova Dumka, Kiev.

Kordyum, E. L., Nedukha, E. M., Sytnik, K. M., and Mashinsky, A. L. (1981). Optical and electron-microscopic studies of the *Funaria hygrometrica* protonema after cultivation for 96 days in space. *Adv. Space Res.* **1**(14), 159–162.

Kordyum, E. L., Nedukha, E. M., Popova, A. F., Sidorenko, P. G., Fomicheva, V. M., and Sytnik, K. M. (1983a). Prospects of autotrophic link functioning in biological life-support systems based on cell biology studies. *Acta Astron.* **10**(4), 225–228.

Kordyum, E. L., Sytnik, K. M., and Chernyaeva, I. I. (1983b). Peculiarities of genital organ formation in Arabidopsis thaliana (L.) Heynh. under space flight conditions. *Adv. Space. Res.* **3**(9), 247–250.

Kordyum, E. L., Belyavskaya, N. A., Nedukha, E. M., Palladina, T. A., and Tarasenko, V. A. (1984a). The role of calcium ions in cytological effects of hypogravity. *Adv. Space Res.* **4**(12), 23–26.

Kordyum, E. L., Popova, A. F., and Mashinsky, A. L. (1984b). Utrastructural organization of *Chlorella vulgaris* Beijer cells (strain LARG-1) grown in autotrophic conditions on board the scientific orbital station "Salyut." *Ukr. Bot. Zh.* **50**(1), 30–32.

Kordyum E. L., Sytnik, K. M., Chernyaeva, I. I., Anikeeva, I. D., and Vaulina, E. N. (1984c). Peculiarities of formation of generative organs in *Arabidopsis thaliana* in space flight conditions. *In* "Biologicheckie Issledovaniya na Orbitalnikh Stanziyakh 'Salyut' " (N. P. Dubinin, Ed.), pp. 81–96. Nauka, Moscow.

Kordyum, E. L., Vaulina, E. N. Grechko, G. M., Zhadko, S. I., Kordyum, V. A., Mashinsky, A. L., Nechitailo, G. S., Popova, A. F., and Sytnik, K. M. (1989). "Alterations in Rate of Biological Processes in Microgravity and Clinostating." Naukova Dumka, Kiev.

Kordyum, E. L., Danevich, L. A., Podluzky, A. G., Chuchkin, V. G., Ivanov, V. B., and Philippenko, V. N. (1992). Ultrastructure of the root cap cells in *Impatiens balsamina* L. seedlings. *In* "Resultati Issledovanii na Biosputnikakh" (O. G. Gazenko, Ed.), pp. 313–316. Nauka, Moscow.

Kordyum, E. L., Sytnik, K. M., Belyavskaya, N. A., Zhadko, S. I., Klimchuk, D. A., and Polulyakh, Yu.A. (1994). Sovremennii problemi kosmicheskoi kletochnoi fitobiologii. *Problemi Kosmicheskoi Biologii* **73,** 5–230.

Kordyum, E. L., Baranenko, V. V., Nedukha, E. M., and Samoilov, V. M. (1995). Formation of *Solanum tuberosum* minitubers in microgravity. *Bot. Zh.* **80**(6), 74–80.

Kordyum, V. A., and Gavrish, T. G. (1983). Investigations of the unicellular alga *Chlorella vulgaris* (strain LARG-1) cultivated under space flight conditions. *In* "11-e Vsesoyuzhoe Rabochee Soveschanie po Voprosam Krugovorota Veschestv v Zamknutikh Sistemakh na Osnove Zhiznedeyatelnosti Nizshikh Organismov" (V. A. Kordyum, Ed.), pp. 63–72. Naukova Dumka, Kiev.

Kordyum, V. A., Polivoda, L. V., Manko, V. G., Mashinsky, A. L., Nechitailo, G. S., and Konshin, N. I. (1974). Investigations of peculiarities of the lag-phase of *Chlorella* growth in extremal conditions. *In* "8-e Vsesoyuznoe Rabochee Soveschanie po Voprosam Krugovorota Veschestv v Zamknutikh Sistemakh na Osnove Zhiznedeyatelnosti Nizshikh Organismov" (V. A. Kordyum, Ed.), pp. 175–180. Naukova Dumka, Kiev.

Kordyum, V. A., Shepelev, E. Ya., and Meleshko, G. I. (1980). Biological studies of *Chlorella pyrenoidosa* cultures grown under space flight conditions. *Life Sci. Space Res.* **18,** 213–218.

Kordyum, V. A., Manko, V. G., Mashinsky, A. L., Nechitailo, G. S., and Popova, A. F. (1983). Changes in symbiotic and associative interrelations in a higher plant–bacteria system during space flight. *Adv. Space Res.* **3**(9), 265–269.

Kostina, L. N., Anikeeva, I. D., and Vaulina, E. N. (1986). Experiments with developing plants on board the Salyut-5, Salyut-6, and Salyut-7 orbital stations. *Kosm. Biol. Aviakosm. Med.* **20,** 49–53.

Kovalenko, E. A. (1974). Patho-morphological analysis of the inluence of weightlessness on the organisms. *In* "Nevesomost" (O. G. Gazenko, Ed.), pp. 237–277. Nauka, Moscow.

Kreis, T., and Vale, R. (1993). "Guidebook to the Cytoskeletal and Motor Proteins." Oxford Univ. Press, London.

Krikorian, A. D. (1990). Chromosomes and plant cell division in space. *28th COSPAR Plenary Meet.,* The Hague, The Netherlands, p. 55.

Krikorian, A. D., and O'Connor, S. A. (1984). Experiments on plants grown in space: Karyological observations. *Ann. Bot.* **54**(Suppl.3), 49–63.

Krikorian, A. D., and Steward, F. C. (1978). Morphogenetic responses of cultured totipotent cells of carrot (*Daucus carota* var. *carota*) at zero gravity. *Sci.* **200,** 67–68.

Krikorian, A. D., Dutcher, F. R., Quinn, C. E., and Steward, F. C. (1981). Growth and development of cultured carrot cells and embryos under spaceflight conditions. *Adv. Space Res.* **1,** 117–127.

Kuang, A., Musgrave, M. E., Matthews, S. W., Cummins, D. B., and Tucker, S. C. (1995). Pollen and ovule development in *Arabidopsis thaliana* under spaceflight conditions. *Am. J. Bot.* **82,** 585–595.

Kuang, A., Musgrave, M. E., and Matthews, S. W. (1996). Modification of reproductive development in *Arabidopsis thaliana* under spaceflight conditions. *Planta* **198,** 588–594.

Kurnosenko, I. A. (1975). Ultrastructural patterns of a heart muscle in conditions of ATP action on general hypoxy of an organism. *In* "Ultrastructurnaya Osnova Narushenii Structur i Funkzii Kletki pp.126–137. Tbilisi, Russia.

Kursanov, A. L., and Paramonova, N. V. (1976). Ultrastructural changes in the leaf mesophyll of *Beta vulgaris* in connection with transport of assimilates. *Fiziol. Rast.* **23**, 286–290.

Kutlakhmedov, Yu. A., Sokirko, G. S., Grodzinsky, D. M., Mashinsky, A. L., Nechitailo, G. S., and Konshin, N. I.(1978). Studies of the influence of space flight factors on awakening of *Spirodela polyrrhiza* turions. *Kosm. Issled. Ukr.* **12**, 49–54.

Laurinavichius, R. S., Yaroschus, A. V., Marchukajtis, A., Shvegzhdene, D. V., and Mashinsky, A. L. (1984). Metabolism of pea plants grown under space flight conditions. *In* "Biologicheskii Issledovaniya na Orbitalnikh Stanziyakh 'Salyut' " (N. P. Dubinin, Ed.), pp. 96–102. Nauka, Moscow.

Laurinavichius, R., Rupainene, O., and Stochkus, A. (1994). The structure of graviperceptive cells in embryonal roots of *Hordeum vulgare* L. in microgravity. *In* "Kosmicheskaya Biologia i Aerokosmicheskaya Medizina" (A. I. Grigorjev, Ed.), pp. 92–94. Slovo, Moscow.

Levine, H. G., and Krikorian, A. D. (1992). Shoot growth in aseptically cultivated daylily and *Haplopappus* plantlets after a 5-day space flight. *Physiol. Plant.* **86**, 349–359.

Lorenzi, G., and Perbal, G. (1990a). Actin filaments responsible for the location of the nucleus in the lentil statocyte are sensitive to gravity. *Biol. Cell.* **68**, 259-263.

Lorenzi, G., and Perbal, G. (1990b). Root growth and statocyte polarity in lentil seedling roots grown in microgravity or on slowly rotating clinostat. *Physiol. Plant.* **78**, 532–537.

Lyon, C. J. (1968). Wheat seedling growth in the absence of gravitational force. *Plant Physiol.* **43**, 1002–1007.

Lyon, C. J. (1971). Growth physiology of the wheat seedling in space. *In* "The experiments of Biosatellite II" ( J. F. Saunders, Ed.), pp. 167–188. NASA, Washington, DC.

Marimuthu, K. M., Schairer, L. A., Sparrow, A. H., and Nawrocky, M. M. (1972). Effects of space flight (Biosatellite II) and radiation on female gametophyte development in *Tradescantia*. *Am. J. Bot.* **59**, 359– 366.

Mashansky, V. F., and Rabinovich, I. M. (1987). "Rannii Reakzii Kletochnikh Organel." Nauka, Leningrad.

Mashinsky, A. L., Panova, S. A., and Izupak, E. A. (1984). Scientific equipments for carrying out biological experiments. *In* "Biologicheskii Issledovaniya na Orbitalnikh Stanziyakh 'Salyut' " (N. P. Dubinin, Ed.), pp. 7–17. Nauka, Moscow.

Mashinsky, A. L., Alekhina, T. P., and Bozhko, A. N. (1991). Peculiarities of *Triticum vulgare* at the first phases of development (ontogeny) in space flight conditions. *Kosm. Biol. Aviakosm. Med.* **1**, 39–41.

Mattoni, R. H., Ebersold W. T., and Eiserling, F. A. (1971). Induction of lysogenic bacteria in the space environment. *In* "The Experiments of Biosatellite II" ( J. F. Saunders, Ed.), pp. 309–324. NASA, Washington, DC.

Mauseth, J. O. (1985). Effect of growth rate, morphogenic activity, and phylogeny on shoot apical ultrastructure in *Opuntia polycantha* (Cactaceae). *Am. J. Bot.* **71**, 1283–1292.

McCuley, M., and Hepler, P. K. (1992). Cortical ultrastructure of freeze substituted protonemata of the moss *Funaria hygrometrica*. *Protoplasma* **169**, 168–178.

Mennigmann, H. D. (1989). Response of unicellular organisms to the conditions in low earth orbit. Preprint of the 40th IAF Congress, Malaga.

Mennigmann, H. D., and Lange, M. (1986). Growth and differentiation of *Bacillus subtilis* under microgravity. *Naturwiss.* **73**, 415–417.

Mergenhagen, D. (1986). The circadian rhythm in *Chlamydomonas reinhardii* in a zeitgeberfree environment. *Naturwiss.* **73**, 410–412.

Mergenhagen, D., and Mergenhagen, E. (1987). The biological clock of *Chlamydomonas reinhardii* in space. *Eur. J. Cell Biol.* **43**, 203–207.

Merkys, A. I. (1973). "Geotropicheskaya Reakziya Rastenii." Mintis, Vilnyus.
Merkys, A. I. (1990). "Gravitaziya v Rostovikh Processakh Rastenii." Nauka, Moscow.
Merkys, A. I., and Laurinavichius, R. S. (1983). Complete cycle of individual development of *Arabidopsis thaliana* (L.) Heynh. plants on board the Salyut-7 orbital station. *Dokl. Akad. Nauk SSSR Ser. Biol.* **271**, 509–512.
Merkys, A. I., and Laurinavichius, R. S. (1990). Plant growth in space. In "Fundamentals of Space Biology" (M. Asashima and G. M. Malacinski, Eds.), pp. 64–89. Springer-Verlag, Berlin.
Merkys, A. I., Mashinsky, A. L., Laurinavichius, R. S., Yaroshius, A. V., and Nechitailo, G. S. (1975). The development of seedling shoots under space flight conditions. *Life Sci. Space Res.* **13**, 53–57.
Merkys, A. I., Laurinavichius, R. S., Mashinsky, A. L., Yaroshius, A. V., Savichene, E. K., and Shvegzhdene, D. V. (1976a). Effect of weightlessness and its simulation on growth and morphology of cells and tissues of pea and lettuce seedlings. In "Organismi i Sila Tyazhesti" (A. I. Merkys, Ed.), pp. 238–246. Akad. Nauk LitSSR, Vilnyus, Russia.
Merkys, A. I., Laurinavichius, R. S., and Rupainene, O. I. (1976b). Growth and development of higher plants in conditions of simulated weightlessness. *Dokl. Akad. Nauk SSSR, Ser. Biol.* **226**, 978–981.
Merkys, A. I., Laurinavichius, R. S., Rupainene, O. I., Shvegzhdene, D. V., and Yaroshius, A. V. (1981). Gravity as an obligatory factor in normal higher plant growth and development. *Adv. Space Res.* **1**(14), 109–116.
Merkys, A. I., Laurinavichius, R. S., and Rupainene, O. I. (1983a). The state of gravity sensors and peculiarities of plant growth during different gravitational loads. *Adv. Space Res.* **3**(9), 211–219.
Merkys, A. I., Laurinavichius, R. S., and Shvegzhdene, D. V. (1983b). Effect of weightlessness on the initial phases of lettuce and *Arabidopsis* seedlings development. In "11-e Vsesoyuznoe Rabochee Soveschanie po Voprosam Krugovorota Veschestv v Zamknutikh Sistemakh na Osnove Zhiznedeyatelnosti Nizshikh Organismov" (V. A. Kordyum, Ed.), pp. 63–72. Naukova Dumka, Kiev.
Merkys, A. I., Rupainene, O. I., and Laurinavichius, R. S. (1983c). About anatomic structure of onion (*Allium cepa* L.) inflorescence formed in space flight conditions. In "11-e Vsesoyuznoe Rabochee Soveschanie po Voprosam Krugovorota Veschestv v Zamknutikh Sistemakh na Osnove Zhiznedeyatelnosti Nizshikh Organismov" (V. A. Kordyum, Ed.), pp. 117–120. Naukova Dumka, Kiev.
Merkys, A. I., Laurinavichius, R. S., and Shvegzhdene, D. V. (1984). Spatial orientation and growth of plants in weightlessness and in the field of artificial gravity. In "Biologicheskie Issledovaniya na Orbitalnikh Stanziyakh 'Salyut'" (N. P. Dubinin, Ed.), pp. 72–81. Nauka, Moscow.
Merkys, A. I., Laurinavichius, R. S., Shveghdene, D. V., and Yaroshius, A. V. (1985). Investigations of higher plants under weightlessness. *Physiologist* **28**(6), 43–46.
Merkys, A. I., Laurinavichius, R. S., Kentaviciene, P. F., and Nechitailo, G. S. (1988). Formation and growth of *Arabidopsis* callus tissue under changed gravity. *27th COSPAR Plenary Meet.*, Helsinki, p. 58.
Mesland, D. A. M. (1992). Possible actions of gravity on the cellular machinery. *Adv. Space Res.* **1**(1), 15–25.
Miquel, J. (1986). An unifying metabolic hypothesis on the effects of microgravity and hypergravity on development and aging. *26th COSPAR Plenary* Meet., Toulouse, France, p. 56.
Miquel, J., Philpott, D. E., Lundgren, P. R., and Binnard, R. (1979). The influence of weightlessness on embryonal development and aging in *Drosophila melanogaster*. In "Biologicheskie Issledovaniya na Biosputnikakh 'Kosmos'" (E. A. Ilyin and G. P. Parfenov, Eds.), pp. 87–92. Nauka, Moscow.

Miroslavov, E. A. (1974). "Structura i Funkziya Epidermisa Lista u Pokritosemennikh Rastenii." Nauka, Leningrad.
Moore, R. (1990). Comparative effectiveness of a clinostat and a slow-turning lateral vessel at mimicking the ultrastructural effects of microgravity in plant cells. *Ann. Bot.* **66,** 541–549.
Moore, R., and Evans, M. L. (1986). How roots perceive and respond to gravity. *Am. J. Bot.* **73,** 574–587.
Moore, R., Fondren, W. M., Koon, E. C., and Wang, C.-L. (1986). The influence of microgravity on the formation of amyloplasts in columella cells of *Zea mays* L. *Plant Physiol.* **82,** 867–868.
Moore, R., Fondren, W. M., McClelen, C. E., and Wang, C.-L. (1987a). Influence of microgravity on cellular differentiation in root caps of *Zea mays*. *Am. J. Bot.* **74,** 1006–1012.
Moore, R., McClelen, C. E., Fondren, W. M., and Wang, C.-L.(1987b). Influence of microgravity on root cap regeneration and the structure of columella cells in *Zea mays*. *Am. J. Bot.* **74,** 218–223.
Musgrave, M. E., Cummins, D. B., Matthews, S. W., Kuang, A., Daugherty, C. J., and Porterfield, D. M. (1993). Growth and flowering of *Arabidopsis thaliana* during space flight. *ASGSB Bull.* **7,** 83.
Nace, G. W. (1983). Gravity and positional homeostasis of the cell. *Adv. Space Res.* **3**(9), 159–168.
Nakagaki, M. (1991). "Physical Chemistry of Membranes." Mir, Moscow.
Nedukha, E. M. (1984). Long clinostation influence on the ultrastructure of *Funaria hygrometrica* moss protonema cells. *Adv. Space Res.* **4**(12), 19–22.
Nedukha, E. M. (1986). Influence of clinostating on the ultrastructure of *Funaria hygrometrica* Hedw. protonema. *Ukr. Bot. Zh.* **43**(2), 20–23.
Nedukha, E. M. (1989). Long clinostation influence on the localization of the free and weakly bound calcium in cell walls of *Funaria hygrometrica* moss protonema. *Adv. Space Res.* **9**(11), 83–86.
Nedukha, E. M. (1992). The role of cellulases in the mechanism of changes of cell walls of *Funaria hygrometrica* protonema at clinostating. *Adv. Space Res.* **12**(1), 99–102.
Nedukha, E. M., Uvarova, S. A., and Tupik, N. D. (1985). Influence of hypogravity on an activity and multiple molecular forms of degydrogenases ln *Funaria hygrometrica*. *Ukr. Bot. Zh.* **42**(1), 52–55.
Nedukha, E. M., Kordyum, E. L., and Danevich, L. A. (1988). Influence of hypogravity on a callose content in *Funaria hygrometrica* Hedw. *Ukr. Bot. Zh.* **45**(6), 58–60.
Nedukha, E. M., Kordyum, E. L., and Nechitailo, G. S. (1991). Influence of the 16-day space flight on an ultrastructure of the leaf cells in wheat plants. Preprint 138-B-91. VINITI, Moscow.
Nedukha, E. M., Chuchkin, V. G., and Philippenko, V. N. (1992). Ultrastructure of cells of a shoot apical meristem, cotyledons and hypocotyles in *Impatiens balsamina* L. seedlings. *In* "Resultati Issledovanii na Biosputnikakh" (O. G. Gazenko, Ed.), pp. 309–312. Nauka, Moscow.
Nedukha, E. M., Sidorov, V. A., and Samoylov, V. M. (1994). Clinostation influence on regeneration of a cell wall in *Solanum tuberosum* L. protoplasts. *Adv. Space Res.* **14**(8), 97–101.
Neubert, J. (1979). Ultrastructural development of the vestibular system under conditions of simulated weightlessness. *Aviat. Space Environ. Med.* **50,** 1058–1061.
Nevruzova, Z. A., Aliev, A. A., and Alekperov, U.K. (1987). Comparative anatomic analysis of *Pisum sativum* seedling organs grown in space flight conditions. *Bot. Zh.* **72,** 657–659.
Nevzgodina, L. V., Grigorjev, Yu. G., and Marenny, A. M. (1990). "Vliyanie Tyazhelikh Ionov na Biologicheskie Objekti." Energoatomizdat, Moscow.
Opsahl, L. R., and Webb, W. W. (1994). Physical mechanisms of tension sensitivity in the channel forming polypeptide alamethicin. *Biophys. J.* **66**(2), 171.

Parfenov, G. P., and Abramova, V. M. (1981). Flowering and maturing of *Arabidopsis* plants in weightlessness: Experiment on the biosatellite Cosmos-1129. *Dokl. Akad. Nauk SSSR Ser. Biol.* **256,** 254–256.
Parfenov, G. P., Platonova, R. N., Karpota, N. I., and Oigenblick, I. A. (1975). Results of biological experiments carried out during Cosmos-69o flight (test with Drosophila and pine seeds). *18th COSPAR Plenary Meet.*, Varna, Bulgaria, p. 375.
Perbal, G. and Driss-Ecole, D. (1994). Sensitivity to gravistimulus of lentil seedling roots grown in space during the IML I mission of Spacelab. *Physiol. Plant.* **90,** 313–318.
Perbal, G., Driss-Ecole, D., Salle, G., and Raffin, J. (1986). Perception of gravity in the lentil root. *Naturwiss.* **73,** 444–446.
Perbal, G., Driss-Ecole, D., and Salle, G. (1987). Graviperception of lentil seedling roots grown in space. *In* "Biorack on Spacelab D-1" (N. Longdon and V. David, Eds.), pp. 109–117. European Space Agency, Paris.
Petrukhin, V. G., Yugalov, E. M., Gaydamakin, N. A., Soloviev, V. I., Antipov, V. V., Petrukhin, S. V., Dolina, L. A., and Mashinsky, A. L. (1976). Morphological investigations of animal organisms in conditions of weightlessness. *Problemi Kosmicheskoi Biol.* **33,** 199–227.
Philippenko, V. N., Chuchkin, V. G., and Ivanov, V. B. (1992). Regeneration and formation de novo of roots in *Impatiens balsamina* seedlings in microgravity. *In* "Resultati Issledovanii na Biosputnikakh" (O. G. Gazenko, Ed.), pp. 316–321. Nauka, Moscow.
Planel, H., Tixador, R., Nefedov, Yu., Grechko, G., and Richoilley, G. (1982). Effects of space flight factors at the cellular level: Results of the Cytos experiment. *Aviat. Space Environ. Med.* **53,** 370–374.
Planel, H., Tixador, R., Richoilley, G., and Gasset, G. (1990). Effects of space flight on a single-cell organism: *Paramecium tetraurelia*. *In* "Fundamentals of Space Biology" (M. Asashima and G. M. Malacinski, Eds.), pp. 85–96. Springer-Verlag, Berlin.
Platonova, R. N., Parfenov, G. P., and Olkhovenko, V. P. (1977). The germination of pine seeds in weightlessness: Investigations on board the "Cosmos- 782." *Izv. Akad. Nauk SSSR Ser. Biol.* **5,** 700–778.
Platonova, R. N., Parfenov, G. P., and Zhvalikovskaya, V. P. (1979). Orientation of plants in weightlessness. *In* "Biologicheskie Issledovaniya na Biosputnikakh 'Kosmos' " (E. A. Ilyin and G. P. Parfenov, Eds.), pp. 149–161. Nauka, Moscow.
Platonova, R. N., Lyubchenko, V. Yu., Devyatko, A. V., Malysheva, G. I., and Tairbekov, M. G.(1980). Orientation of tomato seedlings grown in weightlessness: Investigations on board the satellite "Cosmos-1129." *Izv. Akad. Nauk SSSR. Ser. Biol.* **6,** 91–96.
Platonova, R. N., Lyubchenko, V. Yu., and Devyatko, A. V. (1983). Influence of photo- and chemotropism on the orientation of higher plants in the absence of geotropism. *Izv. Akad. Nauk SSSR Ser. Biol.* **1,** 51–59.
Podluzky, A. G. (1992). Ultrastructural analysis of organization of roots obtained from cell cultures at clinostating and under microgravity. *Adv. Space Res.* **12**(1), 93–98.
Polulyakh, Yu.A. (1988). Phospholipid and fatty acid content of plasma membrane in cells of pea roots under clinostating. *Dokl. Akad. Nauk UkrSSR Ser. Biol.* **10,** 67–69.
Polulyakh, Yu. A., and Volovick, Z. N. (1989). Fluorescent analysis of microviscosity in plant cell membranes under clinostating. *Dokl. Acad. Nauk UkrSSR Ser. Biol.* **12,** 61–63.
Polulyakh, Yu. A., Zhadko, S. I., and Klimchuk, D. A. (1989). Plant cell plasma membrane structure and properties under clinostating. *Adv. Space Res.* **9**(11), 71–74.
Popova, A. F. (1986). Submicroscopic organization of *Anabaena azollae* Strasb. cells under space flight conditions. *In* "Kosmicheskaya Biologiya i Biotekhnologiya" (K. M. Sytnik, Ed.), pp. 18–22. Naukova Dumka, Kiev.
Popova, A. F. (1992). Localization of free and weakly bound calcium ions in *Chlorella* cells under clinostating. *In* "9-i Zyizd Ukrainskogo Botanichnogo Tovaristva" (K. M. Sytnik, Ed.), p. 327. Naukova Dumka, Kiev.

Popova, A. F. (1994). Peculiarities of localization ATFase activated by $Mg^{2+}$ in *Chlorella* cells under clinostating. *Cytol. Genet.* **28**(2), 3–7.

Popova, A. F. and Kordyum, E. L. (1991). Algae. *Problemi Kosmicheskoi Biol.* **69,** 156–228.

Popova, A. F. and Shnyukova, E. I. (1989). Ultrastructure of chloroplasts, fractional composition of amylases and their specific activity in *Chlorella* cells in altered gravity. *Dokl. Akad. Nauk UkrSSR Ser. Biol.* **7,** 78–82.

Popova, A. F., and Sytnik, K. M. (1994). Structural–functional characteristics of *Chlorella* cells after prolonged growth in microgravity. *In* "Kosmicheskaya Biologiya i Aerokosmicheskaya Medizina" (A. I. Grigorjev, Ed.), p. 91. Slovo, Moscow.

Popova, A. F., and Sytnik, K. M. (1996). Peculiarities of ultrastructure of *Chlorella* cells growing aboard the Bion-10 during 12 days. *Adv. Space Res.* **17**(6/7), 99–102.

Popova, A. F., Ivanenko, G. F., and Tupik, N. D. (1987). Ultrastructure of mitochondria and fractional composition of succinate degydrogenase in *Chlorella vulgaris* under clinostating. *Ukr. Bot. Zh.* **43**(3), 11–15.

Popova, A. F., Sytnik, K. M., and Kordyum, E. L. (1989). Ultrastructural and growth indices of *Chlorella* culture in multicomponent aquatic systems under space flight conditions. *Adv. Space Res.* **9**(11), 79–82.

Popova, A. F., Sytnik, K. M., Kordyum, E. L. and Nechitailo, G. S. (1991). Ultrastructural organization of *Chlorella* cells cultivated on the solid medium in microgravity. *Dokl. Akad. Nauk UkrSSR Ser. Biol.* **8,** 161–164.

Popova, A. F., Sytnik, K. M., Nechitailo, G. S., and Mashinsky, G. S.(1992a). The submicroscopic organization of *Chlorella* cells cultivated in solid medium under microgravity. *Adv. Space Res.* **12**(1), 141–146.

Popova, A. F., Sytnik, K. M., Nechitailo, G. S., and Mashinsky, A. L. (1992b). The structural–functional organization of *Chlorella* cells grown over 12 months in microgravity. World Space Congr., Washington, DC, p. 527.

Popova, A. F., Kordyum, E. L., and Shnyukova, E. I. (1993). Ultrastructure of chloroplasts and amylase activity in *Chlorella* under space flight conditions. *Dokl. Akad. Nauk UkrSSR Ser. Biol.* **11,** 152–156.

Popova, A. F., Kordyum, E. L., Shnykova, E. I., and Sytnik, K. M. (1995). Plastid ultrastructure, fractional composition and activity of amylases in *Chlorella* cells in microgravity. *J. Gravit. Physiol.* **2,** 159–160.

Rasmussen, O., Klimchuk, D. A., Kordyum, E. L., Danevich, L. A., Tarnavskaya, E. B., Lozovaya, V. V., Tairbekov, M. G., Baggerud, C., and Iversen, T.-H. (1992). The effect of exposure to microgravity on the development and structural organization of plant protoplasts flown on Biokosmos 9. *Physiol. Plant.* **84,** 162–170.

Rasmussen, O., Baggerud, C., Larsen, H. C., Evjen, K., and Iversen, T.-H. (1994). The effect of 8 days of microgravity on regeneration of intact plants from protoplasts. *Physiol. Plant.* **92,** 404–411.

Richoilley, G., Tixador, R., Gasset, G., Templier, J., and Planel, H.(1986). Preliminary results of the *"Paramecium"* experiment. *Naturwiss.* **73,** 404–406.

Rijken, P. J., de Groot, R. P., Kruijer, W., Verkleij, A. J., Boonstra, J., and de Laat, S. W. (1992). Altered gravity conditions affect early EGF-induced signal transduction in human epidermal A431 cells. *ASGSB Bull.* **35**(2), 77–82.

Rokhlenko, K. D., and Savik, Z. F. (1981). Influence of space flight factors on ultrastructure of skeletal muscles. *Kosm. Biol. Aviakosm. Med.* **15**(1), 72–76.

Rubin, B. A., Ladygina, M. E., Voronkov, L. A., Kartashova, E. R., Taimla, E. A., Sokolovskaya, I. V., Tukeeva, M. I., Zolotukhina, E. Yu., and Zhivopistseva, I. V. (1979). Physiological state of the crown gall tissue induced by *Agrobacterium tumefaciens*. *In* "Biologicheskie Issledovaniya na Biosputnikakh 'Kosmos' " (E. A. Ilyin and G. P. Parfenov, Eds.), pp. 126–136. Nauka, Moscow.

Rumyanzeva, V. B., Merzlyak, M. N., Mashinsky, A. L., and Nechitailo, G. S. (1990). Influence of space flight conditions on the pigment and lipid content in wheat. *Kosm. Biol.* **1,** 53–57.
Sack, F. D. (1991). Plant gravity sensing. *Int. Rev. Cytol.* **127,** 193–251.
Saunders, J. F., Reynolds, O. E., and de Serres, F. J. (1971). The experiments of Biosatellite II. *In* "Gravity and the Organism" (S. A. Glordon and M. J. Cohen, Eds.), pp. 443–450. Univ. of Chicago Press, Chicago.
Schatz, A., Reitstetter, R., Briegleb,W., and Linke-Hommes, A. (1992). Gravity effects on biological systems. *Adv. Space Res.* **12**(1), 51–53.
Schatz, A., Reitstetter, R., Linke-Hommes, A., Briegleb, A., Slenzka, K., and Rahmann, H. (1994). Gravity effects on membrane processes. *Adv. Space Res.* **14**(8), 35–43.
Schneider, E. M., and Sievers, A. (1981). Concanavalin A binds to the endoplasmic reticulum and the starch grain surface of root statocytes. *Planta* **152,** 177–180.
Schnepf, E. (1986). Cellular polarity. *Annu. Rev. Plant Physiol.* **37,** 23–47.
Schönherr, K., Johann, P., and Hampp, R.(1994). Energy and carbohydrate metabolism in regenerating protoplasts grown under different gravity conditions. *30th COSPAR Scientific Assembly,* Hamburg, Germany, p. 280.
Schulze, A., Jensen, P. J., Desrosiers, M., Buta, J. G., and Bandurski, R. S. (1992). Studies on the growth and indole-3-acetic acid content of *Zea mays* seedlings grown in microgravity. *Plant Physiol.* **100,** 692–698.
Schwuchow, J., Sack, F. D., and Hartmann, E. (1990). Microtubule distribution in gravitropic protonemata of the moss Ceratodon. *Protoplasma* **159,** 60–69.
Shen-Miller, J., Hinchman, R., and Gordon, S. A. (1968). Thresholds for georesponse to acceleration in gravity-compensated *Avena* seedlings. *Plant Physiol.* **43,** 338–344.
Shinitzky, M.(1984). "Physiology of Membrane Fluidity," Vols. 1 and 2. CRC Press, Boca Raton, FL.
Shmalgauzen, I. I. (1940). "Puti i Zakonomernosti Evoluzionnogo Prozessa." Izdatelstvo Akad. Nauk SSSR, Moscow.
Shvegzhdene, D. V. (1991). Isuchenie roli gravitazii v Prozessakh Prostranstvennoi Orintazii, Morfogenesa i Rosta Pervichnikh Kornei Salata. Ph. D. thesis, Institute of Botany of the Akad. Nauk LitSSR, Vilnyus, Russia.
Sidorenko, P. G., and Mashinsky, A. L. (1978). Influence of space flight conditions on the higher plant cell cultures grown *in vitro. Kosm. Issled. Ukr.* **12,** 39–42.
Sievers, A., and Hejnowicz, Z. (1992). How well does the clinostat mimic the effect of microgravity on plant cells and organs? *ASGSB Bull.* **5**(2), 69–76.
Sievers, A., and Hensel, W.(1991). Root cap: Structure and function. *In* "Plant Root: The Hidden Half" ( J. Waisel, U. Kafkafi, and A. Eshel, Eds.), pp. 53–74. Dekker, New York.
Sievers, A., Volkmann, D., Hensel, W., Sobick, W., and Briegleb, W. (1976). Cell polarity in root statocytes in spite of simulated weightlessness. *Naturwiss.* **63,** 343.
Sievers, A., Buchen, B., Volkmann, D., and Hejnowicz, Z. (1991). Role of the cytoskeleton in gravity perception. *In* "The Cytoskeletal Basis of Plant Growth and Form" (C. W. Lloyd, Ed.), pp. 169–182. Academic Press, London.
Slocum, R. D., Gaynor, J. J., and Galston, A. W. (1984). Experiments on plants grown in space: Cytological and ultrastructural studies on root tissues. *Ann. Bot.* **54**(Suppl.3), 65–76.
Smith, A. H. (1975). Principles of gravitational biology. *In* "Foundations of Space Biology and Medicine" (M. Calvin and O. Gazenko, Eds.), pp. 129–162. NASA, Washington, DC.
Smith, A. H. (1992). Centrifuges: Their development and use in gravitational biology. *ASGSB Bull.* **5**(2), 33–42.
Sobick, V., and Sievers, A. (1979). Responses of roots to simulated weightlessness on the fast-rotating clinostat. *Life Sci. Space Res.* **17,** 285–290.
Sparrow, A. H., Schairer, L. A., and Marimuthu, K. M. (1968). Genetic and cytologic studies of *Tradescantia* irradiated during orbital flight. *BioScience* **18,** 582–590.

Sparrow, A. H., Schairer, L. A., and Marimuthu, K. M. (1971). Radiobiologic studies of *Tradescantia* plants orbited in Biosatellite II. *In* "The Experiments of Biosatellite II" (J. F. Saunders, Ed.), pp. 99–122. NASA, Washington, DC.

Steinbiss, H. H. (1978). Physiological and cytomorphological changes in leaf segments of *Vicia faba* L. caused by short-time microwave irradiation. *Protoplasma* **94**, 155–166.

Sytnik, K. M. and Musatenko, L. I. (1980). Changes of nucleic acids of wheat seedlings under space flight conditions. *Dokl. Akad. Nauk UkrSSR Ser. Biol.* **2**, 77–80.

Sytnik, K. M., Kordyum, E. L., Mashinsky, A. L., and Popova, A. F. (1979). Ultrastructure of *Chlorella pyrenoidosa* (strain g-11-1) cells grown long time in space flight. *Ukr. Bot. Zh.* **36**(2), 97–105.

Sytnik, K. M., Kordyum, E. L., Belyavskaya, N. A., and Tarasenko, V. A. (1982). Ultrastructure of the root meristem and cap in pea seedlings under space flight conditions. *Dokl. Akad. Nauk UkrSSR, Ser. Biol.* **6**, 78–80.

Sytnik, K. M., Kordyum, E. L., Belyavskaya, N. A., Nedukha, E.M., and Tarasenko, V. A. (1983a). Biological effects of weightlessness and clinostatic conditions registered in cells of the root meristem and cap of higher plants. *Adv. Space Res.* **3**(9), 251–255.

Sytnik, K. M., Kordyum, E. L., Kordyum, V. A., Babsky, V. G., Manko, V. G., and Nedukha, E. M. (1983b). "Mikroorganismi v Kosmicheskom Polete." Naukova Dumka, Kiev.

Sytnik, K. M., Kordyum, E. L., Nedukha, E. M., Sidorenko, P. G., and Fomicheva, V. M. (1984). "Rastitelnaya Kletka pri Izmenenii Geofizicheskikh Faktorov." Naukova Dumka, Kiev.

Sytnik, K. M., Kordyum, E. L., Nedukha, E. M., Nechitailo, G. S., and Samoylov, V. M. (1992). Structural–functional organization of storage parenchyma cells of *Solanum tuberosum* minitubers formed under space flight. *World Space Congr.* Washington, DC. pp. 526.

Tairbekov, M. G. (1991). Studies on plant cell cultures in space. *In* "Biosatellites 'Cosmos' " (A. I. Grigorjev, Ed.), pp. 119–120. Institute of Biomedical Problems, Moscow.

Tairbekov, M. G., and Devyatko, A. V. (1985). Energy exchange of plants in weightlessness. *Dokl. Akad. Nauk SSSR Ser. Biol.* **280**, 509–512.

Tairbekov, M. G., and Parfenov, G. P. (1983). Conduct of cell in the gravity field. *Usp. Sovrem. Biol.* **96**, 426–434.

Tairbekov, M. G., Parfenov, G. P., Platonova, R. N., and Zhvalikovskaya, V. P. (1979). Investigations of plant cells using the "Biofixator-1" instrument. *In* "Biologicheskii Issledovaniya na Biosputnikakh 'Kosmos' " (E. A. Ilyin and G. P. Parfenov, Eds.), pp. 161–169. Nauka, Moscow.

Tairbekov, M. G., Kabitsky, E. N., and Mailyan, E. S. (1980). ATP activity in maize root cap cells grown under changed gravity conditions. *Fiziol. Rast.* **27**, 833–837.

Tairbekov, M. G., Grif, V. G., Barmicheva, E. M., and Valovich, E. M. (1986). Cytomorphology and ultrastructure of the maize root meristem in weightlessness. *Izv. Akad. Nauk SSSR Ser. Biol.* **5**, 680–687.

Tairbekov, M. G., Lozovaya, V. V., and Zabotina, O. A. (1988). Morphological characteristics and composition of maize seedling cell walls formed in weightlessness. *Fiziol. Rast.* **35**, 226–253.

Tarasenko, V. A. (1985). Ultrastructura Kletok Kolumelli v Kornevom Chekhlike Arabidopsis v Usloviyakh Klinostatirovaniya i mikrogravitazii. Ph. D. thesis, Institute of Cytology of the Akad. Nauk SSSR, Leningrad.

Tarasenko, V. A., Kordyum, E. L., and Sytnik, K. M. (1982). Ultrastructure of the *Arabidopsis thaliana* root cap under space flight conditions. *Dokl. Akad. Nauk UkrSSR, Ser. Biol.* **7**, 79–81.

Theimer, R. R., Kudielka, R. A., and Rosch, I. (1986). Induction of somatic embryogenesis in anise in microgravity. *Naturwiss.* **73**, 442–443.

Tixador, R., Richoilley, G., Templier, J., Monrozies, E., Moatti, J.-P., and Planel, H. (1981). Etude de la teneur intra et extracellulaire des electrolytes dans les cultures de paramecies realisees pendant un vol spatial. *Biophys. Biochim. Acta* **649**, 175–178.

Todd, P. (1992). Physical effects on the cellular level under altered gravity conditions. *Adv. Space Res.* **12**(1), 43-49.

Tolbert, N. E., and Essner, E. (1981). Microbodies: Peroxisomes and glycoxisomes. *J. Cell Biol.* **91**, 271-283.

Turkina, M. V., Kulikova, A. L., Koppel, L. A., Akatova, L. Z., and Butenko, R. G. (1995). Actin and thermostabile actin-bound proteins in the wheat cell culture. *Fiziol. Rast.* **42**, 348-355.

Udovenko, G. V. (1979). Mechanisms of plant adaptation to stress. *Fiziol. Biokhim. Kult. Rast.* **11**(2), 99-107.

Van Steveninck, R. F. M. (1976). Cellular differentiation, aging and ion transport. *Encycl. Plant Physiol. New Ser.* **2**, 343-371.

Vaulina, E. N. (1976). Influence of weightlessness on hereditary structures. *Problemi Kosmicheskoi Biol.* **33**, 174-199.

Vaulina, E. N., and Kostina, L. N. (1984). Investigations of air-dry seeds and seedlings of *Crepis capillaris* (L.) Wallr. In "Biologicheskie Issledovaniya na Orbitalnikh Stanziyakh 'Salyut' " (N. P. Dubinin, Ed.), pp. 68-72. Nauka, Moscow.

Vaulina, E. N., Kordyum, V. A., Kordyum, E.L., Konshin, N. I., Mashinsky, A. L., Nechitailo, G. S., and Popova, A. F. (1978). "Vliyanie Kosmicheskogo Poleta na Rasvivayuschiesya Organismi." Naukova Dumka, Kiev.

Vladimirov, Yu. A. (1987). Lipid free-radical oxidation and physical properties of the lipid bilayer in biological membranes. *Biofizika* **32**, 830-844.

Vladimirova, M. G., Markelova, A. G., and Semenenko, V. E. (1982). Identification of localization of RBPC in the pyrenoids of unicellular algae by a cytoimmun-fluorescent method. *Fiziol. Rast.* **29**, 941-951.

Volkmann, D., and Sievers, A. (1979). Graviperception in multicellular organs. *Encycl. Plant Physiol. New Ser.* **7**, 573-600.

Volkmann, D., Behrens, H. M., and Sievers, A. (1986). Development and gravity sensing of cress roots under microgravity. *Naturwiss.* **73**, 438-441.

Volkmann, D., Buchen, B., Tewinkel, M., Hejnowicz, Z., and Sievers, A. (1991). Oriented movement of statoliths studied in a reduced gravitational field during parabolic flights of rockets. *Planta* **185**, 153-161.

Walker, L. M., and Sack, F. D. (1990). Amyloplasts as possible statoliths in gravitropic protonemata of the moss *Ceratodon purpureus*. *Planta* **181**, 71-77.

Ward, C. H., and King, J. M. (1979). Effects of simulated hypogravity on respiration and photosynthesis of higher plants. *Life Sci. Space Res.* **17**, 28-37.

Wells, T., and Baker, R. (1969). Gravity compensation and crown gall development. *Nature* **233**, 734-735.

Whatley, J. M.(1983). The ultrastructure of plastids in roots. *Int. Rev. Cytol.* **85**, 175-220.

Wolf, D., Schulz, M., and Schnabl, H. (1993). Influence of horizontal clinostat rotation on plant proteins: 1. Effects on ubiquitinated polypeptides in the stroma and thylakoid membranes of *Vicia faba* L. chloroplasts. *J. Plant Physiol.* **141**, 304-308.

Wyatt, S. E., and Carpita, N. C. (1993). The plant cytoskeleton-cell wall continuum. *Trends Cell Biol.* **3**, 413-417.

Yuganov, E. M., Kasyan, I. I., and Asyamolov, B. V. (1974). Bioelectric activity of skeleton muscle in conditions of intermittent action of overloadings and weightlessness. *In* "Nevesomost" (O. G. Gazenko, Ed.), pp. 213-219. Nauka, Moscow.

Zabotina, O. A. (1987). Vliyanie Izmenennoi Gravitatii na Sostav Kletochnikh Obolochek Visshikh Rastenii. Ph. D. thesis, Kazanskii Institute Biologii of the Akad. Nauk SSSR, Kazan.

Zaitseva, M. G., Kasumov, E. A., and Kasumova, I. V. (1991). Accumulation of cations and conformational changes of proteins: Their possible role in mitochondrion volume changes. *Fiziol. Rast.* **38**, 708-714.

Zaslavsky, V.A., and Danevich, L.A. (1983). Cytophotometric investigations of nucleus chromatin in the pea root meristem cells grown under clinostating. *Ukr. Bot. Zh.* **40**(2), 63–64.

Zaslavsky, V. A., and Fomicheva, V. M. (1986). Functional state of chromatin and proliferative activity of the meristem cells in pea seedlings under clinostating. *In* "Kosmicheskaya Biologiya i Biotekhnologia" (K. M. Sytnik, Ed.), pp. 23–28. Naukova Dumka, Kiev.

Zhadko, S. I. (1991). Perekisnoe Okislenie Lipidov v Rastitelnikh Kletkakh v Usloviyakh Mikrogravitazii i ee Simulazii. Ph. D. thesis, Institute of Cytology of the Akad. Nauk SSSR, Leningrad.

Zhadko, S. I., Klimchuk, D. A., Kordyum, E. L., Sytnik, K. M., Sidorenko, P. G., and Baraboy, V. A. (1992). Lipid peroxidation and the ultrastructural organization of *Haplopappus* tissue culture cells. *In* "Resultati Issledovanii na Biosputnikakh" (O. G. Gazenko, Ed.), pp. 321–323. Nauka, Moscow.

Zhadko, S. I., Baranenko, V. V., and Kordyum, E. L. (1994). Lipid peroxidation in *Chlorella* cells and pea tissues grown on board the biosatellite Bion-10. *In* "Kosmicheskaya Biologiya i Aerokosmicheskaya Medizina" (A. I. Grigorjev, Ed.), pp. 91–92. Slovo, Moscow.

# The Use of Antibodies to Study the Architecture and Developmental Regulation of Plant Cell Walls

J. Paul Knox
Centre for Plant Biochemistry & Biotechnology, University of Leeds, Leeds, LS2 9JT, United Kingdom

This review covers the generation and use of antibodies to defined components of plant and algal cell walls and how these have contributed to our understanding of the spatial and developmental regulation of cell walls. Particular emphasis is placed upon the generation and characterization of monoclonal antibodies to matrix polysaccharides, extensins, and arabinogalactan-proteins of higher plants, and algal polysaccharides and glycoproteins. Immunolocalization studies are discussed in relation to the identification of molecular domains within cell walls, cell adhesion, cell differentiation, and the establishment of plant interactions with other organisms.
**KEY WORDS:** Arabinogalactan-proteins, Cell wall polysaccharides, Extensins, Monoclonal antibodies, Pectins, Plant cell wall architecture.

## I. Introduction

The primary and secondary cell walls of plants are major contributors to plant form during growth and development. They also provide the bulk of the biomass of plant materials and are the major determinants of the textural properties of plants utilized for food and fiber. Plant cell walls therefore have considerable importance for our understanding of plant growth and utilization, and much attention has been paid to their molecular nature and architecture. We currently have a good broad understanding of the composition of plant cell walls, the structure and function of the major components, and a clear recognition of their molecular and functional complexity. Structural determinations and biochemical analyses of cell wall components are central to this knowledge. There are several recent reviews covering this information and its integration with our understanding of the roles played by plant cell walls (Bolwell, 1993; Carpita and Gibeaut, 1993; Roberts, 1994; Wallace and Fry, 1994; Boudet *et al.*, 1995; Fry, 1995).

It is now firmly established that the primary cell wall is a metabolically dynamic cell component. It is increasingly recognized that there is a need to understand the molecular differences between cell walls and domains of local organization within cell walls. It is also increasingly apparent that the shared regions of the extracellular matrix such as the middle lamella and developing intercellular spaces are likely to be complex in molecular terms, although we are far from having a clear understanding of how these regions develop, or of their biochemistry. Furthermore, we have only a limited knowledge of the mechanisms that exist in plant cells that are responsive to mechanical stresses and that result in cell wall strengthening, or the molecular factors that determine whether cells will break across cell walls or delaminate and separate at the middle lamella when placed under stress. This chapter focuses on the impact that antibodies, and particularly monoclonal antibodies, have had on our understanding of the organization and developmental dynamics of the plant cell wall and its association with the plasma membrane and extracellular secretions.

Twenty-five years ago the first reports of the immunolocalization of plant cell wall antigens concerned a pollen wall component (Knox *et al.*, 1970) and an algal polysaccharide (Vreeland, 1970). An early review covered the use of antibodies in botanical immunocytochemistry and focused on pollen antigens and allergens (Knox *et al.*, 1980). Subsequent reviews on the use of antibodies in the study of plant cell walls include those by Roberts *et al.* (1985), Hoson (1991), Pennell (1992), and Knox (1992). Aspects of antibody methodology and immunochemical techniques in relation to plant cell walls and plant tissues have also been covered in a series of articles (Roberts, 1986; Moore, 1989; VandenBosch, 1992; Pennell and Roberts, 1995).

It has been 20 years since hybridoma technology was developed and allowed the isolation of monoclonal antibodies with fully defined specificities. This technology has made a significant contribution to our understanding of the plant cell surface. The advent of this technology was particularly important for the generation of antibodies to specific epitopes of complex macromolecules and also for the biochemical dissection of complex, biochemically ill-defined antigens such as whole cells or preparations of cells. Table I lists the major monoclonal antibodies that have been generated to plant and algal cell surface antigens. Where appropriate in the following discussions of cell wall components, emphasis will be placed on these probes because their extended availability and capacity to be highly characterized has increased their usefulness over antisera.

USE OF ANTIBODIES IN STUDYING PLANT CELL WALLS 81

TABLE I
Major Monoclonal Antibodies to Components of Plant Cell Surfaces Including the Plant Cell Wall and the Plasma Membrane

| Antibody | Antigen/epitope | References |
|---|---|---|
| Anti-pectin | | |
| JIM5 | Unesterified homogalacturonan | VandenBosch et al. (1989), Knox et al (1990) |
| JIM7 | Methyl-esterified homogalacturonan | Knox et al. (1990) |
| 2F4 | Calcium-requiring configuration of homogalacturonan | Liners et al. (1989, 1992) |
| CCRC-M5[a] | Sycamore RGI | Puhlmann et al. (1994) |
| CCRC-M7 | Arabinogalactan epitope of sycamore and maize RG I | Puhlmann et al. (1994), Steffan et al. (1995) |
| LM5 | (1 → 4)-$\beta$-galactan | L. Jones, G. B. Seymour, and J. P. Knox (unpublished manuscript) |
| Anti-xyloglucan | | |
| CCRC-M1 | (1 → 2)-$\alpha$-linked fucosyl-containing epitope of XG | Puhlmann et al. (1994) |
| Anti-glucan | | |
| LAMP2H12H7 | (1 → 3)-$\beta$-glucan | Meikle et al. (1991) |
| BG1 | (1 → 3,1 → 4)-$\beta$-glucan | Meikle et al. (1994) |
| Anti-extensin | | |
| 11.D2 | Tobacco extensin | Meyer et al. (1988) |
| MC-1[a] | Maize extensin | Fritz et al. (1991) |
| JIM11 | Subset of cells in carrot root | Smallwood et al. (1994) |
| JIM12 | Intercellular spaces and subset of cells in carrot root | Smallwood et al. (1994) |
| JIM20 | Subset of cells in carrot root | Smallwood et al. (1994), Knox et al. (1995) |
| JIM19 | Extensin/ABA perception | Knox et al. (1995), Wang et al. (1995) |
| LM1 | Rice extensin | Smallwood et al. (1995) |
| Anti-arabinogalactan-protein | | |
| PCBC3[a] | Tobacco style AGPs, $\alpha$-arabinofuranoside residues | Anderson et al. (1983) |
| PN16.4B4 | Plasma membrane AGPs | Norman et al. (1986) |
| JIM4 | Plasma membrane AGPs, subset of cells in carrot root | Knox et al. (1989) |
| MAC207 | Plasma membrane AGPs | Pennell et al. (1989) |
| JIM13[a] | Plasma membrane AGPs, subset of cells in carrot root | Knox et al. (1991) |
| JIM15 | Plasma membrane AGPs, subset of cells in carrot root | Knox et al. (1991) |
| JIM8 | Plasma membrane AGPs | Pennell et al. (1991, 1992) |

(*continues*)

TABLE I (continued)

| Antibody | Antigen/epitope | References |
|---|---|---|
| ZUM15 | Carrot AGPs | Kreuger and van Holst (1995) |
| ZUM18 | Carrot AGPs | Kreuger and van Holst (1995) |
| LM2 | Plasma membrane AGPs, β-glucuronosyl residues | Smallwood et al. (1996) |
| **Anti-cell wall enzymes** | | |
| mWP3[a] | Cell wall peroxidases, maize | Kim et al. (1988) |
| αEI[a] | (1 → 3, 1 → 4)-β-glucanase | Høj et al. (1990) |
| **Anti-algal polysaccharides/glycoproteins** | | |
| MAC1[a] | Glycoproteins, *Chlamydomonas* | Smith et al. (1984) |
| 44.2[a] | Glycoproteins, *Chlamydomonas* | Homan et al. (1987) |
| 2-8.7[a] | Polyguluronate *Fucus* | Vreeland et al. (1984), Larsen et al. (1985) |
| FS1[a] | Glycoproteins and polysaccharides, *Fucus* | Jones et al. (1988, 1990) |
| F13 | Fucoidan/ascophyllan, *Fucus* | Green et al. (1993) |
| 3H8[a] | Fucoidan, *Macrocystis* | Eardley et al. (1990) |
| 1C8 | Fucoidan/ascophyllan, *Macrocystis* | Eardley et al. (1990) |
| 3G1 | Lambda carrageenan, *Kappaphycus* | Vreeland et al. (1992) |
| 4D12 | Iota carrageenan, *Kappaphycus* | Vreeland et al. (1992) |
| 6A11[a] | Kappa carrageenan, *Kappaphycus* | Vreeland et al. (1992) |
| **Miscellaneous** | | |
| JIM1 | Plasma membrane, β-galactosyl residues | Knox and Roberts (1990) |
| JIM18 | Plasma membrane, glycosylated phospholipid | Knox et al. (1995), Perotto et al. (1995) |
| G10-4-C3[a] | Guard cell plasma membrane | Key and Weiler (1988) |
| 40.1C2.8 | Membrane and cell wall protein of oat root cap cells | Stout and Griffing (1993) |
| **Anti-peribacteroid membrane** | | |
| MAC64 | Pea nodule peribacteroid membrane glycoprotein | Brewin et al. (1985) |
| MAC209[a] | Peribacteroid membrane glycoprotein/AGPs | Bradley et al. (1988) |
| MAC265[a] | Pea nodule infection thread matrix glycoproteins | Bradley et al. (1988), VandenBosch et al. (1989) |
| MAC254[a] | Peribacteroid membrane and soluble glycoproteins | Bradley et al. (1988) |

[a] Indicates that the monoclonal antibody is a representative of a panel of antibodies of related specificities. In all cases the primary reference of the generation of the antibodies are given, along with references in which the specificity is characterized further. References to subsequent reports describing the use of the antibodies are found in the text.

## II. Antibodies as Molecular Probes for the Plant Cell Surface

Biochemical analyses leading to an understanding of plant cell wall composition require homogenization of large amounts of material, generally including a diversity of cell types. The information on each cell type is lost upon analysis. Clearly, antibodies are powerful tools for complementing this biochemical knowledge when used in immunolocalization studies. There are several approaches in which antibodies can contribute to our knowledge of the structure, organization, and function of the plant cell surface: (1) understanding the organization and architecture of the cell wall and extracellular matrix of an individual cell, (2) determining differences among cell walls associated with the development of cell types, (3) determining the molecular differences in cell walls associated with taxonomic diversity, and (4) probing the function of a cell wall component or of an epitope carried by a polymer. The use of antibodies in the first two of these areas is well established

### A. Spatial Architecture of Cell Walls

The primary plant cell wall is generally in the region of 100 nm thick, and consists of at least two macromolecular networks of polysaccharides and numerous proteins, glycoproteins, and phenolics with immense possibilities for covalent and noncovalent modifications, attachments, and interactions. The primary cell wall creates turgor pressure, accommodates and regulates the direction and extent of cell expansion, has a structural role, is involved in cell-to-cell adhesion and cell separation, and is likely to have a signaling role, of which we have as yet only a vague notion (Berger *et al.,* 1994). Antibodies to cell wall components are extremely important in building up a picture of the wall's organization and changes during cell development. Cell wall components can be visualized with antibody probes in various ways. The first is in sections that cut through cells so that both older and the more recently deposited cellulosic layers are exposed for interaction with antibody probes. In this way, aspects of transmural secretion can also be studied. Second, the outer face of a cell wall at unadhered surfaces can be directly immunolabeled. Such surfaces are readily accessible on intact suspension-cultured plant cells and tip-growing regions of cells such as root hairs and pollen tubes. The inner face of a cell wall (i.e., the cell wall face that is adjacent to the plasma membrane) can also be immunolabeled. This approach has recently been developed to observe patterns in the deposition of a pectin epitope related to the presence of pit fields at the inner face of tomato pericarp cell walls (see later discussion and Casero and Knox, 1995).

A further approach to understanding cell wall architecture using antibodies is the immunolabeling and imaging of isolated polymers, isolated complexes of several polymers, or the residue of partially solubilized walls. An example of this approach is the study at the electron microscope level of a cellulose–xyloglucan complex using an antixyloglucan antiserum (Baba *et al.,* 1994).

## B. Cell Distinctions and Cell Differentiation Antigens

The study of cell surface antigens associated with cell differentiation is well established for animal cells and has revealed extensive molecular changes at the plasma membrane. The generation of anti-plant cell surface antibodies has made it clear that the molecular components of both the cell wall and the plasma membrane are developmentally regulated (Knox, 1992 and later discussion) and therefore to some extent there are two domains of the cell surface that may have so-called differentiation antigens. Both of these locations may also contain components involved in aspects of cell identity or cell recognition events, but what these are or the mechanisms involved are not known. It is currently thought that the upper size limit for passive movement of molecules through a cell wall is about 20,000–30,000 daltons (Baron-Epel *et al.,* 1988). Although molecules at the outer face of a plant cell plasma membrane are unlikely to be able to interact directly with molecular ligands on the plasma membrane of a neighboring cell, there are clearly mechanisms, as yet unknown, for the specific movement of macromolecules (such as polysaccharides and proteins) through cell walls to intercellular spaces, for secretion and possibly to neighboring cells.

Approximately 10 years ago the plant cell plasma membrane was the focus of activity to prepare and characterize both antisera and monoclonal antibodies for the study of its components (Metcalf *et al.,* 1986; Hough *et al.,* 1986; Norman *et al.,* 1986; Villanueva *et al.,* 1986; Fitter *et al.,* 1987; Lynes *et al.,* 1987; Grimes and Breidenbach, 1987; Hahn *et al.,* 1987; Dupont *et al.,* 1988; Key and Weiler, 1988). These antibodies were used to assess plasma membrane purifications, to analyze protoplast properties, and to identify cell surface components that had patterns of expression restricted to tissues or cell types, which in effect was a search for differentiation antigens. A class of plant protoeoglycans, the arabinogalactan-proteins (AGPs), which are present in all plant tissues, emerged as an important component in the generation of the immune response to plant cells in that they appeared to be immunodominant (Anderson *et al.,* 1984; Norman *et al.,* 1986, 1990; Evans *et al.,* 1988; Knox *et al.,* 1989, 1991; Pennell *et al.,*

1989). The anti-AGP antibodies that have been characterized have also indicated extensive developmental regulation of these proteoglycans, and AGPs are now receiving considerable attention in relation to plant cell development; they are discussed in detail in later sections.

### C. Taxonomic Significance of Cell Wall Components

The plant cell wall is a major contributor to the growth and shape of organs, and therefore changes in the cell wall are likely to have been important in the evolution of plant form. The structure of the major polysaccharides such as the pectins and the hemicelluloses is known to vary among taxa (Bacic *et al.*, 1988). The use of antibodies with known specificities, especially those recognizing developmentally regulated cell wall components, in surveys of a range of plant families is likely to be highly instructive, but as yet has not attracted significant attention.

### D. Antibodies as Probes of Function *in Vivo*

With a highly selective probe for up to five amino acids or sugars within a polymer, antibodies are good probes for structural elements of macromolecules in assays related to function. Monoclonal antibodies have been selected by their ability to inhibit a plant plasma membrane ATPase (Chin, 1982), to inhibit the polar transport of auxin (Jacobs and Gilbert, 1983), and to inhibit abscisic acid (ABA) responses of aleurone protoplasts (Wang *et al.*, 1995). Several antisera to cell wall polymers have been shown to inhibit auxin-induced elongation in various systems, including antiglucanase and antiglucan in maize coleoptiles, antixyloglucan in bean epicotyls, and anti-$\beta$-galactosidase in chickpea (Inouhe and Nevins, 1991; Hoson *et al.*, 1991, 1992; Valero and Labrador, 1993). With the advent of high-affinity probes to defined epitopes of cell wall polysaccharides and glycoproteins, and the ability to manipulate them through recombinant DNA technology, this approach is likely to increase.

## III. Generation and Use of Antibodies to the Major Cell Wall Polymers of Higher Plants

### A. Pectin

Pectins are abundant and complex matrix polysaccharides of higher plant primary cell walls and secretions, and are not yet fully defined in structural or functional terms. They contain several structural domains, including the acidic homogalacturonan and rhamnogalacturonans in addition to neutral side chains linked to the acidic backbone through rhamnose (O'Neill et al., 1990). The galacturonic acid residues can be methyl esterified and acetylated, resulting in changes to pectin properties. Much of pectin chemistry has been determined subsequent to enzymatic or chemical release of distinct pectin fractions from cell walls. The details of the fine structure of pectins and their links in the cell wall are far from clear. Our understanding of pectin function is equally unclear; it has probable roles in cell-to-cell adhesion at middle lamellae, the control of primary cell wall ionic status and wall porosity, which are likely to influence enzyme action and access to other cell wall polymers, and pectin oligosaccharide fragments may have a role in development and defense mechanisms. Pectin has received considerable attention in terms of the generation of antibody probes with the primary aim of their use in immunolocalization studies. Three antipectin monoclonal antibodies are now used extensively, predominantly in immunofluorescence and immunogold localization studies. These are

1. JIM5, a monoclonal antibody generated subsequent to immunization with carrot protoplasts that recognizes an epitope of unesterified homogalacturonan (VandenBosch et al., 1989; Knox et al., 1990).

2. JIM7, a monoclonal antibody generated subsequent to immunization with carrot protoplasts that recognizes an epitope containing methyl esterifed homogalacturonan (Knox et al.,1990).

3. 2F4, a monoclonal antibody generated subsequent to immunization with a conjugate of homogalacturonan coupled to methylated bovine serum albumin (BSA). It appears to recognize a calcium-dependent conformation of homogalacturonan that is thought to be a dimer requiring at least nine contiguous galacturonic acid residues for formation (Liners et al., 1989, 1992).

It should be noted that the generation of JIM5 and JIM7 was to a certain extent fortuitous in that they were isolated from a general screen for antibodies recognizing carrot cell surfaces, and their specificities only became apparent during subsequent characterization procedues. The 2F4 antibody recognized homoglacturonan with up to 30% random and 40% blockwise methyl esterification (Liners et al., 1992). The JIM5 antibody bound to pectin with up to 50% methyl esterification and JIM7 to pectin with a 35% methyl esterification and over. The JIM5 and JIM7 epitopes can occur on the same pectin molecule (Knox et al., 1990). The precise structure of the epitope is not known for any of these antibodies in terms of what patterns of methyl esterification in contiguous galacturonic acid residues they will and will not bind to and how a rhamnose insertion will influence binding. This information will require purified oligosaccharides with precisely known patterns of esterification. Nonetheless, all three antibodies are of high affinity and high specificity for the homogalacturonan domains of pectin, and are extremely useful probes. JIM5 has now been used in more than 25 reports. The 2F4 and JIM5 antibodies both recognize unesterified epitopes of homogalacturonan and often give similar localization patterns within cell walls. These two antibodies can also be used to extensively localize methyl esterified galacturonic acid residues after treatment of sections with methyl esterases, and to locate methyl-, acetyl-, and other esterified forms of pectin after tissue sections have been treated with alkali.

The major observation made with these three antibodies has been the heterogeneous distribution of pectin epitopes, indicating spatial variation in pectin and particularly its esterification within cell walls. The use of the antibodies in immunolocalization studies in pea roots (VandenBosch et al., 1989), carrot roots (Knox et al., 1990), suspension-cultured carrot cells (Liners and van Cutsem, 1991, 1992), cultured melon cells (Vian and Roland, 1991), mung bean hypocotyls (Vian et al., 1992), potato stems (Marty et al., 1995), and peach tissues (Wisniewski and Davis, 1995) has indicated that unesterified pectin (as indicated by the presence of JIM5 and 2F4 epitopes) is located in regions of the middle lamella close to cell corners, and abundantly at cell corners, and the linings of intercellular spaces. However, these studies also indicated that a predominantly esterified form of pectin occurs throughout most primary cell walls. In some cases a layer of JIM5-reactive pectin was observed in cell walls close to the plasma membrane. There are variations among species and in some cases cell walls appear evenly unesterified.

The important point that these observations have emphasized is the local regulation of pectin esterification, and that adjacent, adhered cells appear to be capable of independent regulation of the extent of pectin esterification, as indicated by density of antibody labeling (McCann and Roberts, 1991). These observations have generally confirmed and extended those made with an antiserum specific for homogalacturonan and rhamnogalacturonan I (with immunolocalization properties similar to JIM5 and 2F4) in suspension-cultured sycamore cells, and leaves and roots of clover (Moore et al., 1986; Moore and Staehelin, 1988; Lynch and Staehelin, 1992, 1995).

How do these observations relate to our understanding of the role of pectin in primary cell walls and in cell-to-cell adhesion? The molecular domains of differences in pectin structure within cell walls indicated by the antibodies may reflect the local regulation of pH-dependent activities or modifications in cell wall porosity, with implications for growth and developmental processes. To some extent it was surprising that JIM5 and 2F4 did not always specifically recognize the middle lamella between adhered cells but were more often associated with cell separation at middle lamellae.

However, these are just two epitopes, and other structural features or configurations of pectin not recognized by these antibodies may be involved in cross-links. It is of interest that the epitope recognized by 2F4 does not occur abundantly in sugar beet or apple pectins until after treatment resulting in deesterification (both pectins have high levels of methyl esterification and sugar beet also has high levels of acetylation) (Renard *et al.*, 1993). An immunolocalization study of sugar beet calli in conjunction with the use of a methyl esterase and alkaline deesterification, which would also remove acetyl groups, indicated that the 2F4 epitope was most abundant in compact calli and that the friablity of the callus was correlated with an increase in acetylation (Liners *et al.*, 1994). During ripening of tomato fruits, a reduction in the occurrence of the JIM7 epitope and a corresponding increase in the JIM5 epitope in the pericarp cell walls has been reported (Roy *et al.*, 1992). In an interesting study that looked at domains in cell walls of ripening tomato fruit pericarp, the abundant occurrence of the JIM5 epitope at cell corners was shown to correlate with high levels of calcium ions, monitored by secondary ion mass spectrometry on the same tissue. This indicated that pectin-calcium cross-linking at these points may be preventing cell separation (Roy *et al.*, 1994). This may indicate that cell corner calcium cross-links are important in maintaining cell-to-cell contacts (and in fact treatment of tomato pericarp tissue with calcium chelators results in cell separation at the middle lamellae), and the pectin-calcium cross-links may function predominantly in preventing delamination at the corners of intercellular spaces where the stress induced by turgor pressure is greatest (Jarvis, 1992). Other linkage mechanisms may operate in the more firmly adhered middle lamellae of adjacent cell walls. Further analysis of the molecular architecture of intercellular spaces is required. Why, for example, is the JIM5 epitope often observed lining the entire intercellular spaces and not just at the corners where unesterified pectin would be expected to contribute most to the prevention of delamination?

Overall, these observations indicate that the modulation of pectin epitopes is associated with cell separation but the precise details are not yet clear and more information on pectin structure is required. There are even indications of differences in the pectin within areas of mucilages in that the JIM5 and JIM7 epitopes are restricted to regions of the surface mucilage of cap cells of hair roots of *Calluna vulgaris* (Peretto *et al.*, 1990), and pectin epitopes in clover and oat root cap cells and mucilages are subject to high levels of regulation that appear to reflect pectin metabolism (Lynch and Staehelin, 1992, 1995).

Certain cell types do not appear to contain the JIM5 or the JIM7 epitope at all. In particular, the root cap and epidermal cells of grasses do not label with these antibodies and in certain cases the stele and the cortex also show differential patterns of labeling. Often the JIM5 epitope is completely restricted to the lining of intercellular spaces (Knox *et al.*, 1990). Related patterns of epitope occurrence were also observed in the Chenopodiaceae (Knox *et al.*, 1990). Further taxonomic surveys with an extended panel of antipectin antibodies will be of interest. Schindler *et al.* (1995b) have examined the presence of the JIM5, JIM7, and 2F4 epitopes in cell walls of maize coleoptiles at the ultrastructural level in relation to auxin-induced growth. Although they found no evidence supporting a role for pectin in the control of cell expansion, they did demonstrate distinct domains of the pectin epitopes within the cell walls (Schindler *et al.*, 1995b). While pectins seem to occur throughout the outer, thicker, growth-controlling, epidermal cell wall, JIM7 binding in parenchyma cell walls was restricted to the middle lamella and cell corners. It will be interesting to discover if the absence of all three epitopes in a cell wall reflects the absence of pectin or its extensive modification.

The distribution of pectin in cell walls can also be visualized by immunolabeling the inner face of the cell wall of large tomato pericarp cells (which have lost their protoplasts). In such an approach, JIM5 immunofluorescence appeared in ordered, oriented stripes across the face of the cell wall focused at pit fields as shown in Fig. 1 (Casero and Knox, 1995). This pattern may occur because the epitope is masked in the intervening regions by pectin modification, or nonpectin material is present. It will be of great interest to determine the molecular nature of these regions. The significance of these distribution patterns is far from clear. The parameters of growth or cell wall architecture that they reflect, and their direction across the face of the cell wall, are of considerable interest. Comparable immunolabeling has not yet been observed in other cells, such as those undergoing elongation.

In addition to studies focusing on pectin as a component of cell wall architecture, antipectin antibodies are extremely useful for looking at mechanisms and pathways of pectin synthesis in the Golgi apparatus and transport of pection to the surface. Using the anti-RGI antiserum, JIM7, and probes for other cell wall polysaccharides and glycoproteins, epitopes have been located on different subsets of cisternae (Moore *et al.*, 1991; Lynch and Staehelin, 1992, 1995; Sherrier and VandenBosch, 1994a; Zhang and Staehelin, 1992). These observations have demonstrated the flexibility of the Golgi apparatus and how the targeting of vesicles with cell wall polysaccharides changes in relation to cell differentiation. They have also confirmed that pectin is synthesized in a predominantly methyl esterified form. JIM7 has also been used to study of the effects of Brefeldin A and monensin on the Golgi apparatus and on vesicle traffic (Satiat-Jeunemaitre and Hawes, 1993; Satiat-Jeunemaitre *et al.*, 1994).

Antipectin monoclonal antibodies have also been used extensively in studies of floral development and pollen–pistil interactions. Perhaps the

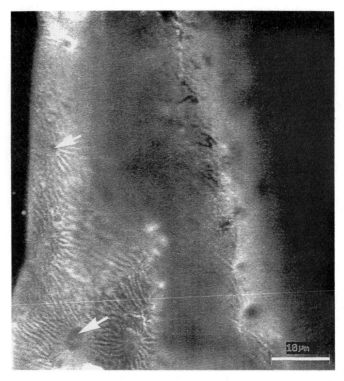

FIG. 1 The deposition of the JIM5 pectin epitope seen on the inner face of tomato pericarp cell walls. The lines of pectin are often parallel and oriented in relation to pit fields (indicated with arrows). Bar = 10 $\mu$m.

most significant observation in this area is JIM5 detection of a periodically arranged annular pattern of pectin deposits along the growing pollen tube in some species. The distance between bands is 4–7 $\mu$m and a similar pattern of arabinogalactan-protein deposition has been detected (Li et al., 1992, 1994). This has been confirmed by immunogold electron microscopy. In tobacco pollen tubes, most of the labeling was found to be in the inner part of the outer layer of the bilayered wall close to the inner callosic layer (Geitmann et al., 1995b). Do these bands of pectin and AGP have a role in the reinforcing the pollen tube? It will be of interest to determine how these domains relate to pollen tube growth. The pollen tubes of some species have a pulsatory growth rate and in several cases (but not all) tube growth with a pulsatory rather than a steady growth rate correlated with periodic pectin depositions (Pierson et al., 1995). Pollen tubes grow through the transmitting tissue of the style with its specialized secretions and controlled cell separations. In the solid style of *Brugmansia suaveolens*, the pectin of the cortex surrounding the transmitting tissue was recognized by JIM7, but that of transmitting tissue by JIM5, indicating the presence of a more acidic pectin in walls of the separated cells. JIM5 also labeled the intercellular substance, and the clustering of gold particles indicated some as yet uncharacterized organization (Vennigerholz et al., 1992, Vennigerholz, 1992). Pectin may be taken up by growing pollen tubes for incorporation into the cell wall, or its presence in the transmitting track may facilitate the passage of a pollen tube. Again, however, there are differences among species since the transmitting track intercellular space of *Smyrnium perfoliatum* labeled extensively with JIM7 (Weber, 1994).

In a recent report, the pectin and callose in cell walls of pollen tubes of *Brugmansia suaveolens* were studied *in vivo* after compatible and incompatible reactions. Pectins appeared more deesterified further away from the plasma membrane, and in the incompatible reaction the movement of pectin through the callosic inner wall apeared to be blocked and the pectin accumulated in inner layers. It was not clear whether this was due to changes in the cell wall or disruption of the secretion of cell wall polysaccharides associated with the disruption of pollen tube growth (Geitmann et al., 1995a). JIM5 and JIM7 have been used to localize pectin in relation to aspects of the development of the pollen of *Arabidospis thaliana* (Van Aelst and Van Went, 1992), *Ledebouria socialis* (Hess and Frosch, 1994), *Ophiorrhiza longiflora* (Weber and Igersheim, 1994), and *Nicotiana tabacum* (Li et al., 1995).

As well as their use in localization studies on cellular development, antipectin antibodies have a diversity of uses, such as the rapid and facile detection of pectin in a range of systems, cells, and culture media (van Engelen et al., 1991; David et al., 1995; Stacey et al., 1995; Stephenson and Hawes, 1994). They also have immense and largely unexplored potential to aid in the structural analysis of pectins by their use in immunoaffinity techniques and in the imaging of isolated molecules in conjunction with such techniques as negative staining electron microscopy (McCann et al., 1992).

Taken as a whole, these observations have clearly indicated that pectin esterification can vary within cell walls and with cell type, tissue, organ, and species. The binding of an antipectin antibody, or lack of it, does not directly correlate with the presence or absence of pectin, but of the epitope, which may be absent or modified by metabolic reactions in the cell wall. With these three antibodies we are just at the beginning of understanding the spatial and developmental regulation of pectin. We need other antipectin probes to develop a greater understanding of pectin within primary cell walls and during cell separations. Defined probes for known methyl esterification patterns, specific oligosaccharide sequences of rhamnogalacturonan backbones, and arabinogalactan side chains are now required. Antibody probes such as these will be essential for the characterization of pectin modulation in response to treatment and manipulation so that we can learn more of the growth and developmental parameters that influence and are influenced by pectin structure.

A series of monoclonal antibodies have been generated subsequent to immunization with a preparation of sycamore rhamnogalacturonan I (Puhlmann et al., 1994). Some of these, such as CCRC-M5, are specific for sycamore RGI and do not recognize maize RGI, whereas CCRC-M7 also recognized maize RGI. CCRC-M7 has subsequently been shown to recognize an epitope of at least three galactosyl residues that are $(1\rightarrow6)$-$\beta$-linked with at least one arabinose residue attached that is common to arabinogalactan-containing plant cell wall polysaccharides and plasma membrane glycoproteins (Puhlmann et al., 1994; Steffan et al., 1995). CCRC-M7 has been used in a study of the compartmentation of the Golgi apparatus (Zhang and Staehelin, 1992). A further monoclonal antibody isolated from this series of fusions is CCRC-M1, which has a high affinity for xyloglucan and is discussed later. The first monoclonal antibody directed against a highly defined pectin epitope and generated using a synthetic neoglycoprotein (a $(1\rightarrow4)$-$\beta$-linked tetragalactoside coupled to BSA) as the immunogen recognizes the neutral side chains of pectin (L. Jones, G. B. Seymour and J. P. Knox, unpublished).

## B. Other Plant Cell Wall Matrix Polysaccharides

As indicated earlier, an important approach to the generation of antipolysaccharide antibodies is the coupling of defined monosaccharides and oligosaccharides to protein to use as immunogens. The fact that several classes of polysaccharides share common compositional features of sugars and linkages means that such antisera are not necessarily specific for one class of polysaccharide. For example, the antiserum resulting from the use of an $\alpha$-L-arabinofuranoside-aminophenyl protein conjugate recognized a range of arabinose-containing polysaccharides, including arabinoxylans and arabinogalactans (Kaku et al., 1986; Misaki et al., 1988). However, this approach has been used to generate specific antisera to xylans, arabinogalactans, and callose, and the antisera indicated the presence of xylans in secondary walls, and callose and arabinogalactans in cell plates of bean and *Zinnia* tissues (Northcote et al., 1989).

Hemicelluloses present similar problems and challenges to the pectins in terms of antibody generation in that they have distinct structural elements but compositional features in common with other polysaccharides. Specific antisera, readily generated to unconjugated hemicelluloses, have generally indicated that hemicelluloses occur throughout the cellulosic regions of cell walls (Moore et al., 1986; Moore and Staehelin, 1988; Barry et al., 1991; Sherrier and VandenBosch, 1994a). A recent report of the use of an antisycamore xyloglucan antiserum (which cross-reacted with a mixed-linkage glucan) has indicated differential labeling of expanded middle lamellae in different cell types and gradients of labelling intensities across root cap cell walls at the oat root apex (Lynch and Staehelin, 1995). The conjugation of xlyoglucan to protein is also effective in producing antisera (Sone et al., 1989a,b) which were demonstrated to have biological activity in inhibiting growth, providing evidence for the role for xyloglucan metabolism in plant growth (Hoson et al., 1991). A xyloglucan heptasaccharide conjugated to BSA was used to produce an antiserum for studying the association of xyloglucan with cellulose microfibrils (Baba et al., 1994).

Although the generation of specific antihemicellulose probes appears to be straightforward by using the unconjugated polysaccharide or conjugated-oligosaccharides, only one antixyloglucan monoclonal antibody has so far been reported and this, CCRC-M1, resulted from a series of immunizations with RGI. CCRC-M1 recognized a $\alpha(1 \rightarrow 2)$-fucosylated epitope common to both RGI and xyloglucan but with a higher affinity to the latter (Puhlmann et al., 1994). This antibody has been used in a study of the compartmentation of the Golgi apparatus (Zhang and Staehelin, 1992). The generation of anti-hemicellulose monoclonal antibodies recognizing defined oligosacharides is clearly difficult and is likely to require a directed carbohydrate synthesis and neoglycoprotein preparation.

The $(1 \rightarrow 3)$-$\beta$-glucan of higher plants, callose, is known to be developmentally regulated and has been localized using fluorochromes based on aniline blue. A $(1 \rightarrow 3)$-$\beta$-glucan-specific monoclonal antibody of high specificity and affinity has been generated using a laminarin–protein conjugate (Meikle et al., 1991). This antibody recognized at least five $(1 \rightarrow 3)$-$\beta$-glucan residues and was used to develop an assay for the quantification of $(1 \rightarrow 3)$-$\beta$-glucans and to immunolocalize $(1 \rightarrow 3)$-$\beta$-glucans to the inner walls of tobacco pollen tubes (Meikle et al., 1991). This antibody has been widely used and is one of the few plant cell surface monoclonal antibodies to be available commercially. It has been used in conjunction with cryofixation technology in a detailed study of cell plate formation in tobacco root tips. In this system callose was observed to be the major lumenal component of the forming cell plates and formed a coat-like structure on the membrane surface, possibly indicating involvement in stabilizing membrane networks (Samuels et al., 1995).

Mixed-linkage $(1 \rightarrow 3,1 \rightarrow 4)$-$\beta$-glucans are unique to grasses, where they can account for up to 75% of cell wall polysaccharides in cereal endosperm and aleurone tissues. Antisera to mixed-linkage $(1 \rightarrow 3,1 \rightarrow 4)$-$\beta$-glucan have been readily obtained and found to inhibit auxin-induced elongation of maize coleoptiles, confirming a role for this hemicellulose in growth regulation (Hoson and Nevins, 1989a,b; Hoson et al., 1992). A monoclonal antibody, BG1, to a $(1 \rightarrow 3, 1 \rightarrow 4)$-$\beta$-glucan epitope has now been generated using a $(1 \rightarrow 3, 1 \rightarrow 4)$-$\beta$-glucan-protein conjugate. BG1 did not cross-react with $(1 \rightarrow 3)$-$\beta$-glucan and showed maximal binding to a heptasaccharide with the structure: Glc$(1 \rightarrow 3)$Glc$(1 \rightarrow 4)$Glc$(1 \rightarrow 4)$Glc$(1 \rightarrow 3)$Glc$(1 \rightarrow 4)$Glc$(1 \rightarrow 4)$Glc (Meikle et al., 1994). The antibody was found to bind strongly to the cell walls of wheat aleurone cells but not to the middle lamella (Meikle et al., 1994).

## C. Lignin

In terms of the characterization of structure, lignin has remained the most intractable of the major plant cell wall polymers and continues to present a considerable challenge for analysis. There appears to be a great need for specific antibody probes in view of the complexity of lignin. Also the detection of lignin is not always unequivocal, especially in lower plants (Lewis and Yamamoto, 1990). Lignin can be an extremely heterogeneous macromolecule with a potential for compositional variety in the proportion of the hydroxyphenyl propane (H), guaiacyl (G), and syringyl (S) residues, and a great diversity of linkages, including those to carbohydrate, resulting from the radical-driven reactions that are the basis of lignification. Furthermore, there are numerous indications of the spatial and developmental regulation of biochemically distinct lignins within cell walls with differences between lignins in middle lamellae, cell corners, and secondary walls being documented. A further aspect is the likelihood of different lignins occurring in different cell types in the same organ, such as xylem elements and phloem fibers, and it is also likely that there are distinctive structural features to the lignin produced due to stress such as that in compression wood or in response to pathogenic microorganisms (Lewis and Yamamoto, 1990; Wallace and Fry, 1994; Boudet et al., 1995). The biological significance (i.e., structural consequences) of these different lignin biochemistries and distributions is largely unknown.

The use of antibodies would make a suitable complementary approach to methods such as microautoradiography, $^{13}$C-nuclear magnetic resonance (NMR) and pyrolysis-mass spectrometry, which have already been used to gain an insight into cellular variations in lignin biochemistry and organization. Ruel et al. (1994) have taken an interesting approach and generated synthetic lignins by preparing three dehydrogenative polymers containing guaiacyl, hydroxyphenyl propane, and both guaiacyl and syringyl residues (GS). These synthetic lignins were used directly as immunogens to generate antisera that are capable of immunolabeling lignin in maize (Ruel et al., 1994). The methods used favor the generation of probes for uncondensed products and may not represent the bulk of natural lignins occurring in plants (Boudet et al., 1995). However, this indicates a promising approach for the development of probes for the qualitative distribution of liginin and the lignification process, especially if the structures of the compounds used as immunogens are more highly defined. Considerable potential remains for the generation of molecular probes to be used to uncover the biological significance of lignin heterogeneity and its formation and function. A further important aspect about which virtually nothing is known is the transfer of the monolignol precursors to the cell wall, and antibodies to monolignols would be important for the study of the transport mechanisms involved.

## D. Cell Wall Structural Proteins

Several groups of proteins known to occur in plant cell walls have no enzyme activity and are thought to have structural roles in cell wall architecture (Showalter, 1993). The most actively studied of these are the extensins, a family of hydroxyproline-rich glycoproteins (HRGPs) with a periodic peptide sequence and extensive glycosylation of hydroxyproline residues by oligoarabinosides and serine with single galactose residues. The precise function of extensins is far from clear. They are known to be insolubilized into the cell wall over time, often by means of oxidative reactions involving cell wall peroxidases and hydrogen peroxide. They appear to have a role in the modulation of cell wall strength or rigidity in relation to both development and defense and thus, although not abundant, they are likely to be key components in the development of cell wall architectures and, in particular, of strengthening mechanisms.

The preparation of antibody probes to extensins has received considerable attention, and antibodies to both carbohydrate and protein epitopes (Kieliszewski and Lamport, 1986; Cassab and Varner, 1987) have been readily generated. In general, the major finding from a series of immunolocalization studies with antiextensin probes involving dicotyledons, monocotyledons, and gymnosperms has been the observation that extensins do not occur in all cell walls but are extensively developmentally regulated. Therefore, what has been thought of as the major structural cell wall protein or family of proteins does not have a fundamental role in plant cell walls in the way, for example, that cellulose does, in that certain plant cell walls can do without them and they appear to be placed in the walls only in certain cells as an aspect of differentiation.

An antiserum to soybean seed coat extensin indicated an abundance of extensin in the two sclereid cell layers of the testa known as palisade and hourglass cells. The extensin was found to accumulate in these cells after differentiation had started and just before drying, indicating that the extensin is a structural protein with a mechanical or protective role in the testa (Cassab and Varner, 1987). This antiserum was also used to pioneer a tissue-printing technique that is particularly applicable to plants, in which soluble antigens are transferred from a cut surface of plant material to an absorptive medium such as nitrocellulose membrane and then probed with antibodies (Cassab and Varner, 1987, Varner and Ye, 1994). The antisoybean antiserum also recognized cells in the region of vascular tissue and cross-reacted widely with extensins of other species. It indicated an abundance of extensin in epidermal cells, subepidermal cortical cells, and vascular tissue of pea epicotyls (Cassab *et al.*, 1988). Further instances of developmental regulation include abundant reaction of this antiserum with the megagametophyte and root cap of the gymnosperm *Araucaria*, which may also indicate a protective role in strengthened walls (Cardemil and Riquelme, 1991). This antiextensin has also been used in a more extensive study in developing soybean tissues, confirming and extending observations on developmental regulation (Ye and Varner, 1991). Antisera to maize extensin have been generated using sequence information to prepare peptide conjugates and purifed maize extensins. Both sets of antisera confirmed observations made in other species that extensins are developmentally regulated and that maize extensin occurred in all cell walls of the embryonic axis but not in the scutellum (Stiefel *et al.*, 1988; Ludevid *et al.*, 1990; Ruiz-Avila *et al.*, 1991). The use of an antiserum to carrot extensin-1 indicated that this extensin was found only in cell walls of secondary root growth; in the phloem parenchyma it appeared to be reduced in the middle lamella and not to occur at the expanded middle lamellae of intercellular spaces. This suggests some control of its movement within the cell wall and that it cannot pass from cell to cell (Stafstrom and Staehelin, 1988). A second antisera to the less abundant extensin-2 of carrot had a distinct pattern of localization in that its binding was restricted to the expanded middle lamellae at cell junctions, suggesting that this extensin is likely to have different properties, interactions, and functions (Swords and Staehelin, 1993). How glycoproteins such as extensin move through cell walls and whether they are always added as cell layers are built up or can be inserted in the wall matrix later are important questions to answer. The appearance of extensin epitopes at the lining of intercellular spaces (see later discussion) seems to indicate that they can move through cell walls and be targeted to specific regions as required.

Antisera have clearly demonstrated the spatial and developmental regulation of extensin. The nature and extent of extensin glycosylation are known to vary both taxonomically (Lamport and Miller, 1971) and developmentally (Klis and Eeltink, 1979), and a first report of the fine details of glycosylational patterns has indicated that contiguous hydroxyproline residues are the major sites of $O$-arabinosylation in a Douglas fir HRGP

(Kieliszewski et al., 1995). Monoclonal antibodies will be needed to explore the occurrence in cell walls and the developmental significance of the glycosylational motifs as they are elucidated. Several antiextensin monoclonal antibodies have been generated as indicated in Table 1. The monoclonal antibody 11D2, generated subsequent to immunization with membranes from tobacco cell suspension cultures, was found to recognize an epitope carried exclusively by extensin, and the antibody did not cross-react with potato lectin (Meyer et al., 1988). Solanaceous lectins are known to contain hydroxyproline-containing and arabinosylated domains similar to those of extensins. The antibody 11D2 recognized extensin in sclerenchyma cells of soybean seed coats and confirmed observations on developmental regulation made with antisera (Meyer et al., 1988). Developmental variations in extensin occurrence were also demonstrated in maize pericarp and maize embryos with a series of monoclonal antibodies. The epitopes recognized by these antibodies were not characterized, but they were not carried by carrot extensin (Fritz et al., 1991).

In a program of work using hybridoma technology to dissect the complexity of the cell surface of suspension-cultured carrot and rice cells and pea guard cells a panel of monoclonal antibodies have been generated that cross-react with both carrot extensin and solanaceous lectins (Smallwood et al., 1994, 1995; Knox et al., 1995). These antibodies have been used to study patterns of extensin epitope regulation within cell walls during tissue differentiation at the carrot root apex. The carrot root apex is currently the system in which cell surface changes associated with plant cell development have been most extensively documented (Knox, 1995).

As seen in a sequential series of transverse sections moving from the root tip toward the root body, the monoclonal antibody JIM11 recognized: central root cap cells, cortical cells, cortical cells plus cells adjacent to phloem elements and then, in addition, the epidermis (Smallwood et al., 1994). A second antiextensin antibody, JIM12, recognized only the expanded middle lamella of intercellular spaces of the double-layered pericycle and the cell walls of developing metaxylem and phloem elements (Smallwood et al., 1994). Figure 2 contains examples of transverse sections of the carrot root apex immunolabeled with JIM11 and JIM12. The restriction of the JIM12 epitope to the lining of intercellular spaces of one tissue was particularly striking and clearly indicated that this extensin must somehow be directed across the cell wall to this location during the formation of these particular spaces. Whether all of the three cells adjacent to the intercellular space are responsible for this particular modification is unknown. The structural requirements connecting the diverse cell types and cell wall regions are not clear. A third monoclonal antibody, JIM20, recognized an epitope common to both the JIM11 and JIM12 sets of antigens, with a corresponding localization pattern at the root apex (Smallwood et al. 1994).

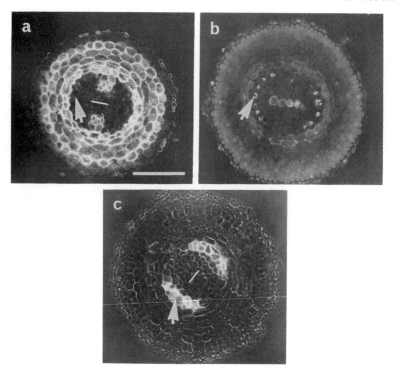

FIG. 2 Developmentally regulated cell surface epitopes at the carrot root apex. Immunofluorescent micrographs of transverse sections of the carrot root apex, in a region before cell differentiation and vacuolation, showing different labeling patterns with the antiextensin monoclonal antibodies JIM11 (a) and JIM12 (b) and the anti-AGP monoclonal antibody JIM4 (c). In each case the arrows indicate the position of an intercellular space between the cells of the double-layered pericycle that is adjacent to the protoxylem poles. The lining of this and related spaces is bound by JIM12 in (b). The position of several adjacent future metaxylem cells is indicated by a line in (a) and (c) and is fluorescently labeled by JIM12 in (b). JIM11 recognized two pairs of pericycle cells and associated stele cells surrounding the phloem sieve elements. JIM4 recognized cells of the two arcs of the double-layered pericycle. Bar = 100 μm.

These observations suggest developmentally related modifications and/or processing of extensin.

JIM20 was generated subsequent to immunization with a preparation of pea guard cell protoplasts and isolated along with a further antiextensin monoclonal antibody, JIM19, because of their patterns of binding related to stomatal architecture in pea and *Commelina* epidermal peels (Knox *et al.*, 1995). A remarkable feature of JIM19 is its biological activity in inhibiting the action of abscisic acid-induced expression of *RAB-16* mRNA in barley aleurone protoplasts, suggesting that the ABA receptor, or an associated molecule at the plasma membrane carries the JIM19 epitope (Wang *et al.,* 1995). Other antiextensin antibodies were not active in this assay. The significance of this is far from clear but it is of interest in relation to this that ABA has itself been immunolocalized to plant cell walls (Bertrand *et al.,* 1992) and that ABA was rapidly released when rose petal cell walls were digested with wall-degrading enzymes (Bianco-Trinchant *et al.,* 1993). Additional evidence that plasma membrane proteins carry extensin-like epitopes has been obtained in rice. The monoclonal antibody LM1 was generated subsequent to immunization with a preparation of the conditioned media of suspension-cultured rice cells and found to recognize an epitope carried by rice extensin (a threonine- and hydroxyproline-rich glycoprotein); however, it also recognized hydrophobic proteins of the plasma membrane and protoplasts of rice cells (Smallwood *et al.,* 1995). At the rice root apex, LM1 bound to a stellate pattern of cells associated with the xylem in addition to the epidermis.

JIM11, JIM20, and LM1 recognized periodate-sensitive epitopes, indicating a carbohydrate component to the epitope and also bound to potato, *Datura,* and tomato lectins, whereas the JIM12 epitope was not periodate sensitive and the antibody bound only tomato lectin (Smallwood *et al.,* 1994, 1995). The fine details of the epitopes bound by these antibodies are not yet known. There is now a need for probes that recognize defined glycopeptide epitopes of extensins so that the developmental regulation and modification of extensins can be understood at the biochemical level and mapped in relation to the strengthening or restructuring of plant cell walls.

Other classes of structural cell wall proteins include the proline-rich proteins (PRPs). The major observation from immunolocalization studies with antisera is that PRPs are also not constitutive cell wall proteins, but are localized to vascular tissues, where their presence correlates with patterns of lignification. In certain cases they also occur specifically in regions of the expanded middle lamellae at intercellular spaces of cortical parenchyma tissues. The patterns of PRP expression are similar in bean and three solanacecous species (Marcus *et al.,* 1991; Ye *et al.,* 1991).

Immunolocalization studies with antisera have also confirmed the nonconstitutive nature of the glycine-rich cell wall proteins (GRPs) that are also thought to be structural proteins. In bean, GRP1.8 was associated specifically with vascular tissue and in particular the protoxylem, where it may have a role in the expansion of the thickened cells or lignification (Keller *et al.,* 1988, 1989). It is of particular interest, in relation to understanding the movement of proteins through cell walls, that GRP1.8 is proposed to be transported to the protoxylem cell walls (and to cell corners of metaxylem elements) from neighboring xylem parenchyma cells. This conclusion arose from immunolabeling experiments that indicated GRP occurred in the Golgi vesicles of the parenchyma cells but not abundantly in the primary cell walls of these cells (Ryser and Keller, 1992). An association of GRP expression with the lignification of xylem vessels and phloem fibers in vascular tissues has been reported for stems of soybean, tomato, petunia, and tobacco (Ye and Varner, 1991; Ye et al., 1991). GRPs appear in these cell types before lignin and therefore may be involved in setting up a framework for the initiation of lignification. In this way GRPs could influence the formation of local domains of lignin biochemistry and lignin archictecture.

Antibodies remain essential tools for demonstrating the unequivocal localization of putative cell wall proteins and for elucidating the patterns of developmental occurrence. As such, they have confirmed the discovery of proteins occurring in secondary cell walls, including a protein isolated from mature wood of loblolly pine (*Pinus taeda* L.) (Bao et al., 1992), a tyrosine and lysine-rich protein occurring in the lignified secondary walls of tomato (Domingo et al., 1994), and a secondary cell wall glycoprotein of bean hypocotyls (Wojtasek and Bolwell, 1995).

### E. Cell Wall Enzymes

Clearly, antibodies are useful for confirming the location of the numerous enzymes that are known to occur in the cell wall (see Fry, 1995 for a recent review of polysaccharide modifying enzymes). Unlike the preparation of antibody probes to cell wall polysaccharides, there are no general problems of cross-reaction in the use of antibodies to proteins or peptide sequences, and antisera to a variety of cell wall enzymes have been generated (Hoson, 1991). There appears to be no extensive observation of restricted domains of enzyme occurrence within cell walls, although such domains may be expected for enzymes that are involved in cross-linking reactions or in modifying cell wall architecture. Monoclonal antibodies to peroxidases have indicated a predominantly cell wall location in maize and were used to quantify changes in response to light; levels of a peroxidase were observed to increase within 10 min of irradiation (Kim et al., 1988, 1989). A series of monoclonal antibodies that recognize and distinguish $(1 \rightarrow 3,1 \rightarrow 4)$-$\beta$-glucanase isoenzymes has been generated (Høj et al., 1990).

## IV. Uncovering Cell Surface Markers of Cell Development with Monoclonal Antibodies

### A. Arabinogalactan-Proteins

Arabinogalactan-proteins are the only known class of plant proteoglycans and occur abundantly in plant secretions. Their function is unknown but they appear to be involved in diverse aspects of plant cell development. AGPs have not been extensively reviewed since Fincher *et al.* (1983), but see Chasan (1994) for a commentary on the recent cloning of cDNAs encoding AGP core proteins. In summary, AGPs are a complex and diverse group of molecules. They can consist of more than 90% carbohydrate that is composed of a branched galactan framework substituted with arabinose, uronic acids, and other sugars. In some cases it has been ascertained that the complex glycan is $O$-linked to hydroxyproline or serine. cDNAs encoding five protein cores have been cloned to date and the protein components of AGPs are varied and developmentally regulated (Mau *et al.*, 1995). Monoclonal antibody technology has also had considerable impact on our understanding of the biology of AGPs, and continues to do so, in that antibodies are helping to put some of the molecular complexity in a developmental context and are also helping with the elucidation of the function of AGPs.

Anti-AGP monoclonal antibodies were developed initially after immunization procedures with complex plant extracts (Anderson *et al.*, 1984; Knox *et al.*, 1989; Pennell *et al.*, 1989; Norman *et al.*, 1990). These antibodies predominantly recognized carbohydrate epitopes, and the arabinogalactan component is now thought to have the capacity to dominate an immune response to a preparation of plant material. Several monoclonal antibodies were generated subsequent to immunization with style extracts from *Nicotiana alata* and the binding of these to style extracts was inhibited by arabinose and/or galactose, or oliogsaccharides and glycoproteins containing these sugars (Anderson *et al.*, 1984). AGPs were the major antigens for these antibodies in the style extracts. The monoclonal antibody PCBC3 was found to preferentially bind to $\alpha$-L-arabinofuranosyl residues and to recognize AGPs, arabinoxylan, and an arabinan (Anderson *et al.*, 1984, 1987). PCBC3 has been found to bind to the surface of growing pollen tubes and used to develop a quantitative assay for *in vitro* pollen tube growth (Harris *et al.*, 1987). It has also been used for the immunogold localization of arabinofuranosyl residues in pollen tubes and for transmitting tissue of *Nicotiana alata* (Anderson *et al.*, 1987; Sedgley *et al.*, 1985). PCBC4 was isolated from the same fusion as PCBC3 and has been characterized as recognizing a style-specific epitope throughout the Solanaceae, and in *Nicotiana alata* recognizes style AGPs and a glycoprotein with properties of both AGPs and the extensins (Lind *et al.*, 1994). The binding of PCBC4 was found to be inhibited by galactose, arabinose, and most effectively, by the disaccharide Gal(1 $\rightarrow$ 6)-$\beta$-Gal (Lind *et al.*, 1994).

Style AGPs are soluble. Several monoclonal antibodies to plasma membrane components have been generated and subsequently found to bind to glycoproteins with the characteristics of AGPs (Norman *et al.*, 1986; Knox *et al.*, 1989; Pennell *et al.*, 1989). In some cases antibodies generated against soluble AGPs were found to bind to plasma membranes (Knox *et al.*, 1991), confirming the association of AGPs with the plasma membrane. PN16.4B4 was generated subsequent to immunization with membranes from tobacco suspension-cultured cells and found to bind to a series of AGP-like glycoproteins appearing at 135–180 kDa on sodium dodecyl sulfate–polyacrylamide gel electrophoresis (SDS-PAGE) analysis and with a 50-kDa core protein (Norman *et al.*, 1986, 1990). The PN16.4B4 epitope has been immunolocated to the external face of the plasma membrane and to intracellular and intravacuolar multivesicular bodies in tobacco callus and leaf cells, which suggests a mechanism of vacuolar-mediated disposal of material from the cell surface (Herman and Lamb, 1992). MAC207, generated subsequent to immunization with a preparation of pea peribacteroid membrane (see later discussion and Bradley *et al.*, 1988), and characterized in relation to suspension-cultured carrot cells, was found to bind to an epitope that was specific to flowering plants. Its binding was inhibited by high levels of arabinose and glucuronic acid (Pennell *et al.*, 1989). MAC207 has been used for the immunoaffinity purification of an extracellular AGP for analysis (Baldwin *et al.*, 1993).

It was with the generation of monclonal antibody JIM4 that AGP epitopes were given a new perspective. JIM4 was generated and characterized in relation to the suspension-cultured carrot cell line that MAC207 had recognized, and had the same broad characteristics in this system. However, in sections of the carrot root apex, MAC207 bound to the plasma membane of all cells whereas JIM4 only recognized two symmetrical sets of cells early in development, centered upon bilayered sectors of the developing pericycle (Knox *et al.*, 1989). The specific immunofluorescent labeling by JIM4 of two sectors of the pericycle in a region of the carrot root apex before differentiation of the vascular tissues is complete is shown in Fig. 2c. Furthermore, in carrot somatic embryos, the appearance of the JIM4 epitope was found to be intimately associated with the formation of the tissue systems (Stacey *et al.*, 1990).

The observations that an AGP epitope was associated in some way with patterning events at a root meristem and in a developing embryo expanded interest in AGPs. Two other, related patterns of AGP epitopes at the carrot root apex that were recognized by antibodies JIM13 and JIM15 were subsequently reported (Knox *et al.*, 1991). JIM13 recognized cells of the region of the developing xylem and epidermis, and JIM15 cells other than these. For a discussion of these observations and how they relate to the patterns of extensin epitopes at the carrot root apex discussed earlier, see Knox (1995). Developmental regulation of AGP epitopes of the plasma membrane has also been reported in floral tissues, where the MAC207 epitope is lost from the generative cells in pea (Pennell and Roberts, 1990) and the JIM8 epitope sequentially appears and disappears in cell layers of the developing flower of oilseed rape (Pennell *et al.*, 1991). Members of this panel of antibodies have been used to study developmental regulation of the AGP epitopes in pollen grains (Van Aelst and Van Went, 1992); the surface of growing pollen tubes, where the JIM8 epitope occurred in a periodic manner equivalent to pectin epitopes, as discussed earlier (Li *et al.*, 1992); pea root nodules (Rae *et al.*, 1991); primary and secondary growth of *Arabidopsis* roots (Dolan *et al.*, 1995; Dolan and Roberts, 1995); and the growth of maize coleoptiles (Schindler *et al.*, 1995a).

In a study of a suspension-cultured rice cells and a comparison with the carrot cell line, no rice-specific or carrot-specific anti-AGP antibodies were isolated. A further anti-AGP monoclonal antibody, LM2, was generated using the rice system, and LM2 bound to two AGPs in the conditioned medium of the cultured rice cells whereas the anti-AGP probes developed in the carrot system bound to just one of these (Smallwood *et al.*, 1996). An important aspect of AGPs is their association with the plasma membrane; the relationship of membrane-bound with soluble extracellular AGPs is not understood. LM2 was also used to indicate that the membrane-associated AGPs in rice were hydrophobic in that they partitioned into the detergent fraction in detergent/aqueous partitions, unlike the situation in carrot where AGPs partition into the aqueous phase. Why rice and carrot AGPs differ in this way is not yet resolved (Smallwood *et al.*, 1996). It is of interest that in some cases the cDNAs encoding core proteins of soluble AGPs indicate probable C-terminal transmembrane helices (Mau *et al.*, 1995).

The immunolocalization studies of the AGP epitopes in different species have indicated that a particular epitope does not occur at the same developmental location in all species. For example, the JIM13 epitope is associated with epidermis and developing metaxylem cells at the carrot root apex (Knox *et al.*, 1991), with a central metaxylem initial cell and endodermal cell files in the *Arabidopsis* primary root (Dolan *et al.*, 1995), with periderm cells and xylem development during the secondary growth of *Arabidopsis* roots (Dolan and Roberts, 1995), with the endodermis of pea root nodules (Rae *et al.*, 1991), and with the plasma membrane of developing sclerenchyma cells and the secondary wall thickenings of developing tracheids in maize coleoptiles (Schindler *et al.*, 1995a). It would therefore be of interest to map the developmental occurrence of AGP epitopes in a wider range of species, including lower plants, in relation to taxonomic diversity. Mapping the presence or absence of the MAC207 and JIM8 epitopes within groups of flowering plants appears to indicate some taxonomic significance (Q. C. B. Cronk, personal communication). The further documentation of locations within the plant body of higher and lower plants would be of considerable interest.

Currently, we have no clear idea of how many developmentally regulated AGP epitopes could be detected within a single plant body or within flowering plants. In the carrot root there are three such epitopes. It is also not clear how such knowledge would be obtained. Our understanding of the structures recognized by the existing anti-AGP antibodies is limited although there are indications from hapten inhibition studies that $\beta$-linked glucuronosyl residues are important for recognition by MAC207, JIM4, JIM13, and LM2 (Yates *et al.,* 1996). A complementary, directed approach to understanding the biochemical basis of the developmental regulation of AGP structure is the use of synthetic neoglycoproteins. An antisera to $(1 \rightarrow 6)$-$\beta$-galactosyl residues has been generated and found to bind widely to cell surfaces in radish (Kikuchi *et al.,* 1993). There is clearly immense scope for synthetic carbohydrate chemists to prepare defined oligosaccharides containing arabinosyl, galactosyl, glucuronosyl, and other residues for the future generation of defined probes for the glycan of AGPs.

The demonstration of biological activity of AGPs when added to embryogenic carrot cell cultures has focused interest on the elucidation of their function. The addition of AGPs derived from carrot seeds has been reported to promote embryo formation (Kreuger and van Holst, 1993) and these AGPs are effective across species boundaries (Kreuger *et al.,* 1995). Two monoclonal antibodies, ZUM15 and ZUM18, directed against carrot seed AGPs, were used to dissect the active AGPs from both carrot and tomato seeds by immunoaffinity chromatography procedures. Several fractions defined by the presence of a single epitope or both epitopes appeared to have different activities (Kreuger and van Holst, 1995). ZUM15-binding AGPs induced vacuolation of cells and reduced the production of embryos, whereas ZUM18-binding AGPs promoted embryo formation.

The embryogenesis-inhibiting and -promoting epitopes were carried on separate as well as the same AGP molecules, indicating a complex biochemistry underlying AGP biosynthesis or processing (Kreuger and van Holst, 1995). These are extremely important observations and warrant further study in association with biochemical studies on the perception of the AGPs by the cells, and molecular responses to uncover mechanisms of action. The activity of adding AGP-binding synthetic phenylazo-glycosides, known as Yariv reagents, to arrest cell proliferation is also extremely interesting in relation to the elucidation of AGP function (Serpe and Nothnagel, 1994).

There is still considerable progress to be made in understanding the fine details of AGP structure, and antibodies are likely to play a role in this. Although it is possible to speculate that AGPs have role in signaling between cells (Dolan *et al.,* 1995; Knox, 1995; Pennell *et al.,* 1995), such a role is far from clear. Many surprises surely lie ahead as we learn more about this remarkable group of molecules in which antibodies have given us such insight about their abundance at the surface of all plant cells and their extensive heterogeneity within the plant body.

## B. Other Cell Wall and Plasma Membrane Antigens

Several other monoclonal antibodies have been derived that specifically recognize the plant cell plasma membrane. The monoclonal antibody JIM1 recognized the outer face of all plasma membranes in all higher plants examined, and this binding was inhibited by $\beta$-linked galactosyl residues but the nature of the antigen was not determined fully (Knox and Roberts, 1989). The monoclonal antibody JIM18 also recognized most plant plasma membranes and bound to a glycosylated form of an inositol-containing phospholipid (Knox *et al.*, 1995; Perotto *et al.*, 1995). A series of monoclonal antibodies to protoplasts prepared from *Vicia faba* guard cells has been generated, some of which recognized protein epitopes of the plasma membrane that were both cell type and species specific (Key and Weiler, 1988). The monoclonal antibody 40.1C2.8 recognized a protein epitope of a 40-kDa protein of the oat root (Stout and Griffing, 1993). A study of the localization of this antigen is very interesting in that in root cap meristem cells it occurred in the vacuole, but as the cells matured it became abundant at the plasma membrane and then in the inner layer of the cell wall (Stout and Griffing, 1993). Whether the absence of the epitope throughout the cell wall of mature root cap cells was due to a barrier to antigen movement or the loss or masking of the epitope is not known.

## V. Monoclonal Antibodies to Algal Cell Walls

Algae display an extensive diversity of morphology and cell wall biochemistry that is generally quite distinct from that of higher plants; therefore specific sets of antibodies to probe cell wall organization are required. Several series of monoclonal antibodies have been developed to aid the analysis of algal cell surfaces and in general have indicated the heterogeneity of algal matrix polysaccharides and glycoproteins, reflecting that observed in higher plant cell walls. Carrageenan is the major cell wall polysaccharide of certain red algae and it occurs in different forms known as kappa, iota, and lambda, which have increasing states of sulfation, but little is known about the distribution of these forms within cell walls or their synthesis. Low states of sulfation and high levels of 3,6-anhydrogalactose are known to correlate with high gel strengths. Eight different monoclonal antibodies have been generated to carrageenans of the red algae *Kappaphycus,* and these were sorted into three groups that preferentially bound to kappa, iota, or lambda epitopes (Vreeland *et al.,* 1992). The epitopes bound by these antibodies were not defined precisely because of a lack of defined carrageenan fragments, but the antibodies were specific in that they did not cross-react with agar-producing algae. These antibodies were used in immunofluorescence studies and demonstrated developmental differences and provided evidence for the intracellular synthesis of carrageenan. Kappa-related epitopes were located throughout the cell wall whereas iota-related epitopes were more abundant in the middle lamella, and lambda epitopes were more abundant at the cuticular surface of the algae where iota and kappa carrageenan epitopes were absent (Vreeland et al., 1992).

The fibrous and acidic matrix cell wall polysaccharides of brown alga have attracted considerable attention in terms of antibody preparation. The principal cell wall polysaccharides are alginates, fucans, and cellulose. The early generation of monoclonal antibodies to alginates and fucans, isolated from the brown alga *Fucus distichus,* demonstrated that it was possible to generate specific probes. However, as in higher plants, the precise structures of the epitopes were not easy to determine, but also as in higher plants, the antibodies have been useful in that they have indicated a variety of labeling patterns that reveal the molecular heterogeneity of algal cell wall matrix polysaccharides (Vreeland, 1970, 1972; Vreeland et al., 1984). One of these antibodies, 2-8.7, was found to be specific for the polyguluronate sequences of alginate, although this specificity appeared to change with antibody incubation conditions, and the fine details of antibody specificity have not been characterized (Larsen et al., 1985). Several monoclonal antibodies directed against the cell walls of the kelp *Macrocystis pyrifera* were found to be reactive with fucoidin and one cross-reacted with the xyloglycur-onan ascophyllan (Eardley et al., 1990). The antibody 3H8 was specific for sulfate determinants on fucoidan that were present on the surface of the developing gametophyte at a very early stage (Eardley et al., 1990). Twelve monoclonal antibodies generated to the sperm of *Fucus serratus* indicated a restricted surface localization for a series of glycoprotein epitopes (Jones et al., 1988, 1990). Several of these same monoclonal antibodies recognized glycoprotein epitopes at the surface of *Fucus* eggs, and the restricted localization patterns at the plasma membrane surface indicated stabilized and exclusive molecular domains (Stafford et al., 1992). These antibodies, plus a further monoclonal antibody F13 (directed against fucoidan and ascophyllan), were also used to look at vegetative tissues (Green et al., 1993). In this case, many of the antibodies again bound to glycoproteins and indicated heterogeneity and molecular domains. The localization of several epitopes of fucose-containing polysaccharides indicated considerable heterogeneity in these polysaccharides in the cell walls of the thallus, the medulla, and the mucilage plug of the ostiole of *Fucus* (Green et al., 1993).

The unicellular green alga *Chlamydomonas reinhardii* has received considerable attention. Its vegetative cell wall is made up of arrays of glycoproteins, the most abundant being 2BII, with compositional similarities to the extensins. Some of the first monoclonal antibodies to plant cell wall components were made against these glycoproteins (Smith et al., 1984). This panel of antibodies was highly specific, showed no cross-reaction with higher plant cell wall proteins, and was split into six groups on the basis of competitive binding assays. These antibodies have been used to study the pathway of biosynthesis of the 2BII glycoprotein (Grief and Shaw, 1987) and in an ultrastructural examination of the tougher and thicker zygote cell wall that has a layered structure (Grief et al., 1987). Monoclonal antibodies have also been generated to the cell surface of *Chlamydomonas eugametos,* and strain-specific epitopes involving *O*-methylated sugars were identified at the surface of both vegetative cells and gametes (Homan et al., 1987).

A novel sulfated glycoprotein, ISG, of *Volvox carteri* is an HRGP that is closer to the extensins of higher plants than other algal HRGPs (Ertl et al., 1992). ISG is synthesized for only a few minutes at the end of embryogenesis when cells forming the surface of a hollow sphere are inverted. Immunolocalization studies with antibodies to ISG indicated that it is under tight developmental control and has an important role in the early stages of the formation and organization of the extracellular matrix architecture (Ertl et al., 1992).

## VI. Cell Walls and Plant–Microbe Interactions

The interaction of plants with other organisms often involves direct interaction with the plant cell wall and the plasma membrane, and the probes developed for cell surface components are directly applicable to studies of modifications of cell surfaces during the development of symbioses or during pathogenic invasion. Many of the monoclonal antibodies to determinants of higher plant cell surfaces have been used to study cell wall changes and also to identify whether the structures that develop at the interface between organisms are plant or nonplant in origin. In some cases the modified plant cell structure has been used as the immunogen and its molecular nature dissected using hybridoma technology. This approach has been taken for the plant plasma membrane-derived peribacteroid membrane (PBM) that envelops bacteroids during the *Rhizobium* invasion of legume root nodules. Using a preparation of PBM as immunogen monoclonal antibody, MAC64 was generated and found to recognize an abundant glycoprotein component of both the PBM and the plasma membrane (Brewin et al., 1985). Continuing and refining this approach, several classes of monoclonal antibody that recognized PBM components were generated and used to study membrane composition and synthesis and interactions with the bacteroid (Bradley et al., 1986, 1988; VandenBosch et al., 1989; Perotto et al., 1991, 1994b). One group of antigens comprises a range of membrane glycoproteins, common to the PBM and the plasma membrane of cells of uninfected nodules and non-nodule tissue, and is typified by those recognized by MAC209 and the anti-AGP MAC207 (Bradley et al., 1988; Perotto et al., 1991).

A second class of PBM and plasma membrane epitopes, typified by that recognized by MAC206, were common to membrane glycoproteins and glycolipids (Perotto et al., 1991). Acidic and neutral forms of a third class of glycoproteins, recognized by MAC236 and MAC265, were found to be components of the infection thread matrix and also the matrix of developing intercellular spaces between cells of uninfected tissues (VandenBosch et al., 1989). The nature of the similarity of the infection thread and intercellular space of uninfected tissues in pea and bean has been explored further using these and antipectin probes (VandenBosch et al., 1989; Rae et al., 1992).

The use of MAC265 has also indicated that the glycoproteins were expressed early in the differentiation of *Rhizobium*-induced nodules and were altered in relation to oxygen tension, indicating a role as a structural component influencing oxygen diffusion resistance in the cortex of legume nodules (James et al., 1991; Rae et al., 1991). In most cases nodule antigens also occurred in uninfected nodules and non-nodule plant tissues, which suggests that much of nodule development involves a rearrangement of components used in normal cell development. It is also of interest that antigens, such as AGPs, are maintained in the membrane as it develops into the PBM, even though this membrane is not an interface with a primary wall but with a bacteroid. In no case has a PBM-specific antigen been identified.

This lack of nodule-specific cell surface components is also exemplified by the observation of the JIM13 AGP epitope during nodule development, where it was an early marker of endodermal cells and was detected before wall thickenings (Rae et al., 1991). JIM18 recognized a widespread, chloroform-soluble plasma membrane antigen and stomatal complexes in pea epidermal peels (Knox et al., 1995). The JIM18 antigen was characterized further in the pea nodule and identified as a glycosylated form of an inositol-containing phospholipid. It was found to be associated with all plasma membranes but was developmentally regulated on the PBM. The antigen occurred in the PBM around actively dividing bacteria, but was abruptly lost at an early point in the maturation of the symbiotic compartment, concomitant with the occurrence of starch (Perotto et al., 1995). Proline-rich proteins also located to secondary walls of xylem elements and phloem fibers of pea root, the primary wall of the nodule endodermis, the casparian strip of the vascular endodermis, and the infection thread matrix of pea nodules (Sherrier and VandenBosch, 1994b).

Clearly this approach and some of these antibodies can be used to study cell walls and membranes in other situations. Several antibodies have been used in a comparative study of the colonization of pea roots by *Rhizobium* bacteria and the mycorrhizal fungus *Glomus versiforme*. This study indicated both common and distinct components of the perisymbiotic membranes of these two systems (Perotto *et al.*, 1994a). In the case of arbuscular mycorrhiza, the fungus is in direct contact with plant cell walls in intercellular regions or it invaginates the plant cell and is separated from the host cytoplasm by interfacial material. This interfacial material has a continuity with the cell wall because wall components, including pectin and HRGPs, have been immunolocalized to it, although callose has not been detected (Bonfante-Fasolo *et al.*, 1990; Bonfante *et al.*, 1991; Balestrini *et al.*, 1994). An isogenic mycorrhiza-resistant mutant interacting with *Glomus mosseae* has been studied using JIM8, MAC207, MAC236, and MAC265 (Gollotte *et al.*, 1995).

Antibodies have also been used to dissect the molecular components at the interface between organisms during the invasion of plant tissues by microbial pathogens. For example, the breakdown of pectins in leaf tissue was followed after invasion by *Erwinia* pathogens (Temsah *et al.*, 1991) and the accumulation of HRGPs at sites of resistance to bacteria and fungi (O'Connell *et al.*, 1990). In this latter case, HRGPs accumulated in cells next to dead cells and in papillae encasing intracellular fungal hyphae. A series of antipolysaccharide probes were used to study cell wall synthesis in cotton roots after infection with *Fusarium* and the induced appositions contained the normal cell wall polysaccharides plus high levels of callose (Rodriguez-Gálvez and Mendgen, 1995). Callose has also been localized to the apposition layers that form during infection of tomato with *Fusarium* (Mueller *et al.*, 1994) and HRGPs, and callose deposition was observed following *Pseudomonas* infection of lettuce (Bestwick *et al.*, 1995). Similarly, of course, the antibody approach can be used to dissect the fungal side of things (eg. Mackie *et al.*, 1991; Pain *et al.*, 1994).

Existing antipectin, extensin, and AGP probes were used as markers of anatomy and to survey cell wall components and changes associated with infection of roots by the parasitic root-knot nematode *Meloidogyne incognita* (Gravato-Nobre *et al.*, 1995). Systems such as this in which nematodes alter root cell morphology offer great potential for elucidating the molecular aspects of the restructuring of plant cells and their walls that are exploited by invading and interacting organisms.

## VII. Conclusions and Prospects

The past 10 years have seen an increased appreciation of the details of plant cell wall architecture, the existence of molecular domains within cell walls, and the changes in cell wall and plasma membrane composition associated with cell differentiation. Antibodies have contributed greatly to this refined understanding of the plant cell surface. Recent developments have extended the ability to isolate and manipulate antibodies from the cell to the gene level. Technologies are now available that can be used to design and select molecular probes by manipulating the DNA that encodes antibody binding sites. These abilities will dramatically increase the flexibility and subtlety of antibody probes in relation to plant cell surfaces. For example, it is now possible to produce smaller, single-chain forms of antibodies that are readily diffusible through the pores of the primary cell walls, or bispecific probes with two distinct binding capacities combined in one molecule, and such probes will facilitate the elucidation of the function of cell wall components.

We still need to know how the domains of cell walls are set up, how proteins are directed to intercellular spaces, and the functions of these local variations in cell wall composition and architecture. Several monoclonal antibodies to cell wall components of higher and lower plants are now widely used and are useful in a range of systems and species. There is a need now to generate probes of more highly defined specificities. In terms of glycans, this will require the preparation of neoglycoproteins using oligosaccharides with 5–7 sugars. The chemical synthesis of such defined oligosaccharides will require dedicated effort. The existing range of monoclonal antibodies and future probes will be most valuable for use in dynamic systems that can be manipulated to see the restructuring of cell walls in response to mechanical and biotic stresses and developmental signals. In addition, the use of defined antibody probes in taxonomic surveys will create a wider perspective on the role that cell walls have played in the development and evolution of plant form.

## References

Anderson, M. A., Sandrin, M. S., and Clarke, A. E. (1984). A high proportion of hybridomas raised to a plant extract secrete antibody to arabinose or galactose. *Plant Physiol.* **75,** 1013–1016.

Anderson, M. A., Harris, P. J., Bonig, I., and Clarke, A.E. (1987). Immunogold localization of α-L-arabinofuranosyl residues in pollen tubes of *Nicotiana alata* Link et Otto. *Planta* **171,** 438–442.

Baba, K., Sone, Y., Misaki, A., and Hayashi, T. (1994). Localization of xyloglucan in the macromolecular complex composed of xyloglucan and cellulose in pea stems. *Plant Cell Physiol.* **35,** 439–444.

Bacic, A., Harris, P. J., and Stone, B.A. (1988). Structure and function of plant cell walls. *In* "The Biochemistry of Plants: A Comprehensive Treatise, Vol 14: Carbohydrates" (J. Preiss, Ed.), pp 297–371, Academic Press, New York.

Baldwin, T. C., McCann, M. C., and Roberts, K. (1993). A novel hydroxyproline-deficient arabinogalactan protein secreted by suspension-cultured cells of *Daucus carota*. *Plant Physiol.* **103**, 115–123.

Balestrini, R., Romera, C., Puigdomenech, P., and Bonfante, P. (1994). Location of cell wall hydroxyproline-rich glycoprotein, cellulose and $\beta$1,3-glucans in apical and differentiated regions of maize mycorrhizal roots. *Planta* **195**, 201–209.

Bao, W., O'Malley, D. M,. and Sederoff, R.R. (1992). Wood contains a cell wall structural protein. *Proc. Natl. Acad. Sci. USA* **89**, 6604–6608.

Baron-Epel, O., Gharyl, P. K., and Schindler, M. (1988). Pectins as mediators of wall porosity in soybean cells. *Planta* **175**, 389–395.

Barry, P., Prensier, G., and Grenet, E. (1991). Immunogold labelling of arabinoxylans in the plant cell walls of normal and bm3 mutant maize. *Biol. Cell* **71**, 307–311.

Berger, F., Taylor, A., and Brownlee, C. (1994). Cell fate determination by the cell wall in early *Fucus* development. *Science* **263**, 1421–1423.

Bertrand, S., Benhamou, N., Nadeau, P., Dostaler, D., and Gosselin. A. (1992). Immunogold localization of free abscisic acid in tomato root cells. *Can. J. Bot.* **70**, 1001–1011.

Bestwick, C. S., Bennett, M. H., and Mansfield, J. W. (1995). Hrp mutant of *Pseudomonas syringae* pv *phaseolicola* induces cell wall alterations but not membrane damage leading to the hypersensitive reaction in lettuce. *Plant Physiol.* **108**, 503–516.

Bianco-Trinchant, J., Guigonis, J. M., and Le Page-Degivry, M. (1993). Early release of ABA from cell walls during Rose petal protoplast isolation. *J. Exp. Bot.* **44**, 957–962.

Bolwell, G. P. (1993). Dynamic aspects of the plant extracellular matrix. *Int. Rev. Cytol.* **146**, 261–324.

Bonfante, P., Tamagnone, L., Peretto, R., Esquerre-Tugaye, M. T., Mazau, D., Mosiniak, M., and Vian, B. (1991). Immunocytochemical localization of hydroxyproline-rich glycoprotein at the interface between a mycorrhizal fungus and its host plants. *Protoplasma* **165**, 127–138.

Bonafante-Fasolo, P., Vian, B., Perotto, S., Faccio, A., and Knox, J. P. (1990). Cellulose and pectin localization in roots of mycorrhizal *Allium porrum*: Labelling continuity between host cell wall and interfacial material. *Planta* **180**, 537–547.

Boudet, A. M., Lapierrre, C., and Grima-Pettenat, J. (1995). Biochemistry and molecular biology of lignification. *New Phytol.* **129**, 203–236.

Bradley, D. J., Butcher, G. W., Galfrè, G., Wood, E. A., and Brewin, N. J. (1986). Physical association between the peribacteroid membrane and lipopolysaccharide from the bacteroid outer membrane in *Rhizobium*-infected pea root nodule cells. *J. Cell Science* **85**, 47–61.

Bradley, D. J., Wood, E. A., Larkins, A. P., Galfrè, G., Butcher, G. W., and Brewin, N. J. (1988). Isolation of monoclonal antibodies reacting with peribacteroid membranes and other components of pea root nodules containing *Rhizobium leguminosarum*. *Planta* **173**, 149–160.

Brewin, N. J., Robertson, J. G., Wood, E. A., Wells, B., Larkins, A. P., Galfrè, G., and Butcher, G. W. (1985). Monoclonal antibodies to antigens in the peribacteroid membrane from *Rhizobium*-induced root nodules of pea cross-react with plasma membranes and Golgi bodies. *EMBO J.* **4**, 605–611.

Cardemil, L., and Riquelme, A. (1991). Expression of cell wall proteins in seeds and during early seedling growth of *Araucaria araucana* is a response to wound stress and is developmentally regulated. *J. Exp. Bot.* **42**, 415–421.

Carpita, N. C., and Gibeaut, D. M. (1993). Structural models of primary cell walls in flowering plants: Consistency of molecular structure with the physical properties of the walls during growth. *Plant J.* **3**, 1–30.

Casero, P. J., and Knox, J. P. (1995). The monoclonal antibody JIM5 indicates patterns of pectin deposition in relation to pit fields at the plasma-membrane-face of tomato pericarp cells. *Protoplasma* **188**, 133–137.

Cassab, G. I., and Varner, J. E. (1987). Immunocytolocalization of extensin in developing soybean seed coats by immunogold-silver staining and by tissue printing on nitrocellulose paper. *J. Cell Biol.* **105**, 2581–2588.

Cassab, G. I., Lin, J. J., Lin, L. S., and Varner, J. E. (1988). Ethylene effect on extensin and peroxidase distribution in the subapical region of pea epicotyls. *Plant Physiol.* **88,** 522–524.
Chasan, R. (1994). Arabinogalactan-proteins: Getting to the core. *Plant Cell* **6,** 1519–1521.
Chin, J. J. C. (1982). Monoclonal antibodies that immunoreact with a cation-stimulated plant membrane ATPase. *Biochem. J.* **203,** 51–54.
David, H., Bade, P., David, A., Savy, C., Demazy, C., and Van Cutsem, P. (1995). Pectins in walls of protoplast-derived cells imbedded in agarose and alginate beads. *Protoplasma* **186,** 122–130.
Dolan, L., and Roberts, K. (1995). Secondary thickening in roots of *Arabidopsis thaliana*: Anatomy and cell surface changes. *New Phytol.* **131,** 121–128.
Dolan, L., Linstead, P., and Roberts, K. (1995). An AGP epitope distinguishes a central metaxylem initial from other vascular initials in the *Arabidopsis* root. *Protoplasma* **189,** 149–155.
Domingo, C., Gómez, M. D., Cañas, L., Hernández-Yago, Conejero, V., and Vera, P. (1994). A novel extracellular matrix protein from tomato associated with lignified secondary cell walls. *Plant Cell* **6,** 1035–1047.
DuPont, F. M., Tanaka, C. K., and Hurkman, W. J. (1988). Separation and immunological characterization of membrane fractions from barley roots. *Plant Physiol.* **86,** 717–724.
Eardley, D. D., Sutton, C. W., Hempel, W. M., Reed, D. C., and Ebeling, A. W. (1990). Monoclonal antibodies specific for sulfated polysaccharides on the surface of *Macrocystis pyrifera* (Phaeophyceae). *J. Phycol.* **26,** 54–62.
Ertl, H., Hallman, A., Wenzl, S., and Sumper, M. (1992). A novel extensin that may organize extracellular matrix biogenesis in *Volvox carteri. EMBO J.* **11,** 2055–2062.
Evans, P. T., Holaway, B. L., and Malmberg, R. L. (1988). Biochemical differentiation in the tobacco flower probed with monoclonal antibodies. *Planta* **175,** 259–269.
Fincher, G. B., Stone, B. A., and Clarke, A. E. (1983). Arabinogalactan-proteins: Structure biosynthesis and function. *Annu. Rev. Plant Physiol.* **34,** 47–70.
Fitter, M. S., Norman, P. M., Hahn, M. G., Wingate, V. P. M., and Lamb, C. J. (1987). Identification of somatic hybrids in plant protoplast fusions with monoclonal antibodies to plasma-membrane antigens. *Planta* **170,** 49–54.
Fritz, S. E., Hood, K. R., and Hood, E. E. (1991). Localization of soluble and insoluble fractions of hydroxyproline-rich glycoproteins during maize kernel development. *J. Cell Sci.* **98,** 545–550.
Fry, S. C. (1995). Polysaccharide-modifying enzymes in the plant cell wall. *Annu. Rev. Plant Physiol. Plant Mol. Biol.* **46,** 497–520.
Geitmann, A., Hudák, J., Vennigerholz, F., and Walles, B. (1995a). Immunogold localization of pectin and callose in pollen grains and pollen tubes of *Brugmansia suaveolens*—Implications for the self-incompatibility reaction. *J. Plant Physiol.* **147,** 225–235.
Geitmann, A., Li, Y-Q., and Cresti, M. (1995b). Ultrastructural immunolocalization of periodic pectin depositions in the cell wall of *Nicotiana tabacum* pollen tubes. *Protoplasma* **187,** 168–171.
Gollotte, A., Gianinazzi-Pearson, V., and Gianinazzi, S. (1995). Immunodetection of infection thread glycoprotein and arabinogalactan-protein in wild type *Pisum sativum* (L.) or an isogenic mycorrhiza-resistant mutant interacting with *Glomus mosseae. Symbiosis* **18,** 69–85.
Gravato-Nobre, M. J., von Mende, N., Dolan, L., Schmidt, K. P., Evans, K., and Mulligan, B. (1995). Immunolabelling of cell surfaces of Arabidopsis thaliana roots following infection by *Meloidogyne incognita* (Nematoda). *J. Exp. Bot.* **46,** 1711–1720.
Green, J. R., Stafford, C. J., Jones, J. L., Wright, P. J., and Callow, J. A. (1993). Binding of monoclonal antibodies to vegetative tissue and fucose-containing polysaccharides of *Fucus serratus* L. *New Phytol.* **124,** 397–408.
Grief, C., and Shaw, P. J. (1987). Assembly of cell-wall glycoproteins of *Chlamydomonas reinhardii*: Oligosaccharides are added in medial and trans Golgi compartments. *Planta* **171,** 302–312.

Grief, C., O'Neill, M. A., and Shaw, P. J. (1987). The zygote cell wall of *Chlamydomonas reinhardii*: A structural, chemical and immunological approach. *Planta* **170,** 433–445.
Grimes, H. D., and Breidenbach, R. W. (1987). Plant plasma membrane proteins. Immunological characterization of a major 75 kilodalton protein group. *Plant Physiol.* **85,** 1048–1054.
Hahn, M. G., Lerner, D. R., Fitter, M. S., Norman, P. M., and Lamb, C. J. (1987). Characterization of monoclonal antibodies to protoplast membranes of *Nicotiana tabacum* identified by an enzyme-linked immunosorbent assay. *Planta* **171,** 453–465.
Harris, P. J., Freed, K., Anderson, M. A., Weinhandl, J. A., and Clarke, A. E. (1987). An enzyme-linked immunosorbent assay (ELISA) for *in vitro* pollen growth based on binding of a monoclonal antibody to the pollen tube surface. *Plant Physiol.* **84,** 851–855.
Herman, E. M., and Lamb, C. J. (1992). Arabinogalactan-rich glycoproteins are localized on the cell surface and in intravacuolar multivesicular bodies. *Plant Physiol.* **98,** 264–272.
Hess, M. W., and Frosch, A. (1994). Subunits of forming pollen exine and Ubisch bodies as seen in freeze substituted *Ledebouria socialis* Roth (Hyacinthaceae). *Protoplasma* **182,** 10–14.
Høj, P. B., Hoogenraad, N. J., Hartman, D. J., Yannakena, H., and Fincher, G. B. (1990). Identification of individual (1 → 3, 1 → 4)-$\beta$-D-glucanase isoenzymes in extracts of germinated barley using specific monoclonal antibodies. *J. Cereal Sci.* **11,** 261–268.
Homan, E. L., van Kalshoven, H., Kolk, A. H. J., Musgrave, A., Schuring, F., and van den Ende, H. (1987). Monoclonal antibodies to surface glycoconjugates in *Chlamydomonas eugametos* recognize strain-specific *O*-methyl sugars. *Planta* **170,** 328–335.
Hoson, T. (1991). Structure and function of plant cell walls: immunological approaches. *Int. Rev. Cytol.* **130,** 233–268.
Hoson, T., and Nevins, D. J. (1989a). Antibodies as probes for the study of location and metabolism of (1 → 3),(1 → 4)-$\beta$-D-glucans. *Physiol. Plant.* **75,** 452–457.
Hoson, T., and Nevins, D. J. (1989b). Effect of anti-wall protein antibodies on auxin-induced elongation, and $\beta$-D-glucan degradation in maize coleoptile segments. *Physiol. Plant.* **77,** 208–215.
Hoson, T., Masuda, Y., Sone, Y., and Misaki, A. (1991). Xyloglucan antibodies inhibit auxin-induced elongation and cell wall loosening of azuki bean epicotyls but not of oat coleoptiles. *Plant Physiol.* **96,** 551–557.
Hoson, T., Masuda, Y., and Nevins, D. J. (1992). Comparison of the outer and inner epidermis. Inhibition of auxin-induced elongation of maize coleoptiles by glucan antibodies. *Plant Physiol.* **98,** 1298–1303.
Hough, T., Singh, M. B., Smart, I. J., and Knox, R. B. (1986). Immunofluorescent screening of monoclonal antibodies to surface antigens of animal and plant cells bound to polycarbonate membranes. *J. Immunol. Methods* **92,** 103–107.
Inouhe, M., and Nevins, D. J. (1991). Inhibition of auxin-induced cell elongation of maize coleoptiles by antibodies specific for cell wall glucanases. *Plant Physiol.* **96,** 426–431.
Jacobs, M., and Gilbert, S. F. (1983). Basal localization of the presumptive auxin transport carrier in pea stems. *Science* **220,** 1297–1300.
James, E. K., Sprent, J. I., Minchin, F. R., and Brewin, N. J. (1991). Intercellular location of glycoprotein in soybean nodules: Effect of altered rhizosphere oxygen concentration. *Plant Cell Environ.* **14,** 467–476.
Jarvis, M. C. (1992). Control of thickness of collenchyma cell walls by pectins. *Planta* **187,** 218–220.
Jones, J. L., Callow, J. A., and Green, J. R. (1988). Monoclonal antibodies to sperm surfac antigens of the brown alga *Fucus serratus* exhibit region-, gamete-, species- and genus preferential binding. *Planta* **176,** 298–306.
Jones, J. L., Callow, J. A., and Green, J. R. (1990). The molecular nature of *Fucus serratus* sperm surface antigens recognised by monoclonal antibodies FS1 to FS12. *Planta* **182,** 64–71.
Kaku, H., Shibata, S., Satsuma, Y., Sone, Y., and Misaki, A. (1986). Interactions of $\alpha$-L-arabinofuranose-specific antibody with plant polysaccharides and its histochemical application. *Phytochemistry* **25,** 2041–2047.

Keller, B., Sauer, N., and Lamb, C. J. (1988). Glycine-rich cell wall proteins in bean: Gene structure and association of the protein with the vascular system. *EMBO J.* **7**, 3625–3633.

Keller, B., Templeton, M.D., and Lamb, C. J. (1989). Specific localization of a plant cell wall glycine-rich protein in protoxylem cells of the vascular system. *Proc. Natl. Acad. Sci. USA* **86**, 1529–1533.

Key, G., and Weiler, E. W. (1988). Monoclonal antibodies identify common and differentiation-specific antigens on the plasma membrane of guard cells of *Vicia faba* L. *Planta* **176**, 472-481.

Kieliszewski, M. J., and Lamport, D. T. A. (1986). Cross-reactivities of polyclonal antibodies against extensin precursors determined via ELISA techniques. *Phytochemistry* **25**, 673–677.

Kieliszewski, M. J., O'Neill, M., Leykam, J., and Orlando, R. (1995). Tandem mass spectrometry and structural elucidation of glycopeptides from a hydroxyproline-rich plant cell wall glycoprotein indicate that contiguous hydroxyproline residues are the major sites of hydroxyproline *O*-arabinosylation. *J. Biol. Chem.* **270**, 2541–2549.

Kikuchi, S., Ohinata, A., Tsumuraya, Y., Hashimoto, Y., Kaneko, Y., and Matsushima, H. (1993). Production and characterization of antibodies to the $\beta$-(1 → 6)-galactotetraosyl group and their interaction with arabinogalactan-proteins. *Planta* **190**, 525–535.

Kim, S-H., Terry, M. E., Hoops, P., Dauwalder, M., and Roux, S. J. (1988). Production and characterization of monoclonal antibodies to wall-localized peroxidases from corn seedlings. *Plant Physiol.* **88**, 1446–1453.

Kim, S-H., Shinkle, J. R., and Roux, S. J. (1989). Phytochrome induces changes in the immunodetectable level of a wall peroxidase that precede growth changes in maize seedlings. *Proc. Natl. Acad. Sci. USA* **86**, 9866–9870.

Klis, F. M., and Eeltink, H. (1979). Changing arabinosylation patterens of wall-bound hydroxyproline in bean cell cultures. *Planta* **144**, 479–484.

Knox, J. P. (1992). Molecular probes for the plant cell surface. *Protoplasma* **167**, 1–9.

Knox, J. P. (1995). Developmentally regulated proteoglycans and glycoproteins of the plant cell surface. *FASEB J.* **9**, 1004–1012.

Knox, J. P., and Roberts, K. (1989). Carbohydrate antigens and lectin receptors of the plasma membrane of carrot cells. *Protoplasma* **152**, 123–129.

Knox, J. P., Day, S., and Roberts, K. (1989). A set of cell surface glycoproteins forms an early marker of cell position, but not cell type, in the root apical meristem of *Daucus carota* L. *Development* **106**, 47–56.

Knox, J. P., Linstead, P. J., King, J., Cooper, C., and Roberts, K. (1990). Pectin esterification is spatially regulated both within cell walls and between developing tissues of root apices. *Planta* **181**, 512–521.

Knox, J. P., Linstead, P. J., Peart, J., Cooper, C., and Roberts, K. (1991). Developmentally regulated epitopes of cell surface arabinogalactan proteins and their relation to root tissue pattern formation. *Plant J.* **1**, 317–326.

Knox, J. P., Peart, J., and Neill, S. J. (1995). Identification of novel cell surface epitopes using a leaf epidermal-strip assay system. *Planta* **196**, 266–270.

Knox, R. B., Heslop-Harrison, J., and Reed, C. (1970). Localization of antigens associated with the pollen grain wall by immunofluorescence. *Nature* **225**, 1067–1068.

Knox, R. B., Vithange, H. I. M. V., and Howlett, B. J. (1980). Botanical immunocytochemistry: A review with special reference to pollen antigens and allergens. *Histochem. J.* **12**, 247–272.

Kreuger, M., and van Holst, G. J. (1993). Arabinogalactan-proteins are essential in somatic embryogenesis of *Daucus carota* L. *Planta* **189**, 243–248.

Kreuger, M., and van Holst, G. J. (1995). Arabinogalactan-protein epitopes in somatic embryogenesis of *Daucus carota* L. *Planta* **197**, 135–141.

Kreuger, M., Postma, E., Brouwer, Y., and van Holst, G. J. (1995). Somatic embryogenesis of *Cyclamen persicum* in liquid-medium. *Physiol. Plant.* **94**, 605–612.

Lamport, D. T. A., and Miller, D. H. (1971). Hydroxyproline arabinosides in the plant kingdom. *Plant Physiol.* **48**, 454–456.
Larsen, B., Vreeland, V., and Laetsch, W. M. (1985). Assay-dependent specificity of a monoclonal antibody with alginate. *Carbohydr. Res.* **143**, 221–227.
Lewis, N. G., and Yamamoto, E. (1990). Lignin: Occurrence, biogenesis and biodegradation. *Annu. Rev. Plant Physiol. Plant Mol. Biol.* **41**, 455–496.
Li, Y. Q., Bruun, L., Pierson, E. S., and Cresti., M. (1992). Periodic deposition of arabinogalactan epitopes in the cell wall of pollen tubes of *Nicotiana tabacum* L. *Planta* **188**, 532–538.
Li, Y. Q., Chen, F., Linskens, H. F., and Cresti, M. (1994). Distribution of unesterified and esterified pectins in cell walls of pollen tubes of flowering plants. *Sex. Plant Reprod.* **7**, 145–152.
Li, Y. Q., Faleri, C., Geitmann, A., Zhang, H. Q., and Cresti, M. (1995). Immunogold localization of arabinogalactan proteins, unesterified and esterified pectins in pollen grains and pollen tubes of *Nicotiana tabacum* L. *Protoplasma* **189**, 26–36.
Lind, J. L., Bacic, A., Clarke, A. E., and Anderson, M. A. (1994). A style-specific hydroxyproline-rich glycoprotein with properties of both extensins and arabinogalactan proteins. *Plant J.* **6**, 491–502.
Liners, F., and van Cutsem, P. (1991). Immunocytochemical localization of homopolygalacturonic acid on plant cell walls. *Micron Micro. Acta* **22**, 265–266.
Liners, F., and van Cutsem, P. (1992). Distribution of pectic polysaccharides throughout walls of suspension-cultured carrrot cells. An immunocytochemical study. *Protoplasma* **170**, 10–21.
Liners, F., Letesson, J-J., Didembourg, C., and van Cutsem, P. (1989). Monoclonal antibodies against pectin. Recognition of a conformation induced by calcium. *Plant Physiol.* **91**, 1419–1424.
Liners, F., Thibault, J-F., and Van Cutsem, P. (1992). Influence of the degree of polymerization of oligogalacturonates and of esterification pattern of pectin on their recognition by monoclonal antibodies. *Plant Physiol.* **99**, 1099–1104.
Liners, F., Gaspar, T., and van Cutsem, P. (1994). Acetyl- and methyl-esterification of pectins of friable and compact sugar-beet calli: Consequences for intercellular adhesion. *Planta* **192**, 545–556.
Liu, Q., and Berry, A. M. (1991). Localization and characterization of pectic polysaccharides in roots and root nodules of *Caenothus* spp. during intercellular infection by *Frankia*. *Protoplasma* **163**, 93–101.
Ludevid, M. D., Ruiz-Avila, L., Vallés, M. P., Stiefel, V., Torrent, M., Torné, J. M., and Puigdomènech, P. (1990). Expression of genes for cell wall proteins in dividing and wounded tissues of *Zea mays* L. *Planta* **180**, 524–529.
Lynch, M. A., and Staehelin, L. A. (1992). Domain-specific and cell type-specific localization of two types of cell wall matrix polysaccharides in the clover root tip. *J. Cell Biol.* **118**, 467–479.
Lynch, M. A., and Staehelin, L. A. (1995). Immunocytochemical localization of cell wall polysaccharides in the root tip of *Avena sativa*. *Protoplasma* **188**, 115–127.
Lynes, M., Lamb, C. A., Napolitano, L. A., and Stout, R. G. (1987). Antibodies to cell surface antigens of plant protoplasts. *Plant Sci.* **50**, 225–232.
Mackie, A. J., Roberts, A. M., Callow, J. A., and Green, J. R. (1991). Molecular differentiation in pea powdery-mildew haustoria. Identification of a 62-kDa N-linked glycoprotein unique to the haustorial plasma membrane. *Planta* **183**, 399–408.
Marcus, A., Greenburg, J., and Averyhart-Fullard, V. (1991). Repetitive proline-rich proteins in the extracellular matrix of the plant cell. *Physiol. Plant.* **81**, 273–279.
Marty, P., Goldberg, R., Liberman, M., Vian, B., Bertheau, Y., and Jouan, B. (1995). Composition and localization of pectic polymers in the stems of two *Solanum tuberosum* genotypes. *Plant Physiol. Biochem.* **33**, 409–417.

Mau, S. L., Chen, C. G., Pu, Z. Y., Moritz, R. L., Simpson, R. J., Bacic, A., and Clarke, A. E. (1995). Molecular cloning of CDNAs encoding the protein backbones of arabinogalactan-proteins from the filtrate of suspension-cultured cells of *Pyrus communis* and *Nicotiana alata*. *Plant J.* **8,** 269–281.

McCann, M. C., and Roberts, K. (1991). Architecture of the primary cell wall. *In* "The Cytoskelatal Basis of Plant Growth" and form. (C. W. Lloyd, Ed.), pp. 109–129. Academic Press, London.

McCann, M. C., Wells, B., and Roberts, K. (1992). Complexity in the spatial localization and length distribution of plant cell wall matrix polysaccharides. *J. Microsc.* **166,** 123–136.

Meikle, P. J., Bonig, I., Hoogenraad, N. J., Clarke, A. E., and Stone, B. A. (1991). The location of $(1 \rightarrow 3)$-$\beta$-glucans in the walls of pollen tubes of *Nicotiana alata* using a $(1 \rightarrow 3)$-$\beta$-glucan-specific monoclonal antibody. *Planta* **185,** 1–8.

Meikle, P. J., Hoogenraad, N. J., Bonig, I., Clarke, A. E., and Stone, B. A. (1994). A $(1 \rightarrow 3,1 \rightarrow 4)$-$\beta$-glucan-specific monoclonal antibody and its use in the quantitation and immunocytochemical location of $(1 \rightarrow 3,1 \rightarrow 4)$-$\beta$-glucans. *Plant J.* **5,** 1–9.

Metcalf, T. N., Villanueva, M. A., Schindler, M., and Wang, J. L. (1986). Monoclonal antibodies directed against protoplasts of soybean cells: Analysis of the lateral mobility of plasma membrane-bound antibody MVS-1. *J. Cell Biol.* **102,** 1350–1357.

Meyer, D. J., Afonso, C. L., and Galbraith, D. W. (1988). Isolation and characterization of monoclonal antibodies directed against plant plasma membrane and cell wall epitopes: Identification of a monoclonal antibody that recognizes extensin and analysis of the process of epitope biosynthesis in plant tissues and cell cultures. *J. Cell Biol.* **107,** 163–175.

Misaki, A., Kaku, H., Sone, Y., and Shibata, S. (1988). Anti-$\alpha$-$L$-arabinofuranose antibodies: Purification, immunochemical characterization, and use in histochemical studies of plant cell wall polysaccharides. *Carbohydr. Res.* **173,** 133–144.

Moore, P. J. (1989). Immunogold localization of specific components of plant cell walls. *In* "Plant Fibers" (H. F. Linskens and J. F. Jackson, Eds). pp 70–88. Springer-Verlag, Berlin.

Moore, P. J., and Staehelin, L. A. (1988). Immunogold localization of the cell wall matrix polysaccharides rhamnogalacturonan I and xyloglucan during cell expansion and cytokinesis in *Trifolium pratense* L.; Implication for secretory pathways. *Planta* **174,** 433–445.

Moore, P. J., Darvill, A. G., Albersheim, P., and Staehelin, L. A. (1986). Immunogold localization of xyloglucan and rhamnogalacturonan I in the cell walls of suspension-cultured Sycamore cells. *Plant Physiol.* **82,** 787–794.

Moore, P. J., Swords, K. M. M., Lynch, M. A, and Staehelin, L. A. (1991). Spatial organization of the assembly pathways of glycoproteins and complex polysaccharides in the Golgi apparatus of plants. *J. Cell Biol.* **112,** 589–602.

Mueller, W. C., Morgham, A. T., and Roberts, E. M. (1994). Immunocytochemical localization of callose in the vascular tissue of tomato and cotton plants infected with *Fusarium oxysporum*. *Can. J. Bot.* **72,** 505–509.

Norman, P. M., Wingate, V. P. M., Fitter, M. S., and Lamb, C. J. (1986). Monoclonal antibodies to plant plasma-membrane antigens. *Planta* **167,** 452–459.

Norman, P. M., Kjellbom, P., Bradley, D. J., Hahn, M. G., and Lamb, C. J. (1990). Immunoaffinity purification and biochemical characterization of plasma membrane arabinogalactan-rich glycoproteins of *Nicotiana glutinosa*. *Planta* **181,** 365–373.

Northcote, D. H., Davey, R., and Lay, J. (1989). Use of antisera to localize callose, xylan and arabinogalactan in the cell plate, primary and secondary walls and plant cells. *Planta* **178,** 353–366.

O'Connell, R. J., Brown, I. R., Mansfield, J. W., Bailey, J. A., Mazau, D., Rumeau, D., and Esquerré-Tugayé, M-T. (1990). Immunocytochemical localization of hydroxyproline-rich glycoproteins accumulating in melon and bean at sites of resistance to bacteria and fungi. *Mol. Plant Microbe Int.* **3,** 33–40.

O'Neill, M., Albersheim, P., and Darvill, A. (1990). The pectic polysaccharides of primary cell walls. *Methods Plant Biochem.* **2**, 415–441.
Pain, N. A., O'Connell, R. J., Mendgen, K., and Green, J. R. (1994). Identification of glycoproteins specific to biotrophic intracellular hyphae formed in the *Colletotrichum lindemuthianum*–bean interaction. *New Phytol.* **127**, 233–242.
Pennell, R. I. (1992). Cell surface arabinogalactan proteins, arabinogalactans and plant development. *In* "Perspectives in Plant Cell Recognition," SEB Seminar Series 48 ( J. A. Callow and J. R. Green, Eds.), pp. 105–121. Cambridge Univ. Press, Cambridge, UK.
Pennell, R. I., and Roberts, K. (1990). Sexual development in the pea is presaged by altered expression of arabinogalactan protein. *Nature* **344**, 547–549.
Pennell, R. I., and Roberts, K. (1995). Monoclonal antibodies to cell-specific cell surface carbohydrates in plant cell biology and development. *Methods Cell Biol.* **49**, 123–141.
Pennell, R. I., Knox, J. P., Scofield, G. N., Selvendran, R. R., and Roberts, K. (1989). A family of abundant plasma membrane-associated glycoproteins related to the arabinogalactan-proteins is unique to flowering plants. *J. Cell Biol.* **108**, 1967–1977.
Pennell, R. I., Janniche, L., Kjellbom, P., Scofield, G. N., Peart, J. M., and Roberts, K. (1991). Developmental regulation of a plasma membrane arabinogalactan protein in oilseed rape flowers. *Plant Cell* **3**, 1317–1326.
Pennell, R. I., Janniche, L., Scofield, G. N., Booij, H., de Vries, S. C., and Roberts, K. (1992). Identification of a transitional cell state in the developmental pathway to carrot somatic embryogenesis. *J. Cell Biol.* **119**, 1371–1380.
Pennell, R. I., Cronk, Q. C. B., Forsberg, L. S., Stöhr, C., Snogerup, L., Kjellbom, P., and McCabe, P. F. (1995). Cell-context signalling. *Philos. Trans. R. Soc. London. B* **350**, 87–93
Peretto, R., Perotto, S., Faccio, A., and Bonfante-Fasolo, P. (1990). Cell surface in *Calluna vulgaris* L. hair roots. In situ localization of polysaccharide components. *Protoplasma* **155**, 1–18.
Perotto, S., VandenBosch, K. A., Butcher, G. W., and Brewin, N. J. (1991). Molecular composition and development of the plant glycocalyx associated with the peribacteroid membrane of pea root nodules. *Development* **112**, 763–773.
Perotto, S., Brewin, N.J., and Bonfante, P. (1994a). Colonization of pea roots by the mycorrhizal fungus *Glomus versiforme* and by *Rhizobium* bacteria: Immunological comparison using monoclonal antibodies as probes for plant cell surface components. *Mol. Plant Microbe Int.* **7**, 91–98.
Perotto, S., Brewin, N.J., and Kannenberg, E. L. (1994b). Cytological evidence for a host defense response that reduces cell and tissue invasion in pea nodules by lipopolysaccharide-defective mutants of *Rhizobium leguminosarum* strain 3841. *Mol. Plant Microbe Int.* **7**, 99–112.
Perotto, S., Donovan, N., Drøbak, B. K., and Brewin, N.J. (1995). Differential expression of a glycosyl inositol phospholipid antigen on the peribacteroid membrane during pea nodule development. *Mol. Plant Microbe Int.* **8**, 560–568.
Pierson, E. S., Li, Q., Zhang, H. Q., Willemse, M. T. M.. Linskens, H. F. and Cresti, M. (1995). Pulsatory growth of pollen tubes: Investigation of a possible relationship with the periodic distribution of cell wall components. *Acta Bot. Neerl.* **44**, 121–128.
Puhlmann, J., Bucheli, E., Swain, M. J., Dunning, N., Albersheim, P., Darvill, A. G., and Hahn, M. G. (1994). Generation of monoclonal antibodies against plant cell wall polysaccharides. I. Characterization of a monoclonal antibody to a terminal $\alpha$-($1 \rightarrow 2$)-linked fucosyl-containing epitope. *Plant Physiol.* **104**, 699–710.
Rae, A. L., Perotto, S., Knox, J. P., Kannenberg, E. L., and Brewin, N. J. (1991). Expression of extracellular glycoproteins in the uninfected cells of developing pea nodule tissue. *Mol. Plant Microbe Int.* **4**, 563–570.
Rae, A. L., Bonfante-Fasolo, P., and Brewin, N. J. (1992). Structure and growth of infection threads in the legume symbiosis with *Rhizobium leguminosarum*. *Plant J.* **2**, 385–395.

Renard, C. M. G. C., Thibault, J-F., Liners, F., and van Cutsem, P. (1993). Immunological probing of pectins isolated or *in situ. Acta Bot. Neerl.* **42,** 199–204.

Roberts, K. (1986). Antibodies and the plant cell surface: practical approaches. In "Immunology in plant science" (T. L. Wang, Ed.), pp. 89–110. Cambridge Univ. Press, Cambridge, UK.

Roberts, K. (1994). The plant extracellular matrix: In a new expansive mood. *Curr. Opin. Cell Biol.* **6,** 688–694.

Roberts, K., Phillips, J., Shaw, P., Grief, C., and Smith, E. (1985). An immunological approach to the plant cell wall. *In* "Biochemistry of Plant Cell Walls" (C. T. Brett and J. R. Hillman, Eds.), pp. 125–154. Cambridge Univ. Press, Cambridge, UK.

Rodriguez-Gálvez, E., and Mendgen, K. (1995). Cell wall synthesis in cotton roots after infection with *Fusarium oxysporum*. The deposition of callose, arabinogalactans, xyloglucans, and pectic components into walls, wall appositions, cell plates and plasmodesmata. *Planta* **197,** 535–545.

Roy, S., Vian, B., and Roland, J-C. (1992). Immunocytochemical study of the deesterification patterns during cell autolysis in the ripening of cherry tomato. *Plant Physiol. Biochem.* **30,** 139–146.

Roy, S., Jauneau, A,. and Vian, B. (1994). Analytical detection of calcium ions and immunocytochemical visualization of homogalacturonic sequences in the ripe cherry tomato. *Plant Physiol. Biochem.* **32,** 633–640.

Ruel, K., Faix, O., and Joseleau, J. P. (1994). New immunogold probes for studying the distribution of the different lignin types during plant cell wall biogenesis. *J. Trace Microprobe Technol.* **12,** 247–265.

Ruiz-Avila, L., Ludevid, D., and Puigdomènech, P. (1991). Differential expression of a hydroxyproline-rich cell-wall protein gene in embryonic tissues of *Zea mays* L. *Planta* **184,** 130–136.

Ryser, U., and Keller, B. (1992). Ultrastructural localization of a bean glycine-rich protein in unlignified primary walls of protoxylem cells. *Plant Cell* **4,** 773–783.

Samuels, A. L., Giddings, T. H., and Staehelin, L. A. (1995). Cytokinesis in tobacco BY-2 and root tip cells: A new model of cell plate formation in higher plants. *J. Cell Biol.* **130,** 1345–1357.

Satiat-Jeunemaitre, B., and Hawes, C. (1993). The distribution of secretory products in plant cells is affected by Brefeldin A. *Cell Biol. Int.* **17,** 183–193.

Satiat-Jeunemaitre, B., Fitchette-Lainé, A. C., Alabouvette, J., Marty-Mazars, D., Hawes, C., Faye, L., and Marty, F. (1994). Differential effects of monesin on the plant secretory pathway. *J. Exp. Bot.* **45,** 685–698.

Schindler, T., Bergfeld, R., and Schopfer, P. (1995a). Arabinogalactan proteins in maize coleoptiles: Developmental relationship to cell death during xylem differentiation but not to extension growth. *Plant J.* **7,** 25–36.

Schindler, T., Bergfeld, R., van Cutsem, P., v. Sengbusch, D., and Schopfer, P. (1995b). Distribution of pectins in cell walls of maize coleoptiles and evidence against their involvement in auxin-induced extension growth. *Protoplasma* **188,** 213–224.

Sedgley, M., Blesing, M. A., Bonig, I., Anderson, M. A., and Clarke, A. E. (1985). Arabinogalactan-proteins are localized extracellularly in the transmitting tissue of *Nicotiana alata* Link and Otto, an ornamental tobacco. *Micron Microsc. Acta* **4,** 247–254.

Serpe, M. D., and Nothnagel, E. A. (1994). Effects of Yariv phenyl glycosides on *Rosa* cell suspensions: Evidence for the involvement of arabinogalactan-proteins in cell proliferation. *Planta* **193,** 542–550.

Sherrier, D. J., and VandenBosch, K. A. (1994a). Secretion of cell wall polysaccharides in *Vicia* root hairs. *Plant J.* **5,** 185–195.

Sherrier, D. J., and VandenBosch, K. A. (1994b). Localization of repetitive proline-rich proteins in the extracellular matrix of pea root nodules. *Protoplasma* **183,** 148–161.

Showalter, A. M. (1993). Structure and function of plant cell wall proteins. *Plant Cell* **5,** 9–23.
Smallwood, M., Beven, A., Donovan, N., Neill, S. J., Peart, J., Roberts, K., and Knox, J. P. (1994). Localization of cell wall proteins in relation to the developmental anatomy of the carrot root apex. *Plant J.* **5,** 237–246.
Smallwood, M., Martin, H., and Knox, J. P. (1995). An epitope of rice threonine- and hydroxyproline-rich glycoprotein is common to cell wall and hydrophobic plasma membrane glycoproteins. *Planta* **196,** 510–522.
Smallwood, M., Yates, E. A., Willats, W. G. T., Martin, H., and Knox, J. P. (1996). Immunochemical comparison of membrane-associated and secreted arabinogalactan-proteins in rice and carrot. *Planta* **198,** 452–459.
Smith, E., Roberts, K., Hutchings, A., and Galfrè, G. (1984). Monoclonal antibodies to the major structural glycoprotein of the *Chlamydomonas* cell wall. *Planta* **161,** 330–338.
Sone, Y., Kuramae, J., Shibata, S., and Misaki, A. (1989a). Immunochemical specificities of antibody to the heptasaccharide unit of plant xyloglucan. *Agric. Biol. Chem.* **53,** 2821–2823.
Sone, Y., Misaki, A., and Shibata, S. (1989b). Preparation and characterization of antibodies against 6-$O$-$\alpha$-D-xylopyranosyl-$\beta$-D-glucopyranose ($\beta$-isoprimeverose), the disaccharide unit of xyloglucan in plant cell walls. *Carbohydr. Res.* **191,** 79–89.
Stacey, N. J., Roberts, K., and Knox, J. P. (1990). Patterns of expression of the JIM4 arabinogalactan-protein epitope in cell cultures and during somatic embryogenesis in *Daucus carota* L. *Planta* **180,** 285–292.
Stacey, N. J., Roberts, K., Carpita, N. C., Wells, B., and McCann, M. C. (1995). Dynamic changes in cell surface molecules are very early events in the differentiation of mesophyll cells from *Zinnia elegans* into tracheary elements. *Plant J.* **8,** 891–906.
Stafford, C. J., Green, J. R., and Callow, J. A. (1992). Organization of glycoproteins into plasma membrane domains on *Fucus serratus* eggs. *J. Cell Sci.* **101,** 437–448.
Stafstrom, J. P., and Staehelin, L.A. (1988). Antibody localization of extensin in cell walls of carrot storage roots. *Planta* **174,** 321–332.
Steffan, W., Kovàc, P., Albersheim, P., Darvill, A. G., and Hahn, M. G. (1995). Characterization of a monoclonal antibody that recognizes an arabinosylated $(1 \rightarrow 6)$-$\beta$-D-galactan epitope in plant complex carbohydrates. *Carbohydr.Res.* **275,** 295–307.
Stephenson, M. B., and Hawes, M. C. (1994). Correlation of pectin metylesterase activity in root caps of pea with root border cell separation. *Plant Physiol.* **106,** 739–745.
Stiefel, V., Pèrez-Grau, L., Alberico, F., Giralt, E., Ruiz-Avila, L., Ludevid, D., and Puigdomènech, P. (1988). Molecular cloning of cDNAs encoding a putative cell wall protein from *Zea mays* and immunological identification of related polypeptides. *Plant Mol. Biol.* **11,** 483–493.
Stout, R. G., and Griffing, L. R. (1993). Transmural secretion of a highly-expressed cell surface antigen of oat root cap cells. *Protoplasma* **172,** 27–37.
Swords, K. M. M., and Staehelin, L. A. (1993). Complementary immunolocalization patterns of cell wall hydroxyproline-rich glycoproteins studied with the use of antibodies directed against different carbohydrate epitopes. *Plant Physiol.* **102,** 891–901.
Temsah, M., Berthau, Y., and Vian, B. (1991). Pectate-lyase fixation and pectate disorganization visualized by immunocytochemistry in *Saintpaulia ionantha* infected by *Erwinia chrysanthemi*. *Cell Biol. Int. Rep.* **15,** 611–620.
Valero, P., and Labrador, E. (1993). Inhibition of cell-wall autolysis and auxin-induced elongation of *Cicer arietinum* epicotyls by $\beta$-galactosidase antibodies. *Physiol. Plant.* **89,** 199–203.
Van Aelst, A. C., and Van Went, J. L. (1992). Ultrastructural immunolocalization of pectins and glycoproteins in *Arabidopsis thaliana* pollen grains. *Protoplasma* **168,** 14–19.
Van Engelen, Sterk, P., Booji, H., Cordewener, J. H. G., Rook, W., van Kammen, A., and de Vries, S. C. (1991). Heterogeneity and cell type-specific localization of a cell wall glycoprotein from carrot suspension cells. *Plant Physiol.* **96,** 705–712.
VandenBosch, K. A. (1992). Localization of proteins and carbohydrates using immunogold labelling in light and electron microscopy. *In* "Molecular Plant Pathology. Vol. II. A Practical

Approach" (S. J. Gurr, M. J. McPherson, and D. J. Bowles, Eds.), pp 31–43. Oxford Univ. Press, Oxford, UK.
VandenBosch, K. A., Bradley, D. J., Knox, J. P., Perotto, S., Butcher, G. W., and Brewin, N. J. (1989). Common components of the infection thread matrix and the intercellular space identified by immunocytochemical analysis of pea nodules and uninfected roots. *EMBO J.* **8**, 335–342.
Varner, J. E., and Ye, Z. (1994). Tissue printing. *FASEB J.* **8**, 378–384.
Vennigerholz, F. (1992). The transmitting tissue in *Brugmansia suaveolens*: Immunocytochemical localization of pectin in the style. *Protoplasma* **171**, 117–122.
Vennigerholz, F., Walles, B., and Geitmann, A. (1992). Immunocytochemical localization of pectin in stylar tissues. *Micron Microsc. Acta* **23**, 125–126.
Vian, B., and Roland, J-C. (1991). Affinodetection of the sites of formation and of the further distribution of polygalacturonans and native cellulose in growing plant cells. *Biol. Cell* **71**, 43–55.
Vian, B., Temsah, M., Reis, D., and Roland, J-C. (1992). Co-localization of the cellulose framework and cell wall matrix in helicoidal constructions. *J. Microsc.* **166**, 111–122.
Villanueva, M. A., Metcalf, T. N., and Wang, J. L. (1986). Monoclonal antibodies directed against protoplasts of soybean cells. Generation of hybridomas and characterization of a monoclonal antibody reactive with the cell surface. *Planta* **168**, 503–511.
Vreeland, V. (1970). Localization of a cell wall polysaccharide in a brown alga with labelled antibody. *J. Histochem. Cytochem.* **18**, 371–373.
Vreeland, V. (1972). Immunocytochemical localization of the extracellular polysaccharide alginic acid in the brown seaweed, *Fucus distichus*. *J. Histochem. Cytochem.* **20**, 358–367.
Vreeland, V., Slomich, M., and Laetsch, W. M. (1984). Monoclonal antibodies as molecular probes for cell wall antigens of the brown alga, *Fucus*. *Planta* **162**, 506–517.
Vreeland, V., Zablackis, E., and Laetsch, W. M. (1992). Monoclonal antibodies as molecular markers for the intracellular and cell wall distribution of carrageenan epitopes in *Kappaphycus* (Rhodophyta) during tissue development. *J. Phycol.* **28**, 328–342.
Wallace, G., and Fry, S. C. (1994). Phenolic components of the plant cell wall. *Int. Rev. Cytol.* **151**, 229–267.
Wang, M., Heimovaara-Dijkstra, S., Van der Meulen, R., Knox, J. P., and Neill, S. J. (1995). The monoclonal antibody JIM19 modulates abscisic acid action in barley aleurone protoplasts. *Planta* **196**, 271–276.
Weber, M. (1994). Stigma, style and pollen tube pathway in *Smyrnium perfoliatum* (Apiaceae). *Int. J. Plant Sci.* **155**, 437–444.
Weber, M., and Igersheim, A. (1994). "Pollen buds" in *Ophiorrhiza* (Rubiaceae) and their role in Pollenkitt release. *Bot. Acta* **107**, 257–262.
Wisniewski, M., and Davis, G. (1995). Immunogold localization of pectins and glycoproteins in tissues of peach with reference to deep supercooling. *Trees Struct. Funct.* **9**, 253–260.
Wojtasek, P., and Bolwell, G. P. (1995). Secondary cell wall-specific glycoprotein(s) from French bean hypocotyls. *Plant Physiol.* **108**, 1001–1012.
Yates, E. A., Valdor, J. F., Haslam, S., Morris, H. R., Dell, A., Mackie, W., and Knox, J. P. (1996). Characterization of carbohydrate structural features recognized by anti-arabinogalactan-protein monoclonal antibodies. *Glycobiology* **6**, 131–139.
Ye, Z. H., and Varner, J. E. (1991). Tissue-specific expression of cell wall proteins in developing soybean tissues. *Plant Cell* **3**, 23–37.
Ye, Z. H., Song, Y. R., Marcus, A., and Varner, J. E. (1991). Comparative localization of three classes of cell wall proteins. *Plant J.* **1**, 175–183.
Zhang, G. F., and Staehelin, L. A. (1992). Functional compartmentation of the Golgi apparatus of plant cells. Immunocytochemical analysis of high-pressure frozen- and freeze-substituted Sycamore Maple suspension culture cells. *Plant Physiol.* **99**, 1070–1083.

# Biophysical Aspects of P-Glycoprotein-Mediated Multidrug Resistance

Randy M. Wadkins[1] and Paul D. Roepe
Molecular Pharmacology & Therapeutics Program, Raymond & Beverly Sackler Foundation Laboratory, Memorial Sloan-Kettering Cancer Center, and Graduate Program in Pharmacology, Cornell University Medical College, New York, New York 10021

In the 45 years since Burchenal's observation of chemotherapeutic drug resistance in tumor cells, many investigators have studied the molecular basis of tumor drug resistance and the phenomenon of tumor multidrug resistance (tumor MDR). Examples of MDR in microorganisms have also become topics of intensive study (e.g., *Plasmodium falciparum* MDR and various types of bacterial MDR) and these emerging fields have, in some cases, borrowed language, techniques, and theories from the tumor MDR field. Serendipitously, the cloning of MDR genes overexpressed in MDR tumor cells has led to elucidation of a large family of membrane proteins [the ATP-binding cassette (ABC) proteins], an important subset of which confer drug resistance in many different cells and microorganisms. In trying to decipher how ABC proteins confer various forms of drug resistance, studies on the structure and function of both murine and human MDR1 protein (also called P-glycoprotein or P-gp) have often led the way. Although various theories of P-gp function have become popular, there is still no precise molecular-level description for how P-gp overexpression lowers intracellular accumulation of chemotherapeutic drugs. In recent years, controversy has developed over whether the protein protects cells by translocating drugs directly (as some type of drug pump) or indirectly (through modulating biophysical parameters of the cell). In this ongoing debate over P-gp function, detailed consideration of biophysical issues is critical but has often been neglected in considering cell biological and pharmacological issues. In particular, P-gp overexpression also changes plasma membrane electrical potential ($\Delta\psi_o$) and intracellular pH ($pH_i$), and these changes will greatly affect the cellular flux of a large number of compounds to which P-gp overexpression confers resistance. In this chapter, we highlight these biophysical issues and describe how $\Delta\psi_o$ and $pH_i$ may in fact be responsible for many MDR-related

[1] Present address: Institute for Drug Development, 14960 Omicron Drive, San Antonio, Texas 78245.

phenomena that have often been hypothesized to be due to direct drug translocation (e.g., drug pumping) by P-gp.

**KEY WORDS:** MDR Protein, Membrane potential, Intracellular pH, Chemotherapeutic drugs, Chemoreversal.

# I. Introduction

## A. Definition of Multidrug Resistance

Multidrug resistance (MDR) is a general term that refers to the phenotype of cells or microorganisms that exhibit resistance to different, chemically dissimilar, cytotoxic compounds. MDR can develop after sequential or simultaneous exposure to all the different drugs to which the cell or microorganism is resistant. Due to promiscuous over-the-counter sale of antibiotics, this was the case in one of the first descriptions of MDR, which involved a strain of *Shigella dysenteriae* discovered in northern Japan in the 1950s (Watanabe, 1963), and is usually the case today with MDR *Mycobacterium tuberculi*. MDR can also develop, surprisingly, *before* exposure to many of the compounds to which the cell or microorganism is ultimately found to be resistant. This is frequently the case with myeloma, non-Hodgkin's lymphomas, and perhaps certain other human tumors; *Escherichia coli* and other gram-negative bacteria exposed to tetracycline; and certain parasites such as *Plasmodium falciparum* (the causative agent in malaria). We refer to the first type of MDR as "additive MDR" because multiple genetic mechanisms are likely induced in cells upon selection with more than one drug to produce a MDR phenotype that is essentially the sum of several drug resistance mechanisms operating together. The second type of MDR is called "pleiotropic MDR" because it is presumably due to only one genetic event. This genetic event in pleiotropic MDR is frequently (but not always) the altered expression and/or mutation of an ATP-binding cassette (ABC) transporter. Transporters in the ABC family share sequence homology and predicted structural features that include six transmembraneous $\alpha$-helices followed by a cytoplasmically or intravesicularly disposed ATP binding site. In prokaryotes, one "six helix core" structure is utilized, but in eukaryotes ABC transporters harbor two highly homologous core structures repeated in tandem.

Over the past 25 years, the most extensive analysis of pleiotropic MDR phenomena has probably been with human tumor cell lines grown *in vitro* (although tetracycline-induced MDR in bacteria has also been a heavily studied topic, as has antimalarial MDR in *P. falciparum*). There are estimated to be a half million deaths annually in the United States due to drug-resistant tumors. Understanding P-glycoprotein (P-gp) at a molecular level will aid design of better chemotherapy; thus, elucidating P-gp function in detail is a key goal of cancer pharmacology. Additionally, because of the similarities in pleiotropic MDR exhibited by a surprising variety of cell types (e.g., various tumor cells) and microorganisms, as well as the recent increased appearance of many pleiotropic MDR microorganisms and the continued failure of certain chemotherapeutic strategies, detailed molecular-level analysis of MDR phenomena and MDR protein has acquired additional significance and urgency.

## B. Tumor MDR

In tumor cell lines, selection with one chemotherapeutic drug (typically an anthracycline, vinca alkaloid, actinomycin-D, or colchicine) frequently results in the overexpression of MDR protein, also called P-gp or gp-170, along with development of pleiotropic MDR. "MDR protein" refers to a family of polytopic, integral membrane proteins (including huMDR1, muMDR1, and muMDR3) whose overexpression is associated with a MDR phenotype. MDR proteins, the multidrug resistance-associated protein MRP; see Cole *et al.*, 1992; Grant *et al.*, 1994) and the cystic fibrosis transmembrane conductance regulator (CFTR) (Riordan *et al.*, 1989) are all members of the ABC superfamily of polytopic integral membrane proteins (Higgins *et al.*, 1990), many of which have been associated with transport phenomena.

Interestingly, MDR protein is also overproduced upon acid or osmotic shock of cells in which the protein is endogenously synthesized (e.g., colon, kidney, and liver cells; Wei and Roepe, 1994). MDR protein is probably the best known and most heavily studied ABC transporter (except perhaps the CFTR, which is mutated in cystic fibrosis patients; Tsui, 1992). Overexpression of the MDR protein must cause some degree of the pleiotropic MDR phenotype exhibited by the drug-selected tumor cells because infection or transfection of cell lines with MDR cDNA, without subsequent growth of the infectants/transfectants in the presence of chemotherapeutic drugs, mimics some aspects of the phenotype (Guild *et al.*, 1988; Devault and Gros, 1990; Hoffman *et al.*, 1996). In transfectants, however, the degree of drug resistance that is unequivocally conferred by MDR protein overexpression alone is substantially less than that exhibited by cells expressing similar levels of MDR protein that have been preselected with drug (Hoffman *et al.*, 1996). Thus, most MDR tumor cell lines developed over the past 25 years exhibit a phenotype that is likely the sum of pleiotropic

MDR (due to MDR protein overexpression) and other drug resistance mechanisms. Only recently have "true" transfectants overexpressing MDR protein, but not in any way exposed to chemotherapeutic drugs, been studied in detail (Hoffman et al., 1996; Weisburg et al., 1996) and several major surprises have already been documented with these cells, including several important phenomena that clearly cannot be explained via conventional hypotheses for MDR protein function. In studying pleiotropic tumor MDR, the distinction between true transfectants and cells exposed to (or selected with) chemotherapeutic drug is critical but often neglected. Several papers (Luz et al., 1994; Wei and Roepe, 1994; Wei et al., 1995; Robinson and Roepe, 1996; Hoffman et al., 1996) have attempted to address this point.

## 1. MDR Genes and Different Hypotheses for P-gp Function

Genes encoding human, mouse, and hamster MDR proteins were isolated and cloned in 1986 (Chen et al., 1986; Gros et al., 1986a,b; Scotto et al., 1986). Upon sequence analysis, it was deduced that these genes encoded proteins homologous to components of periplasmic transport systems from E. coli, such as the HisP and MalK proteins (Ames, 1986). Because of this homology, which includes Walker motifs that imply the binding and perhaps hydrolysis of ATP, as well as the decreased intracellular retention of chemotherapeutic drugs observed for MDR tumor cells (Danø, 1973), it has been suggested (Gros et al., 1986a; Gerlach et al., 1986) that MDR protein (P-gp) is a transport protein and that it actively "pumps" chemotherapeutic drugs out of cells, thereby lowering intracellular drug concentration and conferring drug resistance. The energy for this hypothesized pumping is presumed to be provided by ATP hydrolysis. To some investigators, the P-gp drug pump hypothesis has significant appeal and it has become quite popular in certain fields (see Gottesman and Pastan, 1993; Ruetz and Gros, 1994a,b; Shapiro and Ling, 1995b). The pump hypothesis has been extended to suggest that the ABC transporters involved in multidrug resistance in other systems [e.g., chloroquine resistance protein (CRP) in P. falciparum; Wilson et al., 1989] are also drug pumps.

However, as the MDR field has evolved further in recent years, several difficulties with the P-gp drug pump hypothesis have become apparent. In particular, this model strongly violates the law of enzyme specificity (it proposes pumping of substrates as diverse in structure as actinomycin-D, various gramicidins, progesterone, tetraphenylphosphonium, and Tween detergents), and appears to violate the coupling principle for active transporters (as well as other thermodynamic and kinetic concepts; see within). In addition, biophysical data have been obtained in favor of other models, including the hypothesis that P-gp alters passive cellular drug partitioning through modulating plasma membrane electrical potential ($\Delta\psi_o$) and/or intracellular pH ($pH_i$) (Roepe et al., 1993; Roepe, 1995; Wadkins and Houghton, 1995; this chapter). This model is attractive in that it does not violate fundamental biological, chemical, and physical principles as does the pump hypothesis. On the other hand, some argue that this model cannot explain certain detailed features of the MDR phenotype; however, this criticism is based in part on an incomplete description of the true P-gp-mediated MDR phenotype (see Roepe, 1995). In this chapter, we address this confusion.

Other novel hypotheses for P-gp function that have been introduced include (i) P-gp is an ATP channel (Abraham et al., 1993; Al-Awqati, 1995), (ii) P-gp is a $Cl^-$ channel (Valverde et al., 1992; Altenberg et al., 1994), (iii) P-gp is both a $Cl^-$ channel and a drug transporter (Gill et al., 1992), and (iv) P-gp is a "flippase" (Higgins and Gottesman, 1992) and/or "vacuum cleaner" (Gottesman and Pastan, 1993) that actively translocates drugs in a manner that is quite distinct from that envisioned for all other pumps. During the time these models have been proposed, it has also become clear that homology between MDR proteins and *E. coli* HisP and MalK proteins is the tip of an iceberg and that MDR protein is more homologous to other members of the ABC family of polytopic integral membrane proteins. These include the CFTR and sulfonyl urea receptor (SUR; Aguilar-Bryan et al., 1995), which are not thought of as drug-resistance proteins but whose structure and function are, for other reasons, also important topics in modern biomedical research.

## 2. Homology between P-gp, CFTR, and SUR

CFTR and SUR are expressed in the lung epithelium and the $\beta$ cells of the pancreas, respectively, as well as in other tissues. CFTR functions as a low-conductance ATP and cyclic AMP-dependent $Cl^-$ channel (Bear et al., 1992) that also appears to have additional important functions such as modulation of epithelial $Na^+$ conductance (Stutts et al., 1995) and regulation of outwardly rectified chloride channels (ORCC) (Egan et al., 1992). That is, when CFTR function is absent from polarized airway epithelial cells because of a mutation in the CFTR gene, biophysically distinct 10–15 and 80 pS chloride conductances (due to the separate channel proteins CFTR and ORCC, respectively) are lost and $Na^+$ conductance via the epithelial $Na^+$ channel (ENaC) expressed on the opposite side of the cell is dramatically perturbed. Thus, defects in CFTR function lead to profound alterations in epithelial salt transport and hence altered mucous properties in cystic fibrosis patients, thereby resulting in chronic lung infections that are a hallmark of the disease. Expression of wild-type CFTR in CF cells with the three ion transport defects restores some level of ion transport via

ORCC and ENaC as well as chloride transport via CFTR. Furthermore, cells expressing altered levels of CFTR also exhibit perturbations in $\Delta\psi_o$ and $pH_i$ homeostasis (Elgavish, 1991; Stutts et al., 1993; Wei et al., 1995). Perturbations in these parameters could conceivably contribute to the diverse effects on channel activities and other cellular characteristics caused by loss of CFTR protein in CF cells.

The SUR is somehow triggered by sulfonyl urea drugs (insulin secretogues-like glyburide) to depolarize $\beta$ cells (lower their plasma membrane electrical potential). This depolarization leads to $Ca^{2+}$ influx via a L-type $Ca^{2+}$ channel and hence to $Ca^{2+}$-stimulated fusion of insulin-containing vesicles to the plasma membrane. Depolarization may occur via SUR-mediated inhibition of ATP regulated potassium ($K_{ATP}$) channels in the $\beta$ cell or through some conductance directly mediated by the SUR that appears to involve association of the SUR with a member of the inward rectifier $K^+$ channel family, BIR (Inagaki et al., 1995). A model in which SUR might indirectly mediate $\beta$-cell depolarization is the ATP transporter hypothesis (Al-Aqwati, 1995), wherein hypothesized ATP transport by the SUR activates a purinergic receptor/G-protein signal transduction pathway(s) that then regulates ion channels such as $K_{ATP}$ (Fig. 1). This general idea is reminiscent of the ATP channel hypothesis for CFTR protein introduced by Guggino and colleagues (Schwiebert et al., 1995) and others (Cantiello et al., 1994) as well as a similar model for the MDR protein proposed by Cantiello and colleagues (Abraham et al., 1993). However, note that this model for the CFTR has been challenged (Reddy et al., 1996) and is associated with several difficulties (see Section IV). Nonetheless, although sequence homology is certainly no guarantee

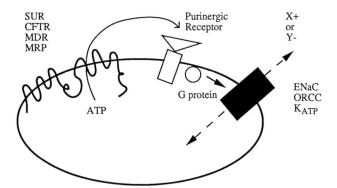

FIG. 1 A schematic diagram of the ATP autocrine pathway that has been proposed for the function of CFTR, MDR, and SUR proteins (see Schwiebert et al., 1995; Al-Awqati, 1995; see also Reddy et al., 1996, for data that contradicts this model).

of functional similarity, the fact that both CFTR and SUR mediate ion conductance (directly or indirectly) and that both are thus somehow intimately involved in plasma membrane electrochemical phenomena lends general support to the idea that MDR protein may be involved in similar phenomena as well.

Indeed, a role for MDR protein overexpression in mediating changes in $\Delta\psi_o$ and $pH_i$ is the hallmark of the altered partitioning model for MDR protein-mediated drug resistance (Roepe et al., 1993; Roepe, 1995), which proposes that MDR protein *indirectly* perturbs intracellular drug retention through altering $\Delta\psi_o$, $pH_i$, and perhaps other biophysical characteristics of the cell but does not *directly* pump, flip, or vacuum drugs. Significant and important perturbations in $\Delta\psi_o$ and the pH gradient across MDR cell plasma membranes harboring overexpressed MDR protein have indeed been observed (Roepe, 1995).

## 3. Direct vs Indirect Models

In summary, current models for MDR protein function essentially fall into direct vs indirect drug transport categories. Either MDR protein directly transports (by pumping, flipping, or vacuuming) the drugs to which MDR cells are resistant, or MDR protein alters drug partitioning and retention through indirect means by modulating $pH_i$, $\Delta\psi_o$, and perhaps other biophysical/biochemical characteristics of the cell. The mechanism for the latter could be direct ion transport by P-gp (Valverde et al., 1992), regulation of ion transport by P-gp (Luckie et al., 1994; Luz et al., 1994; Hardy et al., 1995), involvement of P-gp in an autocrine pathway involving the transport of ATP (Schwiebert et al., 1995; Abraham et al., 1993), or perhaps even lipid transport (Higgins and Gottesman, 1992), which indirectly perturbs ion transporter activity via subtle changes in the physical characteristics of the membrane. There is also the notion that many of the different observations that led to proposal of these various models can be reconciled if P-gp has more than one molecular mechanism of action; for example, that it acts as both a drug pump and a $Cl^-$ channel or $Cl^-$ channel regulator (Gill et al., 1992).

How have so many models for the MDR protein evolved, and can there be a simple way to incorporate the majority of the experimental data into one working model that does not violate the law of enzyme specificity or propose that *one* protein has *two* activities as thermodynamically and kinetically disparate as pump *and* channel function? Can a model for MDR protein be developed that is more consistent with the function of the CFTR and SUR homologs? Although in 1986, homology to MalK and HisP guided our intuition and thus strengthened the appeal of the pump hypothesis, homology to CFTR and SUR, cloned in 1989 and 1995, respectively, is greater. Perhaps comparisons to these proteins are more relevant. We believe the answers to these questions may lie in careful inspection of a variety of interdisciplinary data and concepts that have been gathered and discussed during recent molecular-level analysis of MDR protein function. Study of this provocative polytopic integral membrane protein and its homologs is not simply a pharmacology problem, a membrane biophysics problem, a transport physiology problem, or a molecular biology problem, but is in reality all of these. Because of this, it is virtually impossible to completely satisfy all perspectives in one review article. Many other reviews have focused on the biochemistry and cell biology of MDR (Beck, 1987; Gottesman and Pastan, 1993; Simon and Schindler, 1994; Leveille-Webster and Arias, 1995). In this chapter, it is our goal to analyze P-gp-mediated MDR from a rigorous biophysical perspective and to examine whether available experimental data and biophysical theory agree with various models for P-gp function.

## II. Review of Drug Transport Studies

### A. A Few Definitions and a Little History

The term "transport" has several connotations, and these should be reviewed before proceeding. Several classes of molecules or molecular complexes found in or on various cell membranes "mediate the transport" or "facilitate the translocation" of small molecules, proteins, ions, hormones, etc., but this does not necessarily make them "transporters" by one formal definition of the term. To us, transporters are proteins or protein complexes that directly translocate molecules or ions across membranes—meaning that they specifically bind the substrate and define the actual physical pathway taken by the molecule during transmembraneous translocation. This translocation process should proceed faster than the rate of passive diffusion; otherwise, a cell would waste energy battling to transport a substance that is free to rapidly diffuse down a concentration gradient. In some cases, a transporter may directly transport one or more entities and thereby exert critical indirect effects on the translocation of other molecules that are directly carried by completely different active transport systems or that are not actively transported at all.

As an example of the first case, the $H^+$ ATPase of *E. coli* generates a proton electrochemical potential ($\Delta\mu_{H^+}$), thereby providing the driving force for a variety of $H^+$-coupled sugar and amino acid pumps. Inhibition of the $H^+$ ATPase inhibits sugar or amino acid active transport but not because the ATPase directly translocates sugars and amino acids. One common nomenclature emphasizes the distinction between "primary" active transporters (e.g., $H^+$ ATPase) and "secondary" active transporters (e.g., a $H^+$-coupled sugar pump such as the lactose permease; Roepe *et al.*, 1990).

As an example of the second case, inhibition of the $H^+$-ATPase would also dramatically affect passive distribution of tetraphenylphosphonium ($TPP^+$; Schuldiner and Kaback, 1975) because $TPP^+$ passively equilibrates across the membrane in response to the $\Delta\mu_H^+$ that is generated by the ATPase. However, importantly, there is no known $TPP^+$ pump or transporter in the *E. coli* inner membrane that *directly* catalyzes the transport of $TPP^+$; translocation is passive. If a hydrophobic cation were also a weak base, the situation would be further complicated by any pH gradient influenced by the $H^+$ ATPase (see Section III).

### 1. Pumps, Channels, and Exchangers

More specific definitions for various types of transporters are derived from analyzing the thermodynamics and kinetics of the translocation processes they mediate, as well as biochemical analysis of the transport reactions. Thus, channels (in general) translocate $10^6$ or more ions per second down electrochemical gradients and do not directly require energy to catalyze this translocation; they usually open/close in response to variations in the electrical field in which they operate (i.e., the membrane electrical potential), and they are further regulated in a variety of ways including through ligand binding, phosphorylation, and perhaps even ATP hydrolysis (Gadsby and Nairn, 1994).

Pumps, on the other hand, transport substrate against a concentration gradient (i.e., "uphill" thermodynamically) and they thus require energy, which may be provided in different forms including ATP hydrolysis (Skou and Norby, 1979) or as a stored electrochemical ion gradient (West and Mitchell, 1973). Pumps couple this consumption of energy to direct translocation of substrate at a well-defined stoichiometry. Pumps translocate faster than the rate of passive diffusion of substrate and exhibit exquisite substrate specificity in order to maintain gradients in pumped substrate. Pumps translocate on the order of $10^1$–$10^2$ molecules or ions per second and are regulated in a variety of ways.

Other distinct varieties of transporters include ion exchangers and facilitated diffusion carriers that, in a way, exhibit hybrid kinetic or thermodynamic behavior relative to pumps and channels. For example, the $Cl^-/HCO_3^-$ exchanger isoform 1 (also called red blood cell "band 3" or anion exchanger) can catalyze ion translocation nearly as fast as some channels but translocates $HCO_3^-$ uphill using energy stored in the $Cl^-$ gradient. Under experimental conditions, $HCO_3^-$ transport proceeds at the expense of the Cl⁻ gradient until (neglecting the finer points of regulation) gradients in the two exchanged ions are equal ($[Cl^-]_o/[Cl^-]_i = [HCO_3^-]_i/[HCO_3^-]_o$).

## 2. Coupling

Many pumps have been studied and they are known to exhibit tight coupling to their energy source in either direction, meaning that if, for example, ATP is hydrolyzed inside a bacterial cell by a H⁺ ATPase in order to pump H⁺ out of the cell, "backward" downhill movement of H⁺ through the ATPase back into the cell should synthesize ATP [assuming adenosine disphosphate (ADP) and free inorganic phosphate ($P_i$) are available.] This concept has been experimentally demonstrated for a variety of pumps, usually with vesicles derived from the cells or with reconstituted proteoliposomes harboring pure transporter (Hasselbach, 1978). A frequently used analogy for this "coupling principle" is an old-fashioned bicycle, where a pedal must be pushed forward one rotation to translocate a bicycle uphill a discrete distance and where the pedal spins one rotation in reverse if the bicycle rolls backwards the same distance (assuming constant friction between the tires and the road and no slippage!). Thus, coupling is also stoichiometric in either direction; for example, in the case of the bacterial H⁺ ATPase, one ATP hydrolysis event pumps exactly one H⁺ uphill, and precisely one ATP is synthesized upon downhill translocation of one H⁺ in the other direction.

Therefore, when one uses the definition pump to describe the function of a protein that directly mediates the transmembraneous translocation of an ion, molecule, peptide, etc., there are several implications, and from a biophysical perspective several areas must be investigated before the term can be used correctly: (i) the thermodynamics of the transport process, (ii) the kinetics of the transport process, (iii) the energetics and coupling stoichiometry, and (iv) the substrate specificity. Although via an analysis of most data, P-gp certainly appears to "mediate the translocation of" chemotherapeutic drugs, a variety of recent data argue that it is not a pump in the traditional sense of the term, and thus the four areas above require detailed inspection (Roepe, 1995).

## 3. Historical Context

We should also recall that in the case of P-gp, our current perspective is based in part on hindsight. Scientific fields are frequently born before the theory and language that best describes some relevant phenomena are fully developed, or fields that start out with a focus in one scientific area might undergo significant metamorphosis and evolve into different fields altogether. It was suggested by Danø (1973) that a drug pump existed in MDR tumor cells at about the same time that West and Mitchell (1973) first explored the concept of ion-coupled pump stoichiometry in any detail and that work with membrane-associated ATPases, first isolated in the mid 1960s (Abrams and Baron, 1967), was just beginning to turn toward a detailed focus on active transport (e.g., pumping) mechanisms (Harold *et al.*, 1969; Abrams *et al.*, 1972). It would have been difficult, if not impossible, for Danø to phrase his investigations of drug translocation in the same language that we more routinely use today to describe the details of molecular membrane transport phenomena. Therefore, when analyzing drug transport data and the previous interpretation of these data that bears on the function of the MDR protein, we must consider the long interdisciplinary history of the tumor MDR field that evolved alongside but separate from the discipline of molecular membrane transport.

As work on the molecular mechanisms of transporters proceeded in many laboratories in the 1970s and 1980s, investigators studying the characteristics of drug-resistant tumor cells used a variety of biochemical and molecular biological approaches to explore altered drug translocation in these cells. Out of necessity, and due to their historical context, some of these approaches were not analogous to those that were used by most investigators in the molecular membrane transport fields during the same period. This is in part because interactions between whole tumor cells and hydrophobic compounds were being examined, instead of interactions between bacterial vesicles or proteoliposomes and hydrophillic sugars, amino acids, or ions. Although the concept is frequently underappreciated, theory and experimental practice developed from an analysis of hydrophillic transport is not necessarily directly applicable to the case of hydrophobic compound transport for a variety of reasons (Noy *et al.*, 1986; Deuticke, 1977). Thus, as the possibility of a drug-pumping process took hold in the MDR literature, concepts from the molecular membrane transport fields were borrowed and adapted to the specific case of hydrophobic drug transport but often in a premature fashion. For example, early studies of chemotherapeutic drug transport in MDR cells sometimes emphasized "saturation" and "temperature coefficient" effects as evidence for active drug transport (i.e., drug pumping), in analogy to effects that were being observed in the study of sugar and amino acid pumping, but these arguments and others along these lines cannot always be reliably extended to the case of transport of hydrophobic compounds, and in fact there are very good reasons for why they should not be (Deuticke, 1977; Noy *et al.*, 1986; Wadkins and Houghton, 1993; Roepe, 1995). Physical interactions between hydrophobic drugs and biological membranes that can very easily explain many observed phenomena (see Section III) have often been neglected in favor of interpretations that envision that the mechanism for any altered transmembraneous translocation must be via a process that is directly protein mediated.

A similar pattern to the pertinent literature may explain apparent anomalies in interpretation in the long history of fields that have studied the translocation of other hydrophobic entities such as long chain fatty acids (Noy et al., 1986). As Deuticke (1977) has written, "Data obtained in transport studies differ in their relevance for the characterization of transfer pathways. Kinetic characteristics such as the type of concentration dependency of permeabilities . . . may help to decide whether the number of transport sites is limited but can also provide information on structural aspects if transformed into molecular and physicochemical properties of the permeant."

### B. Arguments in Favor of the Drug Pump Model

**1. ATPase Activity and Stoichiometry**

As mentioned in the Introduction, the drug pump model for MDR protein has been widely championed, but in the MDR literature many of the features of pumps described previously have not been completely explored. Is the best evidence in favor of the pump hypothesis consistent with the features of pumps that we have outlined? First, the protein has indeed been shown to be an ATPase by several laboratories (Ambudkar et al., 1992; Sharom et al., 1993; Al-Shawi et al., 1994; Shapiro and Ling, 1994) and catalyzes the hydrolysis of about 600 ATP/min. Therefore, in theory the protein has the capacity to provide energy for an active transport process. On the other hand, when stoichiometries of ATP hydrolysis:drug translocation are estimated from the best available drug transport data (Sharom et al., 1993; Schlemmer and Sirotnak, 1994), the estimates range from $10^4:1$ to $10^2:1$ (ATP hydrolyzed:drug molecules translocated), which is a very unusual stoichiometry that is not at all consistent with stoichiometries measured for other ATP-driven pumps (they lie between $1:1$ and $1:3$).

Another way to analyze the importance of this concept is to consider the rate of passive diffusion of chemotherapeutic drugs under initial rate conditions. Let us assume that, similar to other ATP-driven pumps, P-gp is capable of hydrolyzing $10^2$ ATP/sec under idealized conditions (not the languid 600/min as described previously) and has a hydrolysis:drug translocation stoichiometry near $1:1$, also similar to other known ATP-driven pumps. If $10^4$–$10^5$ MDR protein molecules are in the membrane (this would be typical, but in some very high-level expressors we might approach 5 × $10^5$ copies/cell), then via these assumptions $10^6$–$10^7$ drug molecules could conceivably be pumped per cell, per second, as "substrate" concentration approaches the $K_m$ for the transport process. Thus, of course we also need to know the $K_m$ for translocated substrate in the pumping process to continue with this analysis in detail, but at 1 $\mu M$ external drug [within the range of 200 n$M$ (Schlemmer and Sirotnak, 1994) to 100 $\mu M$ (F. J. Sharom, personal communication) $K_m$ that have been calculated to date], approximately $10^6$–$10^7$ drug molecules diffuse into a cell per second under initial rate conditions. Thus, via the pump model and using the best available data on ATP hydrolysis by MDR protein (not our idealized assumption of $10^2$ ATP hydrolyzed/sec), we need an extremely high amount of MDR protein and/or an illogical translocation: hydrolysis stoichiometry (and presumably a low $K_m$) to efficiently compete with passive diffusion of hydrophobic drugs under initial rate conditions in order to produce the effects that are seen in MDR cells (see Section III,D). It is not clear if these conditions can be satisfied based on the data that have been reported to date. This is particularly important because it has been suggested, for example, that due to its occurrence in secretory tissue (Yang *et al.*, 1989) and the fact that it can be photolabeled by progesterone (Qian and Beck, 1990), the "normal" (physiologic) role of P-gp might be as a steroid hormone transporter (e.g., a progesterone transporter). However, the rate of passive diffusion of progesterone across phospholipid membranes is among the fastest of any known substance (Koefoed and Brahm, 1994) and rapid stopped-flow techniques are required simply to measure this rate with any reasonable level of accuracy. Progesterone crosses the plasma membranes of cells some six to eight *orders of magnitude* faster than does $H^+$ or $Na^+$ (Magin *et al.*, 1990). Thus, it is virtually unthinkable that any membrane transport system could compete with a gradient-driven passive flux of this compound, and in the absence of detailed rapid-flow data that argue the converse, this possibility and other similar proposals do not appear very likely.

Another related conundrum that is encountered when interpreting data in the context of the pump hypothesis is that linear initial rates of drug accumulation for MDR cells, relative to drug-sensitive parental cells, are reduced by a similar extent over an extremely wide range of drug concentrations. That is, a similar percentage reduction in the linear initial rate of drug accumulation is observed when performing experiments at external drug concentrations ranging from 1 n$M$ (Stein *et al.*, 1994) and even lower (L. J. Robinson and P. D. Roepe, unpublished results), to 100 $\mu M$ (Stow and Warr, 1993) and even higher (L. J. Robinson and P. D. Roepe, unpublished results). Several investigators interpret this to mean that MDR protein pumps intramembraneous drug before it reaches the cytoplasm (Stein *et al.*, 1994). Schlemmer and Sirotnak (1994) calculate an apparent $K_m$ for vinca alkaloid pumping near 200 n$M$, and Stow and Warr (1993) note an 80% reduction in the initial rate of vinca alkaloid accumulation for MDR cells, relative to drug-sensitive parental cells, at that concentration. However, they also note this same percentage reduction in the rate of accumulation at 1, 10, and even 100 $\mu M$ external drug. We know the initial rate of passive diffusion of vinca alkaloids increases dramatically at a concentration of 100 $\mu M$, relative to 10 $\mu M$. Even allowing for some flexibility in the $K_m$ estimate, how can an enzyme with an apparent $K_m$ of 200 n$M$ keep pace with the increase in the rate of passive diffusion under initial rate conditions that occurs between 10 and 100 $\mu M$ vinca alkaloid? At some concentration of substrate, the rate of transport by a pump must plateau; however, if we interpret all the literature summarizing drug accumulation for MDR cells in the context of the pump model, we find that MDR protein does not appear to plateau over a range of substrate concentrations that spans some five or six *orders of magnitude*. We are unaware of any precedent for such behavior.

## 2. Kinetics

The second set of data used to support a pump hypothesis are numerous studies (some reviewed in Roepe, 1995) that have shown that MDR cells exhibit what is frequently called "increased efflux." Distinction between increased efflux vs "decreased retention" has been discussed previously (Roepe, 1995), and it is an important distinction. In the vast majority of examples currently in the literature, increased efflux for MDR cells refers to a phenomenon wherein a greater percentage of intracellularly trapped drug (which is both "bound" and "free" and forced into the cell during an incubation period) is released in a zero-*trans* efflux measurement (i.e., cells preloaded with drug diluted into drug-free media). These data have been interpreted by several investigators over the years to be the manifestation of an outward-directed pumping process. However, as is the case with any other active transporter or carrier (Roepe *et al.*, 1990, and references within) a pumping process should also exhibit a kinetic signature. P-gp does not appear to have an effect on the rate of outward-directed drug transport (Roepe, 1992; Bornmann and Roepe, 1994). Therefore, whether most of the zero-*trans* efflux experiments performed during the past 20 years reveal increased efflux in a kinetic sense (i.e., a heightened rate of outward flux that is not simply due to an increased gradient in freely exchangeable drug and hence increased initial rate of passive diffusion) is not known, and in fact a variety of recent additional data argue that these previous studies do not reveal an increased rate of efflux in the traditional sense of the term (Roepe, 1992, 1995; Wadkins and Houghton, 1995); that is, there is no increase in the rate constant characterizing the outward-directed translocation process (Bornmann and Roepe, 1994).

## 3. Thermodynamics

The best data in favor of a drug pump model are several drug transport studies performed with proteoliposome and vesicle preparations, but their interpretation is not without possible pitfalls. These data have been published and interpreted in terms of a drug pump model by several investigators (Ruetz and Gros, 1993, 1994a; Sharom *et al.*, 1993; Schlemmer and Sirotnak, 1994; Shapiro and Ling, 1995a). Only one of the studies (Ruetz and Gros, 1994a) actually attempts to eliminate both membrane potential and pH gradient effects that could severely complicate interpretation (see below) by measuring drug accumulation in vesicles with inverted orientation before and after collapsing the predicted $\Delta\mu_H+$. In this specific case, membrane potential is calculated for yeast secretory vesicles by measuring equilibrium $TPP^+$ distribution, but it should be noted that this technique has been shown to sometimes be an inaccurate method for measuring potential (Bakker *et al.*, 1976). Addition of nitrate ions to the transport medium is used in this study to dissipate potential, instead of addition of nigericin/valinomycin mixtures, which would be expected to more fully and unequivocally collapse $\Delta\mu_H+$ (Bertl *et al.*, 1993; Ballarin-Denti *et al.*, 1984). Nonetheless, this one paper is, in our opinion, the best evidence in favor of a drug pump model presented to date. However, the above concerns as well as several other unresolved complexities in the proteoliposome and vesicle studies published to date, including very low estimated turnover (drug molecules pumped/MDR protein/minute), unusual predicted stoichiometries, and apparent downhill but not uphill transport in at least one of the studies (Shapiro and Ling, 1995a), leave some doubt. As reviewed previously (Roepe, 1995), these curiosities make it unclear whether active drug pumping is actually revealed in these data, but the reader should note that this is not the interpretation of the authors of those studies (Ruetz and Gros, 1994b; Shapiro and Ling 1995b).

It may also be possible to explain the majority of the observations from these studies and others like them via small perturbations in $pH_i$, $\Delta\psi_o$, volume, and/or other physical characteristics of the preparations (see Section III). If this explanation is neglected, then a less straightforward pumping process that accounts for how low measured turnover could possibly compete with passive diffusion and that also accounts for unusual hydrolysis:translocation stoichiometry and lack of substrate specificity obviously must be invoked if the current data are to be reconciled with some type of direct drug translocation model.

## 4. Photolabeling

Other data that are frequently cited as evidence in favor of a drug-pumping function for the MDR protein include many photolabeling experiments and "altered resistance profiles" for cell lines expressing different mutant MDR proteins. As discussed previously (Tew et al., 1993; Roepe, 1995), photolabeling *per se* is interesting but not strongly suggestive of pumping because no specific photolabeling of MDR protein above background for cells expressing physiologic levels of the protein is observed (Wei et al., 1995). No attempt has been made in the majority of these studies to discern affinity labeling from pseudoaffinity labeling (Tew et al., 1993). Furthermore, no unequivocal equilibrium drug binding, via detailed Scatchard analysis, has yet been demonstrated. Because the compounds used in photolabeling experiments have high lipid partitioning coefficients, these are important caveats. Many membrane proteins are known to be quite efficiently photolabeled by a variety of hydrophobic or amphipathic probes, but they certainly do not transport the compounds; thus, in and of themselves, photolabeling studies do not provide strong evidence for direct transport of drugs by MDR protein (P-gp).

Altered photolabeling and altered resistance profiles have been observed for cells harboring mutant MDR proteins, and some investigtors interpret this to mean that a drug binding site must be mutated in the MDR protein mutant, leading to altered substrate recognition and thus altered drug pumping. However, these observations have been made after the cells harboring the mutant MDR protein were grown in the presence of different drugs or different drug concentrations relative to cells harboring the wild-type protein. Thus, whether the altered photolabeling and resistance pattern are due to the different MDR proteins or different exposure to drug prior to assaying drug resistance is debatable (Roepe, 1995).

In summary, a plethora of data have been published and interpreted exclusively in favor of a drug pump model for the MDR protein, but very reasonable points that are at odds with this interpretation can certainly be raised. If the MDR protein is a drug pump, it violates two fundamental tenets of active transporters—the law of specificity and (apparently) the coupling principle (Roepe, 1994, 1995)—and it would be the first example of a pump to do so. The effects that MDR protein overexpression has on rates of drug accumulation also appear to be completely inconsistent with what is calculated to be the $K_m$ of the pump. Thus, some degree of skepticism is certainly called for. The problems with the drug pump model do not necessarily prove that the model is completely wrong, but they do suggest that other models that are perhaps less frequently discussed but that have less difficulty associated with them should also be seriously considered.

## III. Other Data Important in Analyzing Drug Transport

### A. The Electrical Membrane Potential

As reviewed previously (Roepe, 1995), several studies have shown that significant and stable membrane depolarization occurs in MDR tumor cell lines. To date, every cell line that is MDR solely through overexpression of the MDR protein, and not through growth on medium harboring chemotherapeutic drug, that has been examined for membrane potential has been found to be depolarized relative to the parental cell line from which it was derived (Luz et al., 1994; Hoffman et al., 1996). In addition, depolarization of plasma membrane via overexpression of the CFTR (Stutts et al., 1993; Wei et al., 1995) appears to lead to a MDR phenotype that is similar, but not identical, to the phenotype solely mediated by MDR protein overexpression. Furthermore, a significant collection of model studies using large unilamellar vesicles (LUVs) exhibiting various membrane potential shows that depolarization can have profound effects on the interaction between chemotherapeutic drugs and membranes (Roepe, 1995). In particular, membrane potential greatly affects the ability of LUVs to accumulate drugs that are part of the MDR phenotype (e.g., doxorubicin; (Mayer et al., 1985b; Bally et al., 1985; Praet et al., 1993). Thus, we logically expect that alterations in membrane potential for MDR tumor cells will have critical pharmacologic consequences.

However, confusion regarding this point exists. It has also been argued in the MDR literature (Ruetz and Gros, 1994a,b) that changes in membrane potential should only lead to significant perturbations in the transmembraneous distribution of hydrophobic compounds that are formal cations (or anions). That is, detailed work by Kaback and colleagues in the 1970s (Schuldiner and Kaback, 1975) showed that in some cases the equilibrium transmembraneous distributions (the ratio of inside bulk water phase to outside bulk water phase) of $TPP^+$ and its analogs could be used to estimate the magnitude of membrane potential because the cation has a rapid rate of passive diffusion due to its ability to delocalize charge. Because many of the chemotherapeutic drugs to which MDR cells are resistant are not formal cations, but are actually weak bases, some have argued that changes in membrane potential should not lead to significant changes in the transmembraneous distribution of drugs for MDR cells because a weak base could deprotonate and diffuse down a chemical gradient. This argument follows that if alterations in membrane potential perturbed distribution of the charged weak-base species, the uncharged weak-base species would diffuse down an altered drug chemical gradient and "short circuit" the Nernstian equilibrium. This simplified reasoning raises several detailed issues that have not been completely addressed and that thus merit additional inspection. We address these below (see Sections III,B–III,F).

However, regardless of the precise theoretical explanation, a large amount of direct experimental data show $\Delta\psi_o$ perturbations do indeed alter accumulation and/or distribution of weakly basic chemotherapeutic drugs, and these studies have been previously reviewed (Roepe, 1995). Measured $\Delta\psi_o$ effects are not necessarily due, indirectly, to $\Delta pH$ effects that also occur under some experimental conditions (e.g., LUV $K^+/Na^+$–valinomycin diffusion potential experiments performed in unbuffered media; see de Kroon et al., 1991). These data aside, some investigators also argue that these alterations could not produce ≥50-fold levels of drug resistance, as is frequently seen in many model MDR cell lines. However, when discussing the P-gp-mediated MDR *phenotype*, we first need to define what we mean precisely. It is that phenotype unequivocally mediated by the overexpression of P-gp *alone* and its features include membrane depolarization and alkalinization (Luz et al., 1994; Hoffman et al., 1996). Although the concept is very frequently underappreciated, P-gp overexpression in and of itself mediates relatively low levels of drug resistance (2- to 10-fold; Guild et al., 1988; Devault and Gros, 1990; Hoffman et al., 1996). Shockingly, there has been practically no detailed analysis of drug transport kinetics or altered equilibrium distribution of different chemotherapeutic drugs for true MDR transfectants wherein the phenotype can only be due to MDR protein (P-gp) overexpression (the one exception we are aware of being a paper by Robinson and Roepe, 1996). Thus, we do not really know what precise effects of $\Delta\psi_o$ to expect, and arguments to the effect that observed $\Delta\psi_o$ and $pH_i$ alterations cannot completely explain the P-gp-mediated MDR phenotype are therefore premature and may be inaccurate. In fact, Hoffman et al. (1996) show that $\Delta\psi_o$ and $pH_i$ perturbations due to P-gp overexpression *can* indeed explain the levels of drug resistance mediated solely by P-gp overexpression.

To further explore this concept, we can also compare cells to model systems in which effects of $\Delta pH$ and $\Delta\psi_o$ have been theoretically and experimentally verified. In the following section, we introduce models for drug accumulation that envision P-gp indirectly affecting cellular drug flux through modulating $\Delta\psi_o$ and/or $pH_i$.

Before discussing these models, it is also quite important to remember that investigators do not typically measure the "equilibrium" ratio of *cytoplasmic* drug to extracellular drug for MDR cells. Instead, investigators usually measure relative retention or accumulation of drug over time. Therefore, directly comparing the behavior of model vesicles/liposomes to the behavior of cells is dangerous. Cells are like little chemotherapeutic drug "sponges" that "soak up" drug to very high levels. For example, incubating a cell in 1 $\mu M$ vinblastine for 30–60 min can lead to intracellular concentrations of drug that are 50–100 $\mu M$ or even higher, assuming total intracellular volume is used in the calculation and that extracellular vinblastine is not limiting. Because a MDR cell expressing MDR protein might *retain* anywhere from 90 to 10% of the drug retained by a matched sensitive cell not expressing MDR protein, the perception among some investigators is that there must be huge perturbations in the transmembraneous equilibrium distribution of drug. However, most of the intracellular drug is either tightly or loosely bound. We are not aware of many studies that have rigorously compared equilibrium transmembraneous ratios (that is, cytoplasmic/extracellular bulk water phase ratios) of drug for sensitive vs MDR cell lines. One well-performed study that has attempted this is Simon *et al.* (1994), wherein otherwise drug-sensitive cells were pulsed with $CO_2$ in order to rapidly change $pH_i$. Changes in the (presumably) cytosolic concentration of doxorubicin that are predicted from standard weak-base partitioning theory were indeed observed via rapidly monitoring the change in intracellular doxorubicin fluorescence. It would be interesting to extend these observations with fluorescent derivatives of other drugs [e.g., coumarin–vinblastine (Bornmann and Roepe, 1994), or FITC–colchicine, available from Molecular Probes] and also to perform similar measurements upon perturbation of $\Delta\psi_o$. Without more detailed analysis along these lines, we do not even know how extensively P-gp overexpression, and thus a given change in $\Delta\psi_o$, should alter equilibrium cytosolic drug concentrations. The point is that due to these complexities and unresolved issues, with regard to cellular accumulation of chemotherapeutic drugs the possible effects of membrane potential need to be viewed in a broader perspective.

## B. Effects of Potential on Membrane Flux of Cationic Drugs

A significant body of data has now been published that indicates one aspect of the P-gp-mediated MDR phenotype is altered membrane potential. Indeed, it has also been established that overexpression of CFTR results in cells with lowered plasma membrane potential and a MDR phenotype that is similar, but not identical, to the phenotype mediated by overexpression of P-gp (Wei *et al.*, 1995). It is therefore important to examine the effects of plasma membrane potential on the accumulation and efflux of cationic drugs to understand how alterations in these parameters can result in reduced drug accumulation and less efficient drug retention in cell lines overexpressing P-gp.

When a selectively permeant membrane separates two phases containing a significant concentration of salt ions, an electrochemical potential will be established. When a membrane-permeable substance is introduced to either side of the membrane, it will respond to the electrochemical potential by partitioning into the two phases (and/or the membrane) until both the chemical and the electrical components of the electrochemical system are balanced. In most eukaryotic cells, the selective membrane permeability for $K^+$ generates a voltage across the plasma membrane of 50–60 mV (negative inside). This electrical potential is subsequently capable of affecting the distribution and accumulation of other permeable species within the cell.

The simplest way to view the effect of membrane potential is to invoke the familiar Nernst equation, which describes the effects of internal and external ion concentrations on the membrane potential. Viewed another way, it describes the effect of membrane potential on internal and external cationic drug concentrations. In this application, the Nernst equation in its simplest form is

$$\Delta\psi_o = -60 \log([D_i]/[D_o]), \tag{1}$$

where $\Delta\psi_o$ is the membrane potential in mV, $[D_i]$ is the molar drug concentration within the cell, and $[D_o]$ is the external molar drug concentration.

Compared to the $mM$ concentrations of salt ions in cell incubation solutions, the value of $\Delta\psi_o$ is independent of drug concentration and is generally fixed by the $K^+$ concentration of the medium. Thus, in a simplified way a measurement of $\Delta\psi_o$ should describe what the relative level of drug accumulation will be in a given cell line having a certain membrane potential. In most eukaryotic cells, the membrane potential is approximately $-60$ mV, so the concentration of drug inside the cell is given by

$$[D_i] = 10 \times [D_o].$$

In cells that have been transfected with high levels of CFTR, the membrane potential is decreased to (for example) $-45$ mV. This results in a predicted internal concentration of

$$[D_i] = 5.6 \times [D_o].$$

Consequently, when external drug concentrations are held to similar levels, these CFTR-expressing cells would be predicted to be approximately 1.8-fold resistant to cationic drugs, in good agreement with the 1.1- to 1.5-fold resistance that is observed (Wei *et al.*, 1995). Other CFTR-expressing cells have membrane potentials of approximately $-30$ mV. This then gives an internal concentration of

$$[D_i] = 3.2 \times [D_o],$$

or, simplistically, a predicted 3.1-fold resistance by these cell lines, again in excellent agreement with the 2- to 4-fold resistance that is generally observed (Wei *et al.*, 1995).

The Nernst equation only represents an approximation to the actual situation of cellular drug accumulation. It fails to account for the fact that many hydrophobic drugs are not merely on one side of the membrane or the other but are also found within the membrane. Also, the Nernst equation does not provide any information as to the flux of compounds across the cell membrane because it describes a steady-state situation. This may be important when considering cationic drugs that are also weakly basic with $pK_a$ near physiologic pH and that bind to intracellular targets in a pH-dependent manner (see Section III,E).

## C. A Detailed Steady-State Description of Drug Accumulation

A more appropriate model has been developed by Cafiso and Hubbell (1979, 1982) to describe the effect of membrane potential on steady-state and kinetic accumulation of cationic phosphonium compounds by phosphatidylcholine (PC) vesicles. This model is particularly interesting because similar phosphonium ions have been shown to be excluded from cells upon P-gp overexpression (Gros et al., 1992). We apply a slight modification of the model system to describe the general effects of membrane potential on cationic drug accumulation.

The model is shown in Fig. 2 and is an approximation to the situation occurring when drug is mixed with a cell membrane. The exterior drug is in rapid equilibrium with the outer face of the cell membrane. Drug bound to this outer face diffuses through the membrane to the inner face with a forward rate constant $k_f$ and a reverse rate constant $k_r$. Furthermore, the drug bound to the interior face of the membrane is in rapid equilibrium with the cytoplasm of the cell. However, in contrast to simple PC vesicles, cells contain organelles that may bind the drug internally. We therefore

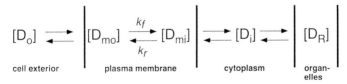

FIG. 2 The model system used to derive the effect of membrane potential on steady-state and kinetic accumulation of cationic drugs. External drug is initially bound to the outside face of the plasma membrane. The limiting step for accumulation is the diffusion from the outside face to the inside face of the membrane and vice versa for efflux. All other equilibria are considered rapid and to be approximated by an effective equilibrium constant $K$.

add an additional feature to the model to describe drug in the cytoplasm that is also in rapid equilibrium with cellular organelles.

When drug is mixed with a cell suspension, an equilibrium situation is ultimately reached. In this steady state, the electrochemical potential for each component of the equilibrium in Fig. 2 is given by

$$\bar{\mu}_j = \mu°_j + RT \ln x_j + ZF\psi_j, \quad (2)$$

Where $\mu°_j$ is the chemical potential of the component in the absence of an electric potential, $x_j$ is the mole fraction of the drug in the component phase, $R$ is the gas constant, $T$ is the absolute temperature, $Z$ is the charge of the drug (+1 for our discussion), $F$ is Faraday's constant, and $\psi_j$ is the electric potential in region $j$. The value for $\psi$ is taken as a reference potential, so $\psi_o = 0$ and $\psi_i = \Delta\psi_o$. The mole fractions are approximated by $n_j/V_j$, where $n_j$ is the number of moles of a species in region $j$ and $V_j$ is the volume of the region. At equilibrium, all electrochemical potentials are equal, and thus $\bar{\mu}_j = \bar{\mu}_k$ for all regions. We also assume that the partition coefficient $K_{mo}$ for the binding of the drug to the outer face of the membrane is the same as $K_{mi}$, the partition coefficient for binding of the drug to the inner face, and also that this partition coefficient describes the affinity of the drug for the organelles ($K_R$). Additionally, for large vesicles, such as cells having a diameter of 12 $\mu$m, we make the approximation that $V_{mi} = V_{mo}$.

The ratio of the number of cell-bound drugs to the number of free drugs can then be written as a function of plasma membrane potential:

$$\frac{n_{cell}}{n_o} = \frac{V_i}{V_o} e^{-ZF\Delta\psi_o/RT} + \frac{K_o V_{mi}}{V_o} e^{-ZF\Delta\psi_o/RT} + \frac{K_o V_{mo}}{V_o} + \frac{K_R V_R}{V_o} e^{-ZF\Delta\psi_o/RT}, \quad (3)$$

where $n_{cell} = n_{mi} + n_{mo} + n_i + n_R$. We may also write the fraction of drug bound to the cell as

$$F_{cell} = \frac{n_{cell}}{n_{cell} + n_o} = \frac{1}{1 + \dfrac{n_o}{n_{cell}}} \quad (4)$$

For drug accumulation assays, a typical experimental condition is that $1 \times 10^6$ cells are mixed with drug in a 2-ml cuvette. We therefore need to make estimates of the parameters shown in Eq. (3) to apply the equation to drug accumulation and efflux experiments. The exterior volume ($V_o$) will be that of the total experiment (2 ml). The interior aqueous volume of cells ($V_i$) is approximately 0.6 $\mu$l for $10^6$ cells (Roepe et al., 1993). For $10^6$ spheres with radii 6 $\mu$m, the total contained volume is 0.9 $\mu$l. We assume the volume not occupied by water is primarily organelle volume, and thus $V_R$ is 0.3 $\mu$l. For $10^6$ spherical shells having a radius of 6 $\mu$m and a thickness of 5 nm (the approximate thickness of a biological membrane), $V_{mo}$ and $V_{mi}$ are 0.001 µl.

The partition coefficient $K_o = K_i = K_R$ will vary depending on the drug in question. For example, actinomycin-D and doxorubicin have partition coefficients (buffer: 1-octanol) on the order of $K \approx 10^1$, whereas vincristine has $K \approx 10^3$ and vinblastine has $K \approx 10^4$ (see Wadkins and Houghton, 1993, and references therein). However, both actinomycin-D and doxorubicin bind tightly to nuclear DNA, increasing $K_R$ to $\approx 10^5$. Therefore, the value for $K$ used here is an effective partition coefficient.

The influence of the partition coefficient on the fraction of cationic drug bound is shown in Fig. 3 at various values of $\Delta\psi_o$. There are two important points to note in Fig. 3. The first is that as the partition coefficient increases, the amount of drug bound becomes less responsive to changes in membrane potential. This result has been observed experimentally for the voltage-sensitive $DiOC_n(3)$ dyes binding to membrane vesicles (Sims et al., 1974) and to P-gp-expressing cell lines (Wadkins and Houghton, 1995). Consequently, this suggests that one reason very hydrophobic cations are able to circumvent MDR (Horichi et al., 1990; Lothstein et al., 1994) is that they are accumulated to the same extent in cells regardless of the plasma membrane potential.

The second important feature pertains to the relative amount of drug accumulated in a cell line having lowered membrane potential. Relative to a sensitive cell ($\Delta\psi_o = -60$ mV), resistant cells may have membrane potentials as low as $-20$ to $-10$ mV (Roepe et al., 1993; Wei et al., 1995). For an effective partition coefficient of $K = 1000$, Fig. 3 indicates that the membrane binding of these resistant cell lines may be reduced to 30% of the

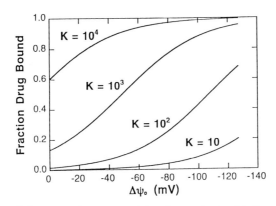

FIG. 3 The effect of plasma membrane potential on the steady-state binding of cationic drugs to $10^6$ cells in 2 ml buffer. The fraction of bound drug is plotted according to Eqs. (3) and (4) at four values of the effective partition coefficient $K$.

sensitive cells. This seems to be the case experimentally in P-gp-expressing cell lines, in which the steady-state accumulation of various drugs may be 10–30% of sensitive cell lines (Wadkins and Houghton, 1995; Roepe, 1992; Wei et al., 1995; Sirotnak et al., 1986). In CFTR transfectants, membrane potentials near $-30$ mV result in a 25% decrease in accumulated vinblastine (Wei et al., 1995), which is more closely modeled by a $K$ of 10,000 (Fig. 3), a value nearer to the true partition coefficient of vinblastine. However, the variety of experimental conditions found in the literature reports of drug accumulation studies, as well as prior drug selection of most model MDR cell lines that induces additional phenomena, makes it difficult to directly compare each result with the predicted values. In general, however, the predicted effects of membrane potential on drug accumulation can be well described for a variety of cationic drugs using a $K$ value of 1000.

## D. Effect of $\Delta\psi_o$ on the Rate of Drug Accumulation

One important aspect of P-gp-expressing cells is that they accumulate less drug than their drug-sensitive counterparts. Another aspect is the apparent "slower" rate of drug accumulation and "faster" efflux of drug. We describe the effects of membrane potential on drug flux kinetics using the model shown in Fig. 2 and the approach of Cafiso and Hubbell (1982), who utilized this model system to explain the kinetic effects of membrane potential on the rates of phosphonium cation accumulation within PC vesicles. We extend this model to the situation at the cell membrane in order to characterize the rates of accumulation of drugs by the cell.

We make the assumption that equilibria at both cell membrane faces and between the cytoplasm and organelles is achieved rapidly with respect to transport through the plasma membrane. Thus, the rate of transport across the plasma membrane is limiting. Then, according to Fig. 2, we can write the kinetic equation for the disappearance of drug bound to the outside interface:

$$-\frac{dn_{mo}}{dt} = k_f n_{mo} - k_r n_{mi}. \tag{5}$$

Again, assuming the partition coefficient for the drugs is the same for both the inside and the outside face of the membrane, and noting that the total number of drugs in the system is given by

$$n_T = n_o + n_i + n_{mo} + n_{mi} + n_R, \tag{6}$$

we may substitute into the rate equation to arrive at an expression that contains only the quantities $n_o$ and $n_T$. Therefore, the change in the number of drugs external to the cell is

$$-\frac{dn_o}{dt} = \left(n_o\gamma - k_r\frac{V_o n_T}{(V_i + K_R V_R)E_i}\right), \quad (7)$$

where

$$\gamma = \left(k_r + k_f\frac{V_o E_o}{(V_i + K_R V_R)E_i}\right) \quad (8)$$

$$E_i = 1 + \frac{K_i V_{mi}}{V_i + K_R V_R} \quad (9)$$

$$E_o = 1 + \frac{K_o V_{mo}}{V_o}. \quad (10)$$

After sufficient time has elapsed, equilibrium will be obtained, and the equality shown in Eq. (7) is zero. Therefore, we find that

$$\gamma n_o(\infty) = k_r\frac{V_o n_T}{(V_i + K_R V_R)E_i}. \quad (11)$$

Substituting Eq. (11) into Eq. (7) and integrating from time $t = 0$, we see that

$$n_o = [n_o(0) - n_o(\infty)]e^{-\gamma t} + n_o(\infty) \quad (12)$$

or

$$F_o = [1 - F_o(\infty)]e^{-\gamma t} + F_o(\infty). \quad (13)$$

Therefore, the number of drugs external to the cell will decrease toward the steady-state concentration $[F_o(\infty)]$ with a rate constant $\gamma$. Conversely, the concentration within the cell will increase toward its steady-state value with an effective rate constant $\gamma$ as given by

$$n_{cell} = n_{cell}(\infty)[1 - e^{-\gamma t}] \quad (14)$$

or

$$F_{cell} = F_{cell}(\infty)[1 - e^{-\gamma t}], \quad (15)$$

where $n_{cell}(\infty)$ and $F_{cell}(\infty)$ are the steady-state values for $n_{cell}$ and $F_{cell}$ as determined from Eqs. (3) and (4).

So far, we have only considered that the membrane potential affects the amplitudes of the accumulation. It is also of some interest to see how the membrane potential affects the rate constant for accumulation of drugs across the membrane. The apparent rate constant, $\gamma$, is given by Eq. (8). To understand the effect of membrane potential on $\gamma$, it is assumed that there is a single barrier to ion flow across the membrane (Fig. 2). This barrier has a maximum height at the center of the membrane and decays exponentially on either side of the center. Cafiso and Hubbell (1982) have successfully described the situation using an Eyring approach to the barrier, such that

$$k_f = k^o_f e^{-nZF\Delta\psi_o/RT} \tag{16}$$

and

$$k_r = k^o_r e^{(1-n)ZF\Delta\psi_o/RT}, \tag{17}$$

where $k^o_r$ and $k^o_f$ are the reverse and forward rate constants in the absence of a membrane potential and $n$ is the fraction of the transmembrane potential that drops from the outer membrane surface to the peak of the barrier. As an approximation, we will assume $n = 0.5$ and that the forward and reverse rate constants are equivalent in the absence of a membrane potential. Substituting Eqs. (16) and (17) into Eq. (8) gives an expression for the apparent rate constant as a function of membrane potential:

$$\gamma = k^o_f \left[ e^{-0.5ZF\Delta\psi_o/RT} + \frac{V_o E_o}{(V_i + K_R V_R)E_i} e^{0.5ZF\Delta\psi_o/RT} \right]. \tag{18}$$

The effects of membrane potential on $\gamma$ are shown in Fig. 4, where $\gamma$ is plotted relative to its value at zero membrane potential for four values of the partition coefficient. These data indicate that $\gamma$ may range from 40 to 275% of its value in the absence of membrane potential, depending on the value of $K$. Using our approximate parameters, the value of $\gamma$ is most sensitive to values of $K$ between 1000 and 10,000.

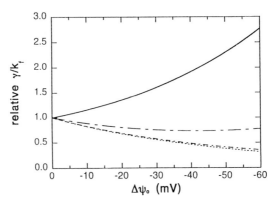

FIG. 4 Effect of plasma membrane potential on the apparent rate constant $\gamma$ at four values of the effective partition coefficient. Data are from Eq. (18) for $K = 10$ ($\cdots$), $K = 100$ ($-\cdot-$), $K = 1000$ ($------$), and $K = 10,000$ ($—$). Values are plotted relative to the value obtained at $\Delta\psi_o = 0$.

Very few reports have attempted to measure the effective rate constants for drug accumulation in sensitive and resistant cell lines, usually only reporting "rates" of accumulation from linear initial rate analysis. However, in the case of the voltage-sensitive $DiOC_n(3)$ dyes, when apparent rate constants for accumulation were measured it was found that only minor differences were present between drug-sensitive and P-gp-expressing drug-resistant cell lines (Wadkins and Houghton, 1995). This is the behavior predicted in Fig. 4 by $K = 1000$, again suggesting this value is a good approximation for partition coefficients for drug accumulation according to the model in Fig. 2.

By using the values of $\gamma$ for $K = 1000$ shown in Fig. 4 and the fraction of drug bound to the cell given by Fig. 3, we can use Eq. (15) to describe the behavior of membrane potential on the accumulation of drugs into sensitive and resistant cell lines. Figure 5 shows the rate of drug accumulation into cells as a function of membrane potential. The predicted drug accumulation is representative of a variety of actual experimental cell accumulation studies (Wei et al., 1995; Wadkins and Houghton, 1995; Robinson and Roepe, 1996; Stow and Warr, 1993; Shalinsky et al., 1993; Dordal et al., 1992). Also note from Eq. (13) that the initial rate of accumulation of drugs (as $t \approx 0$) is simply $\gamma F_{cell}(\infty)$. Quite often (Sirotnak et al., 1986; Stow and Warr, 1993; Shalinsky et al., 1993; Robinson and Roepe, 1996), the rate of drug accumulation into resistant cell lines is plotted for initial accumulation in order to emphasize the difference in resistant vs sensitive cell lines. In Fig. 6, we plot the data shown in Fig. 5 to elaborate the effect of altered membrane potential on these linear initial "rates" of drug accumulation. Note that the effect of membrane potential is immediate and will

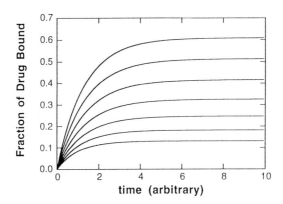

FIG. 5 Simulated accumulation of cationic drugs with variation in plasma membrane potential. Using an effective partition coefficient of 1000, data are (from top to bottom) $-60$, $-50$, $-40$, $-30$, $-20$, $-10$, and $0$ mV membrane potential using Eq. (15) with $\gamma$ taken from Fig. 3.

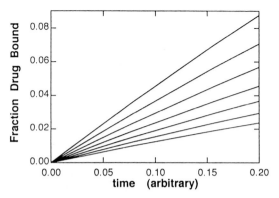

FIG. 6 Data from Fig. 4 to emphasize differences in initial rates. The apparent rate is equal to $\gamma F_{cell}(\infty)$. Using an effective partition coefficient of 1000, data are (from top to bottom) −60, −50, −40, −30, −20, −10, and 0 mV membrane potential.

affect drug accumulation whether or not any "drug pump" is present in the membrane.

The membrane potential also affects the zero-*trans* efflux of drug accumulated into cells. In a typical efflux experiment, drug is accumulated as described previously and the cells are pelleted and resuspended in drug-free media. In this situation, the external drug concentration is zero initially. Using Eq. (12), we can write the fraction of drug bound to the cell during an efflux experiment as

$$F_{cell} = F_{cell}(\infty) + F_o(\infty)e^{-\gamma t}, \qquad (19)$$

where $F_{cell}(\infty)$ and $F_o(\infty)$ are the steady-state fraction of cell bound and external drug, respectively, as determined from Eqs. (3) and (4). Simulated efflux curves using Eq. (19) are given in Fig. 7 for $K = 1000$ and various values of the membrane potential.

Ultimately, the system will come to equilibrium, such that the relative levels of drug given in Fig. 3 are obtained. It will therefore appear that sensitive cells (with $\Delta\psi_o = -60$ mV) efflux less drug than resistant cells (with $\Delta\psi_o = -20$ mV) whether or not any pumping of the drug from the cell occurs. These curves are representative of what is usually observed with P-gp-expressing MDR cells (Roepe, 1992; Ramu *et al.*, 1989; Coley *et al.*, 1993; Sirotnak *et al.*, 1986; Shalinsky *et al.*, 1993; Stow and Warr, 1993). Differentiation of Eq. (19) also shows that the initial rates of efflux will be simply $-\gamma F_o(\infty)$, and thus the rate of drug efflux from resistant cells will appear greater than that for sensitive cells.

In this section, we have applied the approach of Cafiso and Hubbell (1979, 1982) to arrive at a model for the effect of plasma membrane potential

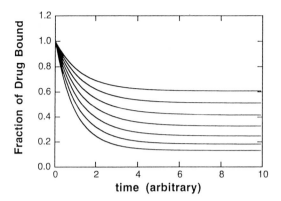

FIG. 7 Simulation of efflux of cationic drugs from cells with respect to membrane potential. Using an effective partition coefficient of 1000, data are (from top to bottom) −60, −50, −40, −30, −20, −10, and 0 mV membrane potential.

on both the rate and steady-state accumulation of cationic drugs into cells. As shown in Figs. 5–7, the membrane potential affects both the accumulation and the efflux of drug and is remarkably similar to the observed drug flux in P-gp-expressing multidrug resistant cell lines. These simulations show that much of the effects previously attributed to drug pumping need not involve any actual direct interaction between drugs and P-gp or pumping of drugs by P-gp. Instead, by regulating membrane potential, much of the clinically relevant level of drug resistance (2- to 10-fold) found in P-gp-expressing cells can be accounted for by physical–chemical effects on drug accumulation that reduce the amount of drug accumulated into cells.

E. Weak Bases, $\Delta\psi_o$, and Alterations in $pH_i$

In addition to alterations in membrane potential, P-gp expression also results in alkalinization of the cytoplasm of resistant cell lines (Roepe, 1995). Many antitumor drugs such as doxorubicin have $pK_a$'s near the pH of the experimental buffer solutions and the cytosolic pH and therefore are extremely sensitive to changes in pH found in resistant cell lines. To understand the combination of effects of $\Delta\psi_o$ and $pH_i$ on the accumulation of antitumor drugs, we modify the model given in Fig. 2 to account for these parameters.

The case for weak bases is described by Fig. 8. We make the assumption that the unprotonated form of the drug is able to penetrate the plasma membrane of the cell and does not bind to a significant extent to the plasma membrane. For doxorubicin, this is a reasonable assumption because the

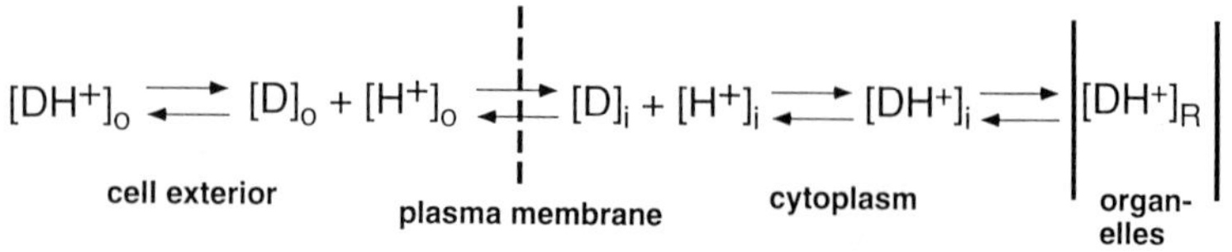

$$[DH^+]_o \rightleftarrows [D]_o + [H^+]_o \rightleftarrows [D]_i + [H^+]_i \rightleftarrows [DH^+]_i \rightleftarrows [DH^+]_R$$

cell exterior | plasma membrane | cytoplasm | organelles

FIG. 8 Model system for the $pH_i$ and $\Delta\psi_o$ dependence of weak base accumulation into cells. Drug in the external solution is dissociated into the neutral form as a function of the $pK_a$ of the drug. The neutral form readily passes through the plasma membrane to the cytoplasm. The protonated drug in the cytoplasm is then capable of binding to organelles (e.g., the nucleus). The $pH_i$ and $pH_o$ are maintained by the buffering capacity of each phase.

partition coefficient ($\sim K_m$) for doxorubicin between buffer and butanol is approximately 0.5. Therefore, the initial entry of drug into the cell is not dependent on the membrane potential. We assume that only the protonated form of the drug in the cytoplasm is affected by membrane potential, in a manner analogous to the "permanent" cationic compounds described previously. Furthermore, the pH of the cell exterior and interior are independent and are fixed by the buffering capacity of both phases, and the rate of protonation of the drug is rapid with respect to diffusion through the membrane. Finally, we assume that the protonated form of the drug is further accumulated into organelles or other binding sites in the cell, which is reasonable considering how substantially cells accumulate these compounds. In the case of doxorubicin, the major organelle to consider is the nucleus.

Based on this model, we can write the fraction of neutral drug molecules outside the cell as

$$F^0_o = \frac{n^0_o}{n^0_o + n^+_o + n^0_i + n^+_i + n^+_R}$$

$$= 1 / \left(1 + \frac{n^+_o}{n^0_o} + \frac{n^0_i}{n^0_o} + \frac{n^+_i}{n^0_o} + \frac{n^+_R}{n^0_o}\right)$$

$$= 1 / \left(1 + \frac{[H^+]_o}{K_a} + \frac{V_i}{V_o} + \frac{V_i[H^+]_i}{V_o K_a} e^{-ZF\Delta\psi_o/RT} + K_R \frac{V_R[H^+]_i}{V_o K_a} e^{-ZF\Delta\psi_o/RT}\right), \quad (20)$$

where $n^+_o$ and $n^+_i$ and $n^+_R$ are numbers of protonated drug molecules outside the cell, in the cell cytoplasm, and in the organelles, respectively, and $n^0_o$ and $n^0_i$ are the unprotonated drugs outside the cell and in the cytoplasm, respectively. $K_a$ is the acid dissociation constant for the weak base, and $[H^+]_i$ and $[H^+]_o$ are the concentration of hydrogen ions inside the cytoplasm and outside the cell, respectively. The other parameters are as described previously.

The fraction of protonated drug external to the cell is related to the fraction of unprotonated drug by

$$F^+_o = F^0_o \left( \frac{[H^+]_o}{K_a} \right). \tag{21}$$

We can therefore use Eqs. (20) and (21) to calculate the effect of both pH$_i$ and $\Delta \psi_o$ on accumulation of drugs into cells. We keep the same experimental parameters as in the cation case, with respect to volumes of each phase. In several experiments with doxorubicin accumulation into P-gp-expressing cell lines (Roepe, 1992; Roepe et al., 1993), the external pH was buffered at 7.3. The internal pH was measured at 7.1 for 8226 drug-sensitive myeloma cell lines, and the membrane potential in these cells is $-52$ mV. The Dox120 cell line was 125-fold resistant to doxorubicin, had an internal pH of 7.6, and membrane potential equal to $-20$ mV. These $\Delta \psi_o$ and pH$_i$ values have been used, along with various values of $K_R$, to model the accumulation of doxorubicin (p$K_a$ = 8.2) into sensitive and resistant cell lines.

Figure 9 shows the simulated accumulation of doxorubicin as a function of pH$_i$ and $\Delta \psi_o$ at three values of $K_R$ [$10^2$, $10^3$, and $10^4$ (Figs. 9A–9C, respectively)]. Depending on the value of $K_R$, the relative accumulation of drug into resistant cells having pH$_i$ of 7.6 may be up to $\frac{1}{4}$ to $\frac{1}{2}$ that observed in the sensitive cell (pH$_i$ = 7.1). The combined effects of $\Delta$pH and $\Delta \psi_o$ may reduce accumulation in the resistant cell to $\frac{1}{6}$ to $\frac{1}{20}$ that of the sensitive cell line. These values would simplistically imply resistance of 2- to 20-fold for these cells, based on reduced drug accumulation alone. They probably do not, however, produce drug resistance in the 40- to 120-fold range, as found for many drug-selected MDR cell lines. It should be noted that highly resistant cell lines (including Dox120 in this example) are selectants or drug-selected cells, which have superimposable mechanisms of resistance (Robinson and Roepe, 1996; Hoffman et al., 1996). We also note that we are neglecting important effects of pH$_i$ on the binding of drugs to target and their subcellular compartmentalization (see Section III,V,F) as well as the role that pH$_i$ probably plays in initiating and/or regulating progression of the apoptotic cascade, which is critical for understanding the cytotoxic function of chemotherapeutic drugs (Pérez-Sala et al., 1995; Barry et al., 1993). Nonetheless, the effects predicted by $\Delta$pH and $\Delta \psi_o$ when $K_R$ = $10^3$–$10^4$ are in excellent agreement with measured steady-state drug accumulation and efflux in some experiments (Roepe, 1992; Roepe et al., 1993) that clearly show the amount of retained doxorubicin in highly resistant cells may be 25–50% of that for the sensitive cell line. Although apparent initial rate would appear different in a simplistic linear initial rate analysis, the rate *constant* for accumulation of doxorubicin is likely independent of

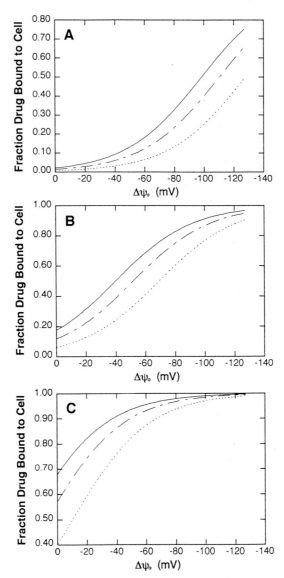

FIG. 9 Dependence of steady-state weak base accumulation on $pH_i$ and $\Delta\psi_o$ at different values for $K_R$. The $pH_o$ is fixed at 7.3, and the $pK_a$ of the base is 8.2. (A) $K_R = 100$, with $pH_i = 7.1$ (—), 7.3 (— — —), and 7.6 (· · · ·). (B) As in A with $K_R = 1000$. (C) As in A with $K_R = 10,000$.

membrane potential, and consequently the rate of accumulation is similar to that shown in Figs. 5–7, with the final amount of drug accumulated as given in Fig. 9.

### F. Summary of Transport Effects Due to $\Delta\psi_o$ and $\Delta$pH

In this section, we have attempted to provide an alternative to the explanations usually offered to describe the role of P-gp in mediating multidrug resistance. As discussed, the more commonly encountered explanations are associated with many significant difficulties, and there is a scarcity of hard physical data showing the detailed thermodynamic and kinetic behavior that would be required for a drug pump model to explain the phenotype. In contrast, experimentally, it is indeed found that both $\Delta$pH and $\Delta\psi_o$ are altered in P-gp-expressing cell lines, and here we have extended established models for drug accumulation into membrane vesicles to the specific case of drug accumulation into cells and show how these physical–chemical phenomena can result in the cellular drug flux that is usually interpreted as direct transport of drugs by P-gp. Clearly, even if P-gp were to actively transport drugs, the electrochemical aspects discussed here are also influencing transport in resistant cells, and they should therefore be considered in any description of the flux of drugs in resistant cell lines. To date, most experiments describing transport of drugs by P-gp have not attempted to account for altered biophysical properties of the cells being examined.

It should also be noted that the physical properties of the drugs being examined will dictate whether $\Delta$pH or $\Delta\psi_o$ is predominant in the resistance mechanism. If we, for the moment, neglect effects of $pH_i$ on binding to intracellular targets, the transport of permanent cations, such as rhodamine 123, TPP$^+$, and DiOC$_n$(3), should be less dependent on $pH_i$, relative to weak bases. However, bases far from their $pK_a$ will be less dependent on changes in pH, and very hydrophobic bases that can bind to membranes in their cationic form will have responses to $\Delta$pH and $\Delta\psi_o$ that fall somewhere in between the analyses given previously. These parameters must be considered for each drug under consideration, and they will provide insight into the "spectrum" of resistance in a given multidrug-resistant cell line with given $\Delta\psi_o$ and $pH_i$. The resistance spectrum or resistance "profile" refers to the rank order of resistance to various compounds to which MDR protein overexpression confers resistance. These differ widely among MDR cell lines, and the molecular basis for this heterogeneity in phenotype is currently not understood. We believe the $\Delta\psi_o/pH_i$ model provides a simple way to explain those phenomena.

In some isolated examples, the biophysical model does not readily explain altered cellular accumulation or retention of certain uncharged drugs (in particular, colchicine). However, it is important to realize that no detailed studies of the transport of colchicine (or any other neutral hydrophobic compound) have been performed with true MDR transfectants not selected with or previously exposed to chemotherapeutic drugs. Thus, it is not even known what level of altered colchicine transport is caused by MDR protein overexpression alone. Second, colchicine's binding target is monomeric (but not dimeric) tubulin, and the dynamic equilibrium between these two is highly $pH_i$ dependent. For this reason, and because of the nature of the colchicine binding site, binding of colchicine to tubulin is very highly pH dependent (Mukhopadhyay et al., 1990). Therefore, subtle changes in $pH_i$ will indeed dramatically affect the level of colchicine that is accumulated in a cell over time, the level of colchicine released from a cell in a zero-*trans* efflux assay, and the percentage of intracellular colchicine that is osmotically sensitive (freely exchangeable). Thus, depending on the assay that is performed, it can also appear as if colchicine is transported in a different way by an MDR cell.

The most recent work available with a series of six individual clones isolated from a mass population of huMDR1/LR73 transfectants not exposed to or selected with chemotherapeutic drug prior to analysis (Hoffman et al., 1996) demonstrates that overexpression of the huMDR1 protein in and of itself leads to cytoplasmic alkalinization in all six cases, and that relative overexpression is related to relative alkalinization. Moreover, all six clones are depolarized and by pulse-elevating $pH_i$ or depolarizing control cells in the absence of MDR protein expression, it is found that the levels of drug resistance exhibited by the transfectants can be completely accounted for by the $pH_i$ and $\Delta \psi_o$ changes that are measured. Notably, the clones are more resistant to doxorubicin and vincristine (both of which have positive charge) and much less resistant to colchicine (which is neutral). Ongoing analysis of the drug transport characteristics of these cells should help to resolve remaining uncertainties and further refine the $\Delta \psi_o$/$pH_i$ model presented previously.

Stable $\Delta \psi_o$ and $pH_i$ alterations likely have important short-term and long-term effects with regard to the cellular pharmacology of chemotherapeutic drugs as well as other drugs (e.g., antimalarials, antibiotics, insulin secretogues, and drugs used in the treatment of cystic fibrosis). That is, these changes affect distribution and passive translocation of drugs relatively quickly via direct electrostatic effects; however, these changes likely also promote (over a longer time period) changes in membrane characteristics (such as leaflet distribution of lipid; (Hope and Cullis, 1987) or target availability (e.g., tubulin structure) that could have additional important effects with regard to conferring drug resistance (see Minotti et al., 1991; Callaghan et al., 1992). For example, there is some indication that significant perturbation of $\Delta \psi_o$ might even affect organization of tubulin near the surface of the plasma membrane (Aszalos *et al.*, 1986), which would be predicted to alter retention of antimitotics, such as colchicine and vinblastine, that bind to tubulin.

## IV. The Recent Controversy over ATP Transport

A. Introduction

In 1993, Cantiello and colleagues (Abraham *et al.*, 1993) reported that cells with increased levels of MDR protein released ATP into cell culture medium at a faster rate relative to cells without significant MDR protein expression. Based on electrophysiologic measurements in the presence of 100 m$M$ ATP, they proposed that MDR protein functioned as an "ATP channel." The measured conductance presumably mediated by the MDR protein that was reported was very small (about 3–5 pS); thus, these data have been viewed skeptically. Nonetheless, these investigators showed that the conductance being measured under these conditions was apparently blocked by an anti-MDR protein antibody, thus strengthening their general conclusion that more ATP can be released from MDR cells under some conditions.

Subsequently, similar observations were made by the same group (Reisin *et al.*, 1994; Cantiello *et al.*, 1994) as well as the Guggino laboratory (Schweibert *et al.*, 1995) for cells expressing different levels of the CFTR. A similar model, that CFTR also acts as an ATP channel, was proposed in these papers. Again, the reported conductance is fairly small and requires high levels of ATP to measure; nonetheless, the results are provocative.

However, Reddy *et al.* (1996) have failed to confirm that CFTR conducts ATP using a variety of preparations including purified and reconstituted CFTR. These data suggest that the ATP conductance measured in previous experiments could be mediated not by the ABC transporter but by some other protein influenced by the ABC transporter. Although negative data with regard to ATP transport by the MDR protein have not yet been published, the fact that strong controversy already exists in the case of the CFTR suggests that the ATP channel hypothesis for MDR protein should be interpreted cautiously.

Although the data are controversial, Al-Awqati (1995) summarized the implications of some of these ATP channel data in terms of a general model for eukaryotic ABC transporter (CFTR, SUR, MDR, etc.) function. In this model (Fig. 1, see p. 126), the proteins exert their important effects on membrane potential, pH$_i$, and perhaps other cell characteristics by stimulating purinergic receptors that then interact with various ion channels (e.g.,

ORCC, $K_{ATP}$, and ENaC) via G-protein-coupled pathways. Although there are clearly conflicting data (Reddy *et al.*, 1996), there remains some interest in this model because it is known that extracellular ATP can exert many effects on a variety of cell types (El-Moatassim *et al.*, 1992) and thus these varied effects might provide an explanation for some of the cell-specific observations that have been made regarding ABC transporters. The model also offers a simple way to explain some of the important effects documented for CF airway epithelia and $\beta$ cell physiology upon treatment with exogenous nucleotides (Knowles *et al.*, 1991).

If some form of ATP transport in MDR cells is found to be physiologically relevant, it may be important to note that Heppel and colleagues have shown that at least in some cell types (including 3T3 and 3T6 mouse fibroblasts, as well as A431 epidermoid carcinoma cells) extracellular ATP leads to elevations in $pH_i$ (Huang *et al.*, 1992). Also, Weiner *et al.* (1986) have suggested that extracellular ATP may regulate $Na^+/H^+$ exchange in Ehrlich ascites tumor cells. Therefore, it is conceivable that earlier observations of $pH_i$ and $\Delta\psi_o$ changes in MDR cells (Roepe *et al.*, 1993) could be connected to recent observations regarding ATP transport.

## B. Connections to Other ABC Transporters

Overexpression of the CFTR has been shown to induce a MDR phenotype (Wei *et al.*, 1995) that is similar to, but not identical to, that mediated by overexpression of the MDR protein in the same cells type. If the ATP transport hypothesis for ABC transporters is correct, then overexpression of CFTR vs MDR1 should lead to similar effects, but perhaps with different magnitude. However, although either depolarization or alkalinization alone is sufficient to cause a low-level MDR phenotype (Hoffman *et al.*, 1996), CFTR overexpressors are depolarized and acid, whereas MDR1 overexpressors are depolarized and alkaline. Thus, if ATP transport mediated directly or indirectly by CFTR is physiologically relevant, there must be additional features to the function of these two proteins that lead to opposite effects with respect to $pH_i$. The difference in $pH_i$ cannot be due to different purinergic receptors, but, via this model, perhaps different levels of ATP are transported by CFTR vs MDR and because different purinergic receptor subtypes bind ATP with different affinities, different effects are found. These models are certainly testable. Also, via this model, similar purinergic receptor pathways must exist in a variety of cell types because similar effects on $\Delta\psi_o$ and $pH_i$ are observed in a variety of cell types overexpressing MDR protein [e.g., murine NIH3T3 and CHO LR73 fibroblasts, K562 myeloid leukemia cells (Choi *et al.*, 1991), and human 293 embryonic kidney cells (M. M. Hoffman and P. D. Roepe, unpublished results)].

## V. Implications for Other Forms of "ABC-MDR"

As mentioned in the beginning of this chapter, one of the reasons for the intense interest in the MDR protein involved in tumor cell chemotherapeutic drug resistance is the tantalizing similarity between MDR protein-mediated tumor MDR and other examples of what we call ABC-MDR; that is, pleiotropic drug resistance apparently mediated by the altered expression and/or mutation of homologues of the MDR protein in the ABC family of transporters. For example, CRP appears to play a major role in some (but not all) examples of pleiotropic drug resistance in the malarial parasite *P. falciparum* (Karcz and Cowman, 1991); pleiotropic drug resistance protein 5 from *Saccharomyces cerevisiae* confers resistance to a wide variety of toxic hydrophobic compounds including ionophores (Balzi and Goffeau, 1991); the EhPgP1 and EhPgP2 proteins from *Entamoeba histolytica* are involved in conferring resistance to emetine and perhaps other drugs (Descoteaux *et al.*, 1992); and the *sapABCDF* complex from *Salmonella typhimurium* appears to play a role in resistance to defensins (Parra-Lopez *et al.*, 1993). Detailed information on the structure and function of P-gp will aid further investigations of these forms of ABC-MDR. Thus, the implications for correct elucidation of the P-gp mechanism are enormous.

Importantly, a case can be made that at least several of these ABC proteins contribute to pleiotropic drug resistance by altering membrane potential and/or pH gradients, in particular the CRP protein (which appears to be involved in regulating *P. falciparum* digestive vacuolar pH; (Van Es *et al.*, 1994), the *sapABCDF* complex (which is required for *sapJ* $K^+$ channel function), and another homolog encoded by the *ami* locus of *Streptococcus* (Alloing *et al.*, 1990) that is clearly involved in regulating membrane potential and also in conferring pleiotropic drug resistance (Trombe, 1993; Alloing *et al.*, 1990). In addition, as mentioned, forced overexpression of another MDR homolog, the CFTR, has also been found to decrease membrane potential and confer a low-level pleiotropic drug resistance phenotype (Wei *et al.*, 1995), although the CFTR is not generally thought of as a drug resistance protein. The implications of this, in terms of searching for additional proteins involved in drug resistance, are significant.

Indeed, interestingly, in pursuing the isolation and sequencing of genetic loci known to be involved in conferring drug resistance of several types in other systems, genes encoding other proteins that are clearly involved in ion transport or ion transport regulation have been found, and these proteins almost certainly impact upon $pH_i$, $\Delta\psi_o$, or both. For example, in cloning the *tet* B(L) gene of *Bacillus subtilis,* which has been a key goal of investigators interested in bacterial tetracycline resistance for quite some time, Krulwich and colleagues discovered that this locus encodes a $Na^+/$

H⁺ antiporter that is clearly involved in regulating cellular pH (Cheng *et al.*, 1994). As we try to understand drug resistance mediated by ABC proteins, if we exclusively focus on nonspecific drug pumping mechanisms alone, without considering other more reasonable possibilities (Lewis, 1994), we will likely slow the rate at which the ABC-MDR field capitalizes on the significant concepts that are inherent in the above observations.

## VI. Conclusions

During the past 20 years, investigators interested in designing better chemotherapeutic treatment for cancer have offered several theories for how overexpression of ABC proteins [in particular the MDR and MRP (Grant *et al.*, 1994) proteins] leads to decreased intracellular accumulation of chemotherapeutic drugs. In the tumor MDR community, a popular model has been the drug pump hypothesis or some derivative of this model (e.g., the flippase or vacuum cleaner model, or the dual function channel/pump model of Higgins and colleagues). However, the MDR protein-mediated MDR phenotype, removed from complications due to drug selection or drug exposure, has only recently been properly defined (Hoffman *et al.*, 1996), and study of drug translocation/drug accumulation in true MDR transfectants is just beginning (Robinson and Roepe, 1996). Because of a deluge of selective interpretation in the MDR literature, there is the perception among many that somehow key issues regarding MDR protein function are settled. This situation has been exacerbated by underappreciation of the important role of $pH_i$ and $\Delta\psi_o$ in altering chemotherapeutic drug transport and retention and confusion regarding the difference between the MDR phenotypes exhibited by drug-selected cells vs true transfectants. All this considered, the investigator familiar with concepts in transport physiology and membrane biophysics finds the notion that MDR protein may confer pleiotropic drug resistance through modulating $pH_i$, $\Delta\psi_o$, or both, is quite attractive. This does not necessarily make the model completely correct, but it does suggest that it deserves much more serious attention.

### Acknowledgments

Some of the experimental work reviewed here was performed in the Sackler Laboratory of Membrane Biophysics at Memorial Sloan-Kettering Cancer Center. Work in our laboratory has been supported by the Society of Sloan-Kettering, the Cystic Fibrosis Foundation, a Cancer Center Support Grant (No. NCI-P30-CA-08748), NIH Grant Nos. RO1 GM55349 and RO1 GM54516, and the Wendy Will Case Fund. P. D. R. thanks his colleagues in the Sackler laboratory for their hard work, enthusiasm, and many helpful discussions. We also thank Drs. Jim O'Brien, Frank Sirotnak, Kathy Scotto, and Joe Bertino (Sloan-Kettering Institute), David Gadsby (Rockefeller University), Olaf Andersen (Cornell University Medical College), M. Jackson Stutts (University of North Carolina, Chapel Hill), Peter Houghton (St. Jude Children's Research Hospital), and Wilfred Stein (Hebrew University) for helpful advice and discussions. This chapter is dedicated to a very courageous woman, Shelly Cronin.

## References

Abraham, E. H., Prat, A. G., Gerweck, L., Seneveratne, T., Arceci, R. J., Kramer, R., Guidotti, G., and Cantiello, H. F. (1993). The multidrug resistance (*mdr1*) gene product functions as an ATP channel. *Proc. Natl. Acad. Sci. USA* **90,** 312–316.

Abrams, A., and Baron, C. (1967). The isolation and subunit structure of streptococcal membrane adenosine triphosphatase. *Biochemistry* **6,** 225–229.

Abrams, A., Smith, J. B., and Baron, C. (1972). Carbodiimide-resistant membrane adenosine triphosphatase in mutants of *Streptococcus faecalis*. I. Studies of the mechanism of resistance. *J. Biol. Chem.* **247,** 1484–1488.

Aguilar-Bryan, L., Nichols, C. G., Wechsler, S. W., Clement, J. P., Boyd, A. E., Gonzalez, G., Herrera-Sosa, H., Nguy, K., Bryan, J., and Nelson, D. A. (1995). Cloning of the β cell high-affinity sulfonylurea receptor: A regulator of insulin secretion. *Science* **268,** 423–426.

Al-Awqati, Q. (1995). Regulation of ion channels by ABC transporters that secrete ATP. *Science* **269,** 805–806.

Alloing, G., Trombe, M. C., and Claverys, J. P. (1990). The *ami* locus of the gram-positive bacterium *Streptococcus pneumoniae* is similar to binding protein-dependent transport operons of gram-negative bacteria. *Mol. Microbiol.* **4,** 633–644.

Al-Shawi, M. K., Urbatsch, I. L., and Senior, A. E. (1994). Covalent inhibitors of P-glycoprotein ATPase activity. *J. Biol. Chem.* **269,** 8986–8992.

Altenberg, G. A., Young, G., Horton, J. K., Glass, D., Belli, J. A., and Reuss, L. (1993). Changes in intra- or extracellular pH do not mediate P-glycoprotein-dependent multidrug resistance. *Proc. Natl. Acad. Sci. USA* **90,** 9735–9738.

Altenberg, G. A., Deitmer, J. W., Glass, D. C., and Reuss, L. (1994). P-glycoprotein-associated Cl⁻ currents are activated by cell swelling but do not contribute to cell volume regulation. *Cancer Res.* **54,** 618–622.

Ambudkar, S. V., Lelong, I. H., Zhang, J., Cardarelli, C. O., Gottesman, M. M., and Pastan, I. (1992). Partial purification and reconstitution of the human multidrug-resistance pump: Characterization of the drug-stimulatable ATP hydrolysis. *Proc. Natl. Acad. Sci. USA* **89,** 8472–8476.

Ames, G. F.-L. (1986). The basis of multidrug resistance in mammalian cells: Homology with bacterial transport. *Cell* **47,** 323–324.

Aszalos, A., Damjanovich, S., and Gottesman, M. M. (1986). Depolymerization of microtubules alters membrane potential and affects the motional freedom of membrane proteins. *Biochemistry* **25,** 5804–5809.

Bakker, E. P., Rottenberg, H., and Caplan, S. R. (1976). An estimation of the light-induced electrochemical potential difference of protons across the membrane of *Halobacterium halobium*. *Biochim. Biophys. Acta* **440,** 557–564.

Ballarin-Denti, A., Den Hollander, J. A., Sanders, D., Slayman, C. W., and Slayman, C. L. (1984). Kinetics and pH-dependence of glycine–proton symport in *Saccharomyces cerevisiae*. *Biochim. Biophys. Acta* **778,** 1–16.

Bally, M. B., Hope, M. J., Van Echteld, C. J. A., and Cullis, P. R. (1985). Uptake of safranine and other lipophillic cations into model membrane systems in response to membrane potential. *Biochim. Biophys. Acta* **812,** 66–76.

Balzi, E., and Goffeau, A. (1991). Multiple or pleiotropic drug resistance in yeast. *Biochim. Biophys. Acta* **1073,** 241–252.

Barry, M. A., Reynolds, J. E., and Eastman, A. (1993). Etoposide-induced apoptosis in human HL-60 cells is associated with intracellular acidification. *Cancer Res.* **53,** 2349–2357.

Bear, C. E., Li, C., Kartner, N., Bridges, R. J., Jensen, T. J., Ramjeesingh, M., and Riordan, J. R. (1992). Purification and functional reconstitution of the cystic fibrosis transmembrane conductance regulator (CFTR). *Cell* **68,** 809–818.

Beck, W. T. (1987). The cell biology of multiple drug resistance. *Biochem. Pharmacol.* **36,** 2879–2887.

Bertl, A., Slayman, C. L., and Gradmann, D. (1993). Gating and conductance in an outwardly-rectifying K+ channel from the plasma membrane of *Saccharomyces cerevisiae*. *J. Membrane Biol.* **132,** 183–199.

Biedler, J. L., and Riehm, H. (1970). Cellular resistance to actinomycin D in chinese hamster cells in vitro: Cross resistance, radioautographic, and cytogenetic studies. *Cancer Res.* **30,** 1174–1184.

Bornmann, W., and Roepe, P. D. (1994). Analysis of drug transport kinetics in multidrug-resistant cells using a novel coumarin–vinblastine compound. *Biochemistry* **33,** 12665–12675.

Burchenal, J. H., Robinson, E., Johnston, S. F., and Kushida, M. M. (1950). Induction of resistance to aminopterin in a strain of transmitted mouse leukemia. *Science* **111,** 116–121.

Cafiso, D. S., and Hubbell, W. L. (1979). Estimation of transmembrane potentials from the phase equilibria of hydrophobic paramagnetic ions. *Biochemistry* **17,** 187–195.

Cafiso, D. S., and Hubbell, W. L. (1982). Transmembrane electrical curents of spin-labeled hydrophobic ions. *Biophys. J.* **39,** 263–272.

Callaghan, R., van Gorkom, L. C. M., and Epand, R. M. (1992). A comparison of membrane properties and composition between cell lines selected and transfected for multi-drug resistance. *Br. J. Cancer* **66,** 781–786.

Cantiello, H. F., Prat, A. G., Reisin, I. L., Ercole, L. B., Abraham, E. H., Amara, J. F., Gregory, R. J., and Ausiello, D. A. (1994). External ATP and its analogs activate the cystic fibrosis transmembrane conductance regulator by a cyclic AMP-independent mechanism. *J. Biol. Chem.* **269.** 11224–11232.

Chen, C.-J., Chin, J. E., Ueda, K., Clark, D. P., Pastan, I., Gottesman, M. M., and Roninson, I. B. (1986). Internal duplication and homology with bacterial transport proteins in the *mdr1* (P-glycoprotein) gene from multidrug-resistant human cells. *Cell* **47,** 381–389.

Cheng, J., Guffanti, A. A., and Krulwich, T. A. (1994). The chromosomal tetracycline resistance locus of *Bacillus subtilis* encodes a $Na^+/H^+$ antiporter that is physiologically important at elevated pH. *J. Biol. Chem.* **269,** 27365–27371.

Choi, K., Frommel, T. O., Stern, R. K., Perez, C. F., Kriegler, M., Tsuruo, T., and Roninson, I. B. (1991). Multidrug resistance after retroviral transfer of the human MDR1 gene correlates with P-glycoprotein density in the plasma membrane and is not affected by cytotoxic selection. *Proc. Natl. Acad. Sci. USA* **88,** 7386–7390.

Cole, S. P. C., Bhardwaj, G., Gerlach, J. H., Mackie, J. E., Grant, C. E., Almquist, K. C., Stewart, A. J., Kurz, E. U., Duncan, A. M. V., and Deeley, R. G. (1992). Overexpression of a transporter gene in a multidrug-resistant human lung cancer cell line. *Science* **258,** 1650–1654.

Coley, H. M., Twentyman, P. R., and Workman, P. (1993). The efflux of anthracyclines in multidrug-resistant cell lines. *Biochem. Pharmacol.* **46,** 1317–1326.

Danø, K. (1973). Active outward transport of daunomycin in resistant Ehrlich ascites tumor cells. *Biochim. Biophys. Acta* **323,** 466–483.

de Kroon, A. I. P. M., Vogt, B., van't Hof, R., de Kruijff, B., and de Gier, J. (1991). Ion gradient-induced membrane translocation of model peptides. *Biophys. J.* **60,** 525–537.

Descoteaux, S., Shen, P. S., Ayala, P., Orozco, E., and Samuelson, J. (1992). P-glycoprotein genes of *Entamoeba Histolytica*. *Arch. Med. Res.* **23,** 23–25.

Deuticke, B. (1977). Properties and structural basis of simple diffusion pathways in the erythrocyte membrane. *Rev. Physiol. Biochem. Pharmacol.* **78,** 1–97.

Devault, A., and Gros, P. (1990). Two members of the Mouse mdr gene family confer multidrug resistance with overlapping but distinct drug specificities. *Mol. Cell. Biol.* **10,** 1652–1663.

Dordal, M. S., Winter, J. N., and Atkinson, A. J., Jr. (1992). Kinetic analysis of p-glycoprotein-mediated doxorubicin efflux. *J. Pharmacol. Exp. Ther.* **263,** 762–766.

Egan, M., Flotte, T., Afione, S., Solow, R., Zeitlin, P. L., Carter, B. J., and Guggino, W. B. (1992). Defective regulation of outwardly rectifying Cl⁻ channels by protein kinase A corrected by insertion of CFTR. *Nature* **358,** 581–584.

Elgavish, A. (1991). High intracellular pH in CFPAC: A pancreas cell line from a patient with cystic fibrosis is lowered by retrovirus-mediated CFTR gene transfer. *Biochem. Biophys. Res. Commun.* **180,** 342–348.

El-Moatassim, C., Dornand, J., and Mani, J.-C. (1992). Extracellular ATP and cell signalling. *Biochim. Biophys. Acta* **1134,** 31–45.

Gadsby, D. C., and Nairn, A. C. (1994). Regulation of CFTR channel gating. *Trends Biochem. Sci.* **19,** 513–518.

Gerlach, J. H., Endicott, J. A., Juranka, P. F., Henderson, G., Sarangi, F., Deuchars, K. L., and Ling, V. (1986). Homology between P-glycoprotein and a bacterial haemolysin transport protein suggests a model for multidrug resistance. *Nature (London)* **324,** 485–489.

Gill, D. R., Hyde, S., Higgins, C. F., Valverde M. A., Mintenig, G. M., and Sepúlveda, F. V. (1992). Seperation of drug transport and chloride channel functions of the human multidrug resistance P-glycoprotein. *Cell* **71,** 23–32.

Gottesman, M. M., and Pastan, I. (1993). Biochemistry of multidrug resistance mediated by the multidrug transporter. *Annu. Rev. Biochem.* **62,** 385–427.

Grant, C. E., Valdimarsson, G., Hipfner, D. R., Almquist, K. C., Cole, S. P. C., and Deeley, R. G. (1994). Overexpression of multidrug resistance-associated protein (MRP) increases resistance to natural product drugs. *Cancer Res.* **54,** 357–361.

Gros, P., Croop, J., and Housman, D. (1986a). Mammalian multidrug-resistance gene: Complete cDNA sequence indicates a strong homology to bacterial transport proteins. *Cell* **47,** 371–380.

Gros, P., Neriah, Y. B., Croop, J. M., and Housman, D. E. (1986b). Isolation and expression of a complementary DNA that confers multidrug resistance. *Nature (London)* **323,** 728–731.

Gros, P., Talbot, F., Tang-Wai, D., Bibi, E., and Kaback, H. R. (1992). Lipophillic cations: A group of model substrates for the multidrug-resistance transporter. *Biochemistry* **31,** 1992–1998.

Guild, B. C., Mulligan, R. C., Gros, P., and Housman, D. E. (1988). Retroviral transfer of a murine cDNA for multidrug resistance confers pleiotropic drug resistance to cells without prior drug selection. *Proc. Natl. Acad. Sci. USA* **85,** 1595–1599.

Hardy, S. P., Goodfellow, H. R., Valverde, M. A., Gill, D. R., Sépulveda, F. V., and Higgins, C. F. (1995). Protein kinase C-mediated phosphorylation of the human multidrug resistance P-glycoprotein regulates cell volume-activated chloride channels. *EMBO J.* **14,** 68–75.

Harold, F. M., Baarda, J. R., Baron, C., and Abrams, A. (1969). Inhibition of membrane-bound adenosine triphosphatase and of cation transport in *Streptococcus faecalis* by N,N′-dicyclohexylcarbodiimide. *J. Biol. Chem.* **244,** 2261.

Hasselbach, W. (1978). The reversibility of the SR pump. *Biochim. Biophys. Acta* **515,** 23–53.

Higgins, C. F., and Gottesman, M. M. (1992). Is the multidrug transporter a flippase? *Trends Biochem. Sci.* **17,** 18–19.

Higgins, C. F., Gallagher, M. P., Hyde, S. C., Mimmack, M. L., and Pearce, S. R. (1990). ABC transporters: From micro-organisms to man. *Phil. Trans. R. Soc. London* **326,** 353–365.

Hoffman, M. M., Wei, L.-Y., and Roepe, P. D. (1996). Are altered $pH_i$ and membrane potential in hu MDR 1 transfectants sufficient to cause MDR protein-mediated multidrug resistance? *J. Gen. Physiol.* **108,** 295–313.

Hope, M. J., and Cullis, P. R. (1987). Lipid asymmetry induced by transmembrane pH gradients in large unilamellar vesicles. *J. Biol. Chem.* **262,** 4360–4366.

Horichi, N., Tapiero, H., Sugimoto, Y., Bungo, M., Nishiyama, M., Fourcade, A., Lampidis, T. J., Kasahara, K., Sasaki, Y., Takahashi, T., and Saijo, N. (1990). 3′-Deamino-3′-morpholino-13-deoxo-10-hydroxycarminomycin conquers multidrug resistance by rapid influx following higher frequency of formation of DNA single- and double-strand breaks. *Cancer Res.* **50,** 4698–4701.

Huang, N.-N., Ahmed, A. H., Wang, D.-J., and Heppel, L. A. (1992). Extracellular ATP stimulates increases in Na+/K+ pump activity, intracellular pH and uridine uptake in cultures of mammalian cells. *Biochem. Biophys. Res. Commun.* **182,** 836–843.

Inagaki, N., Gonoi, T., Clement, J. P., Namba, N., Inazawa, J., Gonzalez, G., Aguilar-Bryan, L., Seino, S., and Bryan, J. (1995). Reconstitution of IKATP: An inward rectifier subunit plus the sulfonyl urea receptor. *Science* **270,** 1166–1170.

Karcz, S., and Cowman, A. F. (1991). Similarities and differences between the multidrug resistance phenotype of mammalian tumor cells and chloroquine resistance in *Plasmodium falciparum*. *Exp. Parasitol.* **73,** 233–240.

Knowles, M. R., Clarke, L. L., and Boucher, R. C. (1991). Activation by extracellular nucleotides of chloride secretion in the airway epithelia of patients with cystic fibrosis. *N. Engl. J. Med.* **325,** 533–538.

Koefoed, P., and Brahm, J. (1994). The permeability of the human red cell membrane to steroid sex hormones. *Biochim. Biophys. Acta* **1195,** 55–62.

Leveille-Webster, C. R., and Arias, I. M. (1995). The biology of the P-glycoproteins. *J. Membrane Biol.* **143,** 89–102.

Lewis, K. (1994). Multidrug resistance pumps in bacteria: Variations on a theme. *Trends Biol. Sci.* **19,** 119–123.

Lothstein, L., Rodrigues, P. J., Sweatman, T. W., and Israel, M. (1994). Effects of 14-acyl congeners of N-benzyladriamycin-14-valerate (AD 198) on cytotoxicity and subcellular drug distribution in AD 198-resistant mouse J774.2 cells. *Proc. Am. Assoc. Cancer Res.* **35,** 342. [Abstract No. 2037]

Luckie, D. B., Krouse, M. E., Harper, K. L., Law, T. C., and Wine, J. J. (1994). Selection for MDR1/P-glycoprotein enhances swelling-activated K+ and Cl-currents in NIH/3T3 cells. *Am. J. Physiol. C* **267,** 650–658.

Luz, J. G., Wei, L. Y., Basu, S., and Roepe, P. D. (1994). Transfection of mu MDR 1 inhibits Na+-independent Cl−/HCO3− exchange in chinese hamster ovary cells. *Biochemistry* **33,** 7239–7249.

Magin, R. L., Niesman, M. R., and Bacic, G. (1990). Influence of fluidity on membrane permeability: Correspondence between studies of membrane models and simple biological systems. *Adv. Membrane Fluidity* **4,** 221–237.

Mayer, L. D., Bally, M. B., Hope, M. J., and Cullis, P. R. (1985a). Uptake of dibucaine into large unilamellar vesicles in response to a membrane potential. *J. Biol. Chem.* **260,** 802–808.

Mayer, L. D., Bally, M. B., Hope, M. J., and Cullis, P. R. (1985b). Uptake of antineoplastic agents into large unilamellar vesicles in response to a membrane potential. *Biochim. Biophys. Acta* **816,** 294–302.

Minotti, A. M., Barlow, S. B., and Cabral, F. (1991). Resistance to antimitotic drugs in chinese hamster ovary cells correlates with changes in the level of polymerized tubulin. *J. Biol. Chem.* **266,** 3987–3994.

Mukhopadhyay, K., Parrack, P. K., and Bhattacharyya, B. (1990). The carboxy terminus of the α subunit of tubulin regulates its interaction with colchicine. *Biochemistry* **29,** 6845–6850.
Noy, N., Donnelly, T. M., and Zakim, D. (1986). Physical–chemical model studies for the entry of water-insoluble compounds into cells. Studies of fatty acid uptake by the liver. *Biochemistry* **25,** 2013–2021.
Parra-Lopez, C., Baer, M. T., and Groisman, E. A. (1993). Molecular genetic analysis of a locus required for resistance to antimicrobial peptides in *Salmonella typhimurium. EMBO J.* **12,** 4053–4062.
Pérez-Sala, D., Collado-Escobar, D., and Mollinedo, F. (1995). Intracellular alkalinization suppresses lovastatin-induced apoptosis in HL-60 cells through the inactivation of a pH-dependent endonuclease. *J. Biol. Chem.* **270,** 6235–6242.
Praet, M., Defrise-Quertain, F., and Ruysschaert, J. M. (1993). Comparison of adriamycin and derivatives uptake into large unilamellar lipid vesicles in response to a membrane potential. *Biochim. Biophys. Acta* **1148,** 342–350.
Qian, X.-D., and Beck, W. T. (1990). Progesterone photoaffinity labels p-glycoprotein in multidrug-resistant human leukemic lymphoblasts. *J. Biol. Chem.* **265,** 18753–18756.
Ramu, A., Pollard, H. B., and Rosario, L. M. (1989). Doxorubicin resistance in P388 leukemia—Evidence for reduced drug influx. *Int. J. Cancer* **44,** 539–547.
Reddy, M. M., Quinton, P. M., Haws, C., Wine, J. J., Grygorczyk, R., Tabcharani, J. A., Hanrahan, J. W., Gunderson, K. L., and Kopito, R. R. (1996). Failure of the cystic fibrosis transmembrane conductance regulator to conduct ATP. *Science* **271,** 1876–1879.
Reisin, I. L., Prat, A. G., Abraham, E. H., Amara, J. F., Gregory, R. J., Ausiello, D. A., and Cantiello, H. F. (1994). The cystic fibrosis transmembrane conductance regulator is a dual ATP and chloride channel. *J. Biol. Chem.* **269,** 20584–20591.
Riordan, J. R., Rommens, J. M., Kerem, B.-S., Alon, N., Rozmahel, R., Grzelczak, Z., Zielenski, J., Lok, S., Plavsic N., Chou, J.-L., Drumm, M. L., Iannuzzi, M. C., Collins, F. S., and Tsui, L.-C. (1989). Identification of the cystic fibrosis gene: Cloning and characterization of complementary DNA. *Science* **245,** 1066–1072.
Robinson, L. J., and Roepe, P. D. (1996). Effects of membrane potential vs. $pH_i$ on the cellular retention of doxorubicin analyzed via a comparison between CFTR and MDR transfectants. *Biochem. Pharmacol.* **52,** 1081–1095.
Roepe, P. D. (1992). Analysis of the steady-state and initial rate of doxorubicin efflux from a series of multidrug-resistant cells expressing different levels of p-glycoprotein. *Biochemistry* **31,** 12555–12564.
Roepe, P. D. (1994). Indirect mechanism of drug transport by P-glycoprotein. Drug transport mediated by P-glycoprotein may be secondary to electrochemical perturbations of the plasma membrane. *Trends Pharmacol. Sci.* **15,** 445–446.
Roepe, P. D. (1995). The role of the MDR protein in altered drug translocation across tumor cell membranes. *Biochim. Biophys. Acta* **1241,** 385–406.
Roepe, P. D., Consler, T. G., Menezes, M. E., and Kaback, H. R. (1990). The *lac* permease of *Escherichia coli:* Site-directed mutagenesis studies on the mechanism of β-galactoside/ $H^+$ symport. *Res. Microbiol. (Ann. Pasteur Inst.)* **141,** 290–308.
Roepe, P. D., Wei, L. Y., Cruz, J., and Carlson, D. (1993). Lower electrical membrane potential and altered $pH_i$ homeostasis in multidrug-restant (MDR) cells: Further characterization of a series of MDR cell lines expressing different levels of p-glycoprotein. *Biochemistry* **32,** 11042–11056.
Ruetz, S., and Gros, P. (1994a). Functional expression of P-glycoprotein in secretory vesicles. *J. Biol. Chem.* **269,** 12277–12284.
Ruetz, S., and Gros, P. (1994b). A mechanism for P-glycoprotein action in multidrug resistance: Are we there yet? *Trends Pharmacol. Sci.* **15,** 260–263.
Ruetz, S., Raymond, M., and Gros, P. (1993). Functional expression of P-glycoprotein encoded by the mouse *mdr3* gene in yeast cells. *Proc. Natl. Acad. Sci. USA* **90,** 11588–11592.

Schlemmer, S. R., and Sirotnak, F. M. (1994). Functional studies of P-glycoprotein in inside-out plasma membrane vesicles derived from murine erythroleukemia cells overexpressing MDR3. *J. Biol. Chem.* **269**, 31059–31066.
Schuldiner, S., and Kaback, H. R. (1975). Membrane potentials and active transport in membrane vesicles from *Escherichia coli*. *Biochemistry*, **14**, 5451–5461.
Schwiebert, E. M., Egan, M. E., Hwang, T.-H., Fulmer, S. B., Allen, S. S., Cutting, G. R., and Guggino, W. B. (1995). CFTR regulates outwardly rectifying chloride channels through an autocrine mechanism involving ATP. *Cell* **81**, 1063–1073.
Scotto, K. W., Biedler, J., and Melera, P. W. (1986). Amplification and expression of genes associated with multidrug resistance in mammalian cells. *Science* **232**, 751–755.
Shalinsky, D. R., Jekunen, A. P., Alcaraz, J. E., Christen, R. D., Kim, S., Khatibi, S., and Howell, S. B. (1993). Regulation of initial vinblastine influx by p-glycoprotein. *Br. J. Cancer* **67**, 37–46.
Shapiro, A. B., and Ling, V. (1994). ATPase activity of purified and reconstituted P-glycoprotein from chinese hamster ovary cells. *J. Biol. Chem.* **269**, 3745–3754.
Shapiro, A., and Ling, V. (1995a). Reconstitution of drug transport by purified P-glycoprotein. *J. Biol. Chem.* **270**, 16167–16175.
Shapiro, A., and Ling, V. (1995b). Using purified P-glycoprotein to understand multidrug resistance. *J. Bioenergetics Biomembranes* **27**, 7–13.
Sharom, F. J., Yu, X., and Doige, C. A. (1993). Functional reconstitution of drug transport and ATPase activity in proteoliposomes containing partially purified P-glycoprotein. *J. Biol. Chem.* **268**, 24197–24202.
Simon, S. M., and Schindler, M. (1994). Cell biological mechanisms of multidrug resistance in tumors. *Proc. Natl. Acad. Sci. USA* **91**, 3497–3504.
Simon, S. M., Roy, D., and Schindler, M. (1994). Intracellular pH and the control of multidrug resistance. *Proc. Natl. Acad. Sci. USA* **91**, 1128–1132.
Sims, P. J., Waggoner, A. S., Wang, C.-H., and Hoffman, J. F. (1974). Studies on the mechanism by which cyanine dyes measure membrane potential in red blood cells and phosphatidylcholine vesicles. *Biochemistry* **13**, 3315–3330.
Sirotnak, F. M., Yang, C.-H., Mines, L. S., Oribe, E., and Biedler, J. L. (1986). Markedly altered membrane transport and intracellular binding of vincristine in multidrug-resistant chinese hamster cells selected for resistance to vinca alkaloids. *J. Cell. Physiol.* **126**, 266–274.
Skou, J. C., and Norby, J. G. (1979). "(Na.K)ATPase Structure and Kinetics." Academic Press, New York.
Stein, W. D., Cardarelli, C., Pastan, I., and Gottesman, M. M. (1994). Kinetic evidence suggesting that the multidrug transporter differentially handles influx and efflux of its substrates. *Mol. Pharmacol.* **45**, 763–772.
Stow, M. W., and Warr, J. R. (1993). Reduced influx is a factor in accounting for reduced vincristine accumulation in certain verapamil-hypersensitive multidrug-resistant CHO cell lines. *FEBS Lett.* **320**, 87–91.
Stutts, M. J., Gabriel, S. E., Olsen, J. C., Gatzy, J. T., O'Connell, T. L., Price, E. M., and Boucher, R. C. (1993). Functional consequences of heterologous expression of the cystic fibrosis transmembrane conductance regulator in fibroblasts. *J. Biol. Chem.* **268**, 20653–20658.
Stutts, M. J., Canessa, C. M., Olsen, J. C., Hamrick, M., Cohn, J. A., Rossier, B. C., and Boucher, R. C. (1995). CFTR as a c-AMP dependent regulator of sodium channels. *Science* **269**, 847–850.
Tew, K. D., Houghton, P. J., and Houghton, J. A. (1993). "Preclinical and clinical Modulation of Anticancer Drugs," pp. 156–160. CRC Press, Boca Raton, FL.
Trombe, M. C. (1993). Pleiotropic drug resistance in *Streptococcus pneumoniae:* The cross-chat hypothesis. *Mol. Microbiol.* **8**, 199–200.

Tsui, L.-C. (1992). The spectrum of cystic fibrosis mutations. *Trends Genet.* **8,** 392–398.
Valverde, M., Diaz, M., Sepulveda, F. V., Gill, D. R., Hyde, S. C., and Higgins, C. F. (1992). Volume-regulated chloride channels associated with the human multidrug-resistance P-glycoprotein. *Nature* **355,** 830–833.
Van Es, H. H. G., Renkema, H., Aerts, H., and Schurr, E. (1994). Enhanced lysosomal acidification leads to increased chloroquine accumulation in CHO cells expressing the *pfmdr 1* gene. *Mol. Biol. Parasitol.* **68,** 209–215.
Wadkins, R. M., and Houghton, P. J. (1993). The role of drug–lipid interactions in the biological activity of modulators of multi-drug resistance. *Biochem. Biophys. Acta* **1153,** 225–236.
Wadkins, R. M., and Houghton, P. J. (1995). Kinetics of transport of dialkyloxacarbocyanines in multidrug-resistant cell lines overexpressing p-glycoprotein: Interrelationship of dye alkyl chain length, cellular flux, and drug resistance. *Biochemistry* **34,** 3858–3872.
Watanabe, T. (1963). Infective heredity of multiple drug resistance in bacteria. *Bacteriol. Rev.* **27,** 87–115.
Wei, L. Y., and Roepe, P. D. (1994). Low external pH and osmotic shock increase the expression of human MDR protein. *Biochemistry* **33,** 7229–7238.
Wei, L. Y., Stutts, M. J., Hoffman, M. M., and Roepe, P. D. (1995). Overexpression of the cystic fibrosis transmembrane conductance regulator in NIH 3T3 cells lowers membrane potential and intracellular pH and confers a multidrug resistance phenotype. *Biophys. J.* **69,** 883–895.
Weisburg, J. H., Curcio, M., Caron, P. C., Raghu, G., Mechetner, E., Roepe, P. D., and Scheinberg, D. A. (1996). The multidrug resistance (MDR) phenotype confers immunological resistance. *J. Exp. Med.* **183,** 2699–2704.
West, I. C., and Mitchell, P. (1973). Stoichiometry of lactose—$H^+$ symport across the plasma membrane. *Escherichia coli. Biochem. J.* **132,** 587–592.
Wiener, E., Dubyak, G., and Scarpa, A. (1986). $Na^+/H^+$ exchange in Ehrlich ascites tumor cells. Regulation by extracellular ATP and 12-O-tetradecanoylphorbol 13-acetate. *J. Biol. Chem.* **261,** 4529–4534.
Wilson, C. M., Serrano, A. E., Wasley, A., Bogenschutz, M. P., Shankar, A. H., and Wirth, D. F. (1989). Amplification of a gene related to mammalian mdr genes in drug-resistant *Plasmodium falciparum*. *Science* **244,** 1184–1186.
Yang, C.-P. H., DePinho, S. G., Greenberger, L. M., Arceci, R. J., and Horwitz, S. B. (1989). Progesterone interacts with p-glycoprotein in multidrug-resistant cells and in the endometrium of Gravid Uterus. *J. Biol. Chem.* **264,** 782–788.

# Normal and Pathological Tau Proteins as Factors for Microtubule Assembly

André Delacourte and Luc Buée
Unité INSERM 422, Place de Verdun, 59045 Lille Cedex, France

Tau proteins are microtubule-associated proteins. They regulate the dynamics of the microtubule network, especially involved in the axonal transport and neuronal plasticity. Tau proteins belong to a family of developmentally regulated isoforms generated by alternative splicing and phosphorylation. This generates several Tau variants that interact with tubulin and other proteins. Therefore, Tau proteins are influenced by many physiological regulations. Tau proteins are also powerful markers of the neuronal physiological state. Their degree of phosphorylation is a good marker of cell integrity. It is heavily disturbed in numerous neurodegenerative disorders, leading to a collapse of the microtubule network and the presence of intraneuronal lesions resulting from Tau aggregation. However, different biochemical and immunological patterns of pathological Tau proteins found among neurodegenerative disorders are useful markers for the understanding of the role of Tau protein isoforms and the diagnosis of these pathological conditions.

**KEY WORDS:** Tau proteins, Microtubules, MAPs, Phosphorylation, Neurofibrillary degeneration, PHF, Alzheimer's disease, Corticobasal degeneration, Pick disease.

## I. Introduction

The interest in Tau proteins has dramatically increased after the discovery that these microtubule-associated proteins (MAPs) aggregate in nerve cells during Alzheimer's disease and many other neurodegenerative disorders. In fact, Tau proteins polymerize into intraneuronal lesions with different shapes and biochemical characteristics, according to the type of neurodegeneration involved. Also, Tau lesions in the isocortex are always strongly associated with cognitive impairment. Aside from these observations, the main question remains to determine the role of Tau proteins during normal life and how they participate to various degenerating processes. After 10 years of intense research, the most valuable and straightforward results have been obtained from studies *in vivo* and especially from the study of human brain diseases. In this chapter, we will concentrate on lessons given by several neurodegenerative disorders, after the presentation of the molecular aspects of Tau–microtubule (MT) interactions.

First, we will describe what is known about the Tau structure at both gene and protein levels. We will emphasize how phosphorylation is the main posttranslational modification under a delicate balance of kinases/phosphatases. Second, we will focus on tau–MT interactions: the role of the different Tau regions and how Tau phosphorylation modulates MT assembly in all cells and especially in neurons. Third, we will ask why Tau aggregation is a common feature to numerous neurodegenerative disorders and how Tau phosphorylation is involved in this process. Finally, we will indicate why we think that pathological Tau proteins present an opportunity to understand the physiological role of Tau proteins.

## II. Microtubules and Microtubule-Associated Proteins

A. Microtubules

Cytoplasmic MTs are cytoskeletal filaments found in all eukaryotic cells. These filaments play an important role in maintenance of cell shape, cell division, axonal transport, secretion, and receptor activity (Burgoyne, 1991). MTs are fragile and dynamic structures constituted by the polymerization of tubulin dimers. They are stabilized by accessory proteins that bind along the MT and form long filamentous projections on the surface of the polymer, which are named MAPs.

**1. Neuronal Microtubules**

*a. Morphology of Neurons* Microtubules are ubiquitous cytoskeleton filaments formed of tubulin polymers. They are more abundant in nerve cells, where they are involved in the mechanisms of intraneuronal transport, especially of axonal transport. In fact, the physiology of neurons is quite different from that of other cells and the MT network is well adapted to neuronal features. First, neurons are highly polarized cells with long cytoplasmic extensions. For example, the cytoplasmic mass of the axon can be 100–1000 times as large as that of the perikaryon. The second point is the limited protein synthesis that takes place in the axon and the long distance between the cell nucleus and axonal endings. Therefore, the neuron develops a unique transport through the axon in order to supply macromolecules necessary for neurophysiological function and allows a repair role of the damaged structures. Lastly, neurons do not divide and multiply and they have to work properly during the possible 120 years of human life's expectancy (Jeanne Calmant, a French citizen of Nimes, is the oldest human being at 121 years of age).

*b. Challenge of the Axonal Transport* MTs are abundant in axons. They are tracks that convey vesicles and macromolecules to the axon terminals, constituting the anterograde axonal transport. Conversely, damaged vesicles and biochemical messages backwards to the perikaryon constitute the retrograde axonal transport. Two types of MAPs interact with axonal MTs. One consists of force-producing MAPs, the main components of which are kinesin and cytoplasmic dynein (MAP Ic). These molecular motors produce energy for the transport by ATP hydrolysis. The other is composed of fibrous MAPs, which include Tau and MAP2 (Mandelkow and Mandelkow, 1994; Mercer *et al.*, 1994; Schoenfeld and Obar, 1994).

At least three types of protein interactions are found on MTs: one is linked to MT assembly and stability, whereas the others are related to MT position in the axoplasm and interactions with other organelles, such as microfilaments and neurofilaments, and associated with molecular motors. Since its characterization, Tau has been known to be a potent promoter of tubulin polymerization *in vitro* (Weingarten *et al.*, 1975). The absence of ATPase activity on Tau proteins demonstrates that they are not molecular motors. Their binding sites on tubulin dimers are different from the kinesin one (Marya *et al.*, 1994).

## 2. Microtubules in Nonneuronal Cells

MTs are also found in nonneuronal cells. They have a role for maintaining the cytoarchitecture, they are the main part of the mitotic spindle that guides chromosomes during cell division, and, in a general manner, they are involved in the intracellular transport (Dustin, 1984).

## B. Tau Proteins as Microtubule-Associated Proteins

The role of Tau proteins is directly related to that of MTs because they copurify with brain MTs through repeated cycles of warm polymerization and cold disassembly. Therefore, they belong to the MAPs family and are likely to play a major role in the regulation of MT assembly.

## 1. Gene Organization

Tau proteins arise from alternative splicing of a single gene located on the long arm of chromosome 17. The human *Tau* gene is located over 100 kilobases (kb) at band position 17q21 (Neve *et al.*, 1986) and contains 16 exons (Andreadis *et al.*, 1992, 1995) (Fig. 1). The restriction analysis and sequencing of the gene shows that it contains two CpG islands; one associated with the promoter region and the other with exon 9 (Andreadis *et al.*, 1992, 1995). The CpG island in the putative *Tau* promoter region resembles that of previously described neuron-specific promoters. Two regions homologous to mouse Alu-like sequence are present. The sequence of the promoter region also reveals a TATA-less that is likely to be related to the presence of multiple initiation sites, typical of housekeeping genes. Three SP1-binding sites that are important in directing transcription initiation in other TATA-less promoters are also found in the proximity of the first transcription initiation site. A 335-bp sequence located immediately upstream of exon -1 is sufficient and necessary to activate *Tau* gene expression in a neuron-specific manner (Sadot *et al.*, 1996a). Conversely, it was also reported that this promoter region is not neuron specific (Andreadis *et al.*, 1996).

## 2. Splicing

The *Tau* primary transcript contains 16 exons but 3 are not present in human brain (exon 4A, 6, and 8). They are specific to peripheral Tau proteins. Exon 4A is found in bovine, human, and rodent with a high degree of homology. Exons 6 and 8 are also highly conservative and may modify binding to MTs. To our knowledge, *Tau* mRNA with either exons 6 or 8 have not been described in human. Exon 8 is only found in bovine. Exon -1

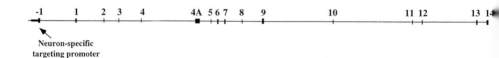

FIG. 1 Organization of the human *Tau* gene. The genomic organization is based on the work of Andreadis *et al.* (1992). The human *Tau* gene is composed of 16 exons. Exons are black boxes. Exon 4A is the largest one, whereas exon 8 is the shortest one. Exons −1, 1, 4, 5, 7, 9, 11, 12, and 13 are constitutive. Exons 2, 3, 4A, 8, and 10 are cassette exons. The 335-bp promoter region (bold line) located immediately upstream of exon −1 is sufficient and necessary to activate *Tau* expression in neurons (Sadot *et al.*, 1996a). The relative sizes and positions of exons and introns are correct.

is part of the promoter. It is transcribed but not translated. Exons 1, 4, 5, 7, 9, 11, 12, and 13 are constitutive exons. Exon 14 is found in messenger RNA, but it is not translated in protein. However, because a combination of splice sites is found between exons 13 and 14 in human, mice, and rat, distinct molecular diversity at the carboxy terminus of *Tau* is likely to occur (Goedert *et al.*, 1989a,b; Andreadis *et al.*, 1992; Sawa *et al.*, 1994). Exons 2, 3, and 10 are alternatively spliced and are adult brain specific (Andreadis *et al.*, 1992). Alternative splicing of these 3 exons allows six combinations (2−3−10−, 2+3−10−, 2+3+10−, 2−3−10+, 2+3−10+, and 2+3+10+) (Goedert *et al.*, 1989a,b; Kosik *et al.*, 1989). Thus, in human brain, the *Tau* primary transcript gives rise to six mRNAs (Goedert *et al.*, 1989a,b; Himmler, 1989) (Fig. 2). The distribution of *Tau* mRNAs with or without exon 10 may be different according to neuronal cell types (Goedert *et al.*, 1989a).

## 3. Structure

In the central nervous system, (CNS), Tau proteins constitute a family of six isoforms that range from 352 to 441 amino acids and a molecular weight (MW) ranging from 45 to 62 kDa. Tau proteins are organized into several regions that have different chemical and physiological properties. The amino-terminus part corresponds to the projection domain, whereas the carboxy-terminus part is a MT assembly domain. Tau protein primary structure is organized into three regions. The amino terminus is highly acidic and projects from the MT surface where it may interact with other cytoskeletal elements (Hirokawa *et al.*, 1988). The central region is proline rich and contains target sites for different types of kinases. The carboxy terminus is highly basic and possesses MT-binding activity. It contains either three or four repeated domains that interact with MTs (Fig. 2).

***a. Projection Domain (Exons 2, 3, and 4A)*** Exons 2 and 3 of the gene are alternatively spliced cassettes in which exon 3 never appears independently of exon 2 (Andreadis *et al.*, 1995). These exons are expressed in adult life. They are localized in the projection domain and give different lengths to the projection domain.

Exon 4A is a long exon that generates an additional sequence of 253 amino acids (Andreadis *et al.*, 1992) (Fig. 3). Tau isoforms with exon 4A, named Big Tau, have an apparent molecular weight of 130 kDa. They are found mostly in the peripheral nervous tissue (Georgieff *et al.*, 1993).

***b. Microtubule Assembly Domain (Exons 9–12)*** The carboxy terminus contains three (3R) or four copies (4R) of a highly conserved 18-amino acid repeat (Lee *et al.*, 1988, 1989; Goedert *et al.*, 1989a; Himmler, 1989)

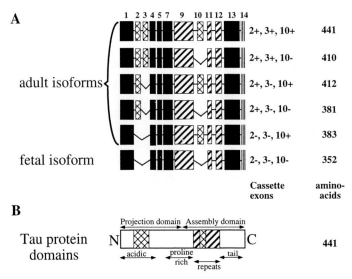

FIG. 2 Schematic representation of the CNS six Tau isoforms. The six Tau isoforms differ from each other by the addition of one or two inserts corresponding to exons 2 and/or 3 in the amino terminus in combination with both three and four (addition of the exon 10) MT-binding domains. Multiple splice acceptor sites between exons 13 and 14 lead to a striped carboxy terminus. The shortest Tau isoform (without exons 2, 3, and 10) is made of 352 amino acids and only found in fetal brain. It is referred to as fetal isoform. Adult Tau isoforms include the 383-amino acid form (2−,3−,10+), the 381-amino acid form ((2+,3−,10−), the 412-amino acid form (2+,3−,10+), the 410-amino acid form (2+,3+,10−), and the 441-amino acid form (2+,3+,10+). With the exception of exon 14, the relative sizes of exons are correct. (B) Protein domains of the longest Tau isoform. Tau proteins are organized into several regions that have different chemical and physiological properties. The amino-terminus part (N) is highly acidic and corresponds to the projection domain, whereas the carboxy-terminus part (C) is highly basic and possesses MT-binding activity and is referred to as the assembly domain. The central region is proline rich and contains target sites for different types of kinases.

separated from each other by less conserved 13- or 14-amino acid interrepeat domains (IR) (Fig. 4). Most notably from a developmental perspective, all of the 4R Tau isoforms possessing a 18+13 amino acid insert are expressed specifically in adult neurons (Kosik et al., 1989).

## C. Tau as a Phosphoprotein

### 1. Sites of Phosphorylation

Many sites of phosphorylation on normal Tau have been discovered while studying the biochemistry of neurofibrillary tangles (NFTs) of Alzheimer's

FIG. 3 Schematic representation of the Big Tau found in peripheral tissues. Big Tau differs from CNS Tau isoforms by the addition of exon 4A encoding for a 253-amino acid sequence. The relative sizes of exons are correct.

disease (AD). These brain lesions are intraneuronal bundles of paired helical filaments (PHFs) composed of hyperphosphorylated Tau proteins, referred to as PHF-Tau or pathological Tau. Phosphorylation-dependent monoclonal antibodies against PHF-Tau were developed in a number of laboratories. They allowed for the identification of the so-called "abnormally phosphorylated sites" on PHF-Tau: AT8 on Ser 202 and Thr 205 (Mercken et al., 1992; Goedert et al., 1995), PHF-1 (Greenberg et al., 1992; Otvos et al., 1994) and AD2 (Buée-Scherrer et al., 1996a) on Ser 396 and 404, C5 on Ser 396 and M4 on Thr 231 and Ser 235 (Hasegawa et al., 1993, 1996), 12E8 on Ser 262 (Seubert et al., 1995), AT180 on Thr 181, AT270 on Thr231 (Mercken et al., 1992; Rosner et al., 1995), and AP422 on Ser 422 (Hasegawa et al., 1996) (Fig. 5). The analysis was completed by ion spray mass spectrometry and direct sequencing of phosphorylated peptides (Hasegawa et al., 1992).

At least 20 hyperphosphorylated sites on PHF-Tau have been described, including Thr 181; Ser 198, 199, 202; Thr 205, 208, 210; Thr 212; Ser 214;

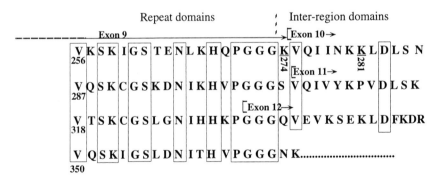

FIG. 4 Sequence of the MT-binding domains. Tau proteins can exist with three or four MT-binding domains resulting from the alternative splicing of exon 10. The first domain is encoded by the end of exon 9; the second one by exon 10; the third one by exon 11 and the beginning of exon 12; and the fourth one by the remaining part of exon 12. Consensus sequences among the four MT-binding domains are boxed. MT-binding domains can be divided in repeat domain and interregion. In the first interregion domain, there is an important contribution of Lys 274 (K) and 281 (K) implicating strong ionic interactions with tubulin.

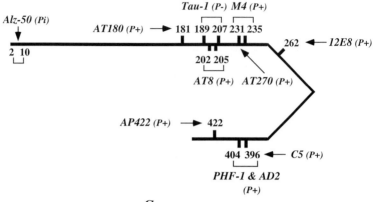

FIG. 5 The binding sites of Tau antibodies. The different well-known antibodies and their binding sites are represented on this schematic map of the longest human tau isoform. Antibodies (AD2, AP422, AT8, AT180, AT270, C5, M4, PHF-1, and 12E8) recognize their epitope when it is phosphorylated (P+), and Tau-1 recognizes a sequence (amino acids 189–207) when it is nonphosphorylated (P−). Alz-50 is a phosphorylation-independent antibody (Pi).

Thr 217; 231; Ser 235, 262, 396, 400, 403, 404, 409, 412, 413, and 422. Most of these phosphorylation sites are on Ser-Pro and Thr-Pro motives. A number of sites on non-Ser/Thr-Pro sites have also been identified (Morishima-Kawashima et al., 1995).

With the exception of Ser 422, these phosphorylated sites that were supposed to be specific to PHF-Tau are also found on native Tau from biopsy-derived samples (Matsuo et al., 1994; Seubert et al., 1995). All these sites are localized outside the MT-binding domains and are Ser-Pro or Thr-Pro motives (Gustke et al., 1992), with the exception of Ser 262 (numbering according to the largest Tau isoform; Goedert et al., 1989a).

D. Other Posttranslational Modifications

**1. O-glycosylation**

O-glycosylation is characterized by the addition of a GlcNAc residue on Ser or Thr at proximity of Pro (Haltiwanger et al., 1992). It is found in neurofilaments (Dong et al., 1993) and on Tau proteins (Arnold et al., 1994).

**2. Others**

Modifications other than phosphorylation and glycosylation are also known, including removal of the first methionine residue, N-acetylation at the amino terminus, and deamidation at two asparaginyl residues (Hasegawa et al., 1992).

## III. Role of Tau Proteins in Microtubule Assembly

The role of Tau proteins is likely to be fundamental in MT asssembly because they are found in many species. For instance, a Tau-like gene referred to as *ptl-1* was found in *Caenorhabditis elegans* (Dermott et al., 1996). The *ptl-1* transcript is also alternatively spliced. The predicted *ptl-1* products have strong sequence homology to the repeat regions and have several protential phosphorylation sites. They bind to MTs *in vitro*. In mammalians, Tau protein functions include stabilization of the axonal architecture (Drubin et al., 1988) Caceres and Kosik, 1990; Shea et al., 1992) and formation of bundles (Kanai et al., 1989, 1992; Lee and Rook, 1992). Tau functions are modulated by the presence of different sets of isoforms as well as by phosphorylation and other posttranslational modifications. Splicing and phosphorylation are developmentally regulated.

### A. Different Sets of Tau Isoforms for Different Functions

**1. Projection Domain**

As described previously, exons 2 and 3 could also play a role by interacting with other organelles. Transfection studies on fibroblasts indicate that Tau isoforms having exons 2 and 3 may allow a more efficient MT bundling than those lacking this N-terminal region (Kanai et al., 1992). Indeed, although bundling of MTs is observed in the cells transfected with each isoform, Tau proteins with a longer N-terminal region induce both thicker MT bundles and higher bundling percentages (Kanai et al., 1992). Tau amino-terminal projection domain also has an important role in neuritic development and allows interactions with the neural plasma membrane (Brandt et al., 1995). Thus, Tau may act as a mediator between MTs and plasma membrane.

In cell culture, exons 2 and 3 are not present in mRNA from SKNSH-SY5Y human neuroblastoma cells (Dupont-Wallois et al., 1995). Their presence in differentiated adult tissue is possibly related to the stabilization of the MT network, and their main role could be, as for exon 4A, to allow MTs spacing and large axonal diameter.

In PNS, Big Tau resulting from the transcription of exon 4A is distributed and has a limited and specific distribution in all central neurons with an extension in the peripheral nervous system. This neuronal specificity is consistent with a role in stabilizing MTs in axons that are subjected to great shear forces (Boyne et al., 1995). One other possible role would be the maintenance of a wide spacing between MTs in peripheral axons. On average, spacing between MTs in axons of the CNS is 1–1.5 MT diameters, whereas the distance increases to 2–2.5 MT diameters in the peripheral nervous system (PNS) (Georgieff et al., 1993).

### 2. Microtubule Assembly Domain

Tau proteins promote MT assembly (Weingarten et al., 1975). They bind MTs through repetitive regions in their carboxy-terminus part. These repetitive regions are the repeat domains encoded by exons 9–12 (Lee et al., 1989) (Figs. 2 and 4). Tau synthetic peptides corresponding to single repeats as well as four repeats (4R) can induce MT assembly (Butner and Kirschner, 1991; Gustke et al., 1994; Lee and Rook, 1992). 4R Tau is more efficient than 3R Tau (Goedert and Jakes, 1990). The 18-amino acid repeats bind to MTs through a flexible array of distributed weak sites (Lee et al., 1989; Butner and Kirschner, 1991). Interestingly, the most potent part to induce MT polymerization is the interregion between repeats 1 and 2 (R1–R2 IR), and more specifically peptide KVQIINKK within this sequence (Figs. 4 and 6). R1–R2 IR binding to tubulin involves different molecular mechanisms when compared to that of the 18-amino acids repeats. For instance, there is an important contribution of Lys 274 and 281 implicating strong ionic interactions with tubulin (Fig. 4). This R1–R2 interregion is unique to 4R Tau, adult-specific, and responsible for a 40-fold difference in the binding affinities between 3R and 4R Tau (Fig. 6; Goode and Feinstein, 1994; Panda et al., 1995).

This provides significant support for a mechanism by which the binding of Tau to individual tubulin subunits in MTs induces a conformational change that strengthens intertubulin bonding (Fig. 6). The R1–R2 IR and the presence of four repeats increase binding properties toward tubulin dimers and generate stiffer or longer MTs (Figs. 4 and 6). These properties are well adapted to an adult neuronal cytoskeleton that must be more stable than in a developing neuron, which needs a cytoskeletal plasticity for neuronal migration and synapse formation. It is noteworthy that other MAPs also have similar repeats, with up to five repeats for MAP4 (Chapin et al., 1995).

### B. Tau–Tubulin Interactions

Regulation of MT assembly may also be brought about by the structure of tubulin itself. Interactions between subtilisin-treated MTs and MAPs

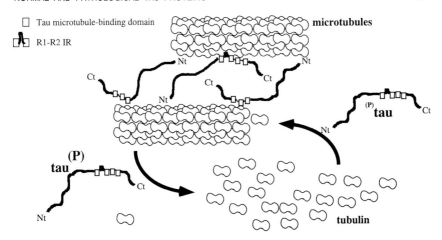

FIG. 6  Tau phosphorylation and MT polymerization. Polymerization of tubulin into MTs is regulated by the state of phosphorylation of Tau proteins. Hypophosphorylated forms of Tau proteins [($^P$) tau, on the right] induce tubulin polymerization, whereas hyperphosphorylated forms of Tau proteins [tau($^P$), on the left] lead to the depolymerization of MTs into tubulin. Interactions with tubulin and MTs is mediated through the carboxy terminus (Ct) of Tau proteins. Tau proteins with either three or four MT-binding domains bind to MTs through a range of reversible weak interactions (Butner and Kirschner, 1991). However, Tau proteins with four repeats bind to MTs with high affinity. It is likely to be related to the presence of an extra repeat and the interregion encoded by exon 10 between repeats 1 and 2 (R1–R2 IR) that also binds tubulin. This binding site on tubulin is different from that of the repeats (see Fig. 4). According to Panda et al. (1995), it may allow to anchor the interrepeat within the MT and thus increase the interactions between tubulin and the two first MT-binding domains. Distance between MTs is determined by the amino-terminal part (Nt) of Tau proteins corresponding to the projection domain.

suggest that the carboxy terminals of $\alpha$- and $\beta$-tubulins are not primarily involved in the binding of MAPs onto MTs. However, interactions between tubulin carboxy terminals and MAPs remain possible and might be involved in the regulation of MAP-induced MT bundling (Saoudi et al., 1995). In fact, Tau interacts differentially with the diverse posttranslationally modified isotubulins: The recently discovered polyglutamylation at the end of tubulin C-termini, which consists of the sequential, posttranslational addition of one to six glutamyl units to both $\alpha$- and $\beta$-tubulin subunits, regulates the binding of Tau as a function of its chain length. The relative affinity of Tau, which is very low for unmodified tubulin, increases progressively for isotubulins carrying from one to three glutamyl units, reaches an optimal value, and then decreases progressively when the polyglutamyl chain lengthens up to six residues (Boucher et al., 1994).

## C. Tau Phosphorylation

Phosphorylation results from the actions of kinases and phosphatases. Both can act directly and indirectly through a cascade of enzyme actions. Kinases can also phosphorylate and, therefore, modulate activities of other kinases and/or phosphatases. Protein phosphorylation undergoes a complex control including intra- and extracellular modulating factors and specificity of enzymes (kinases and phosphatases) toward a specific site. This regulation is developmentally regulated.

### 1. Kinases

***a. Kinases Involved in Tau Phosphorylation*** Most of the kinases involved in Tau phosphorylation are part of the proline-directed protein kinases (PDPK), which include MAP kinase (Drewes *et al.*, 1992; Vulliet *et al.*, 1992), glycogen synthase kinase 3 (GSK3) (Hanger *et al.*, 1992), and cyclin-dependent kinases including cdc2 and cdk5 (Baumann *et al.*, 1993; Liu *et al.*, 1995). These kinases have been shown to phosphorylate Tau *in vitro* at a number of Ser-Pro and Thr-Pro sites. For example, GSK3-$\beta$ after heparin potentiation phosphorylates Tau on sites of Ser 199, Thr 231, Ser 235, Ser 262, Ser 396, and Ser 400 (Song and Yang, 1995).

Non-Ser/Thr-Pro sites can be phosphorylated by many other protein kinases, including $Ca^{2+}$/calmodulin-dependent protein kinase II (Baudier and Cole, 1987; Johnson, 1992), cyclic-AMP-dependent kinase (Litersky and Johnson, 1992), and casein kinase II (Greenwood *et al.*, 1994).

***b. Kinase Activity and Specificity*** Kinases can be more or less specific for the different potential phosphorylation sites on Tau proteins. Ser 262 is specifically phosphorylated *in vitro* by a 110-kDa protein kinase named p110 (mark) (Drewes *et al.*, 1995). However, other kinases, including protein kinase A and $Ca^{2+}$/calmodulin-dependent kinase, were shown to phosphorylate Ser 262 and other sites *in vitro* (Drewes *et al.*, 1995; Litersky and Johnson, 1992).

Mitogen-activated protein kinases have a broad action on Ser-Pro or Thr-Pro motives. Interestingly, significant proportions of mitogen-activated protein kinase are associated with MTs in culture cells but it is not clear which protein kinases act on Tau in the human brain (Reszka *et al.*, 1995).

Transient transfection of human GSK3-$\beta$ into Chinese hamster ovary (CHO) cells stably transfected with individual human Tau isoforms leads to hyperphosphorylation of Tau at all sites investigated using phosphorylation-dependent anti-Tau antibodies. Thus, GSK3-$\beta$ is a protein kinase that phosphorylates Tau protein at multiple sites in intact cells (Sperber *et al.*, 1995). Also, numerous kinases, proline directed and nonproline directed, must work in tandem in order to have a complete phosphorylation of recombinant Tau and may be positively modulated at the substrate level by non-PDPK-catalyzed phosphorylations (Singh et al., 1996).

Hyperphosphorylation of Tau proteins is also observed using kinase activators such as phorbol ester for protein kinase C activation (Couratier et al., 1996) and heparin (Song and Yang, 1995; Hasegawa et al., 1996).

Together, these results show that Tau phosphorylation is likely the result of numerous and complex cascades of kinases.

## 2. Phosphatases

There are four types of Ser/Thr phosphatase proteins that are present in the brain: 1, 2A, 2B (calcineurin), and 2C (Ingebristen and Cohen, 1983; Cohen and Cohen, 1989).

*a. Phosphatases Involved in Tau Dephosphorylation* Purified phosphatase proteins 1, 2A, and 2B can dephosphorylate Tau proteins *in vitro* (Goto et al., 1985; Yamamoto et al., 1988, 1995). Phosphatase 2A is highly efficient for dephosphorylating recombinant Tau proteins that were previously phosphorylated using PDPK (Goedert et al., 1995). Other studies indicate that phosphatase 2B can dephosphorylate Tau proteins *in situ* (Fleming and Johnson, 1995). Recent data have shown that Tau proteins from brain tissue or neuroblastoma cells are rapidly dephosphorylated by endogenous phosphatases (Garver et al., 1994; Matsuo et al., 1994; Buée-Scherrer et al., 1996a; Soulié et al., 1996).

*b. Phosphatase Regulations* Like kinases, phosphatases have many direct or indirect physiological effects. They counterbalance the action of kinases. Phosphatases acting on Tau are developmentally regulated (Pope et al., 1993; Dudek and Johnson, 1995) and are associated or not with MTs (Dudek and Johnson, 1995; Sontag et al., 1995). Activity levels and association of phosphatases with MTs are independently regulated during postnatal development. In fact, activities and levels of phosphatase 1 decrease significantly from P1 to adult, but association with MTs does not change with postnatal age. Furthermore, activities and levels of phosphatase 2A decrease significantly from P1 to adult, whereas association with MTs is increased in the adult compared to the neonate (P3). Finally, activities and levels of phosphatase 2B increase significantly from P1 to P21 and remain elevated in adult. Association with MTs is also increased in the adult compared to the neonate (P3) (Dudek and Johnson, 1995).

It is likely that phosphatases 2A and 2B are differentially involved in dephosphorylation of Tau proteins in neurons. In fact, in fetal rat primary cultured neurons, the use of phosphatase 2A inhibitors induces phosphorylation of Tau proteins on some sites, whereas phosphatase 2B inhibitors allow phosphorylation on other sites (Saito *et al.*, 1995; Ono *et al.*, 1995). Phosphatases inhibitors used to study phosphatase activities are okadaic acid, which specially inhibits phosphatase 2B, and others such as orthovanadate, tautaumycin, microcystin, sodium fluoride, and calyculin A (Lobert *et al.*, 1994; Baum *et al.*, 1995; Shinozaki *et al.*, 1995).

### 3. Phosphorylation Regulations

The different states of Tau phosphorylation result from activities of both kinases and phosphatases. Phosphorylation of Tau proteins is developmentally regulated. It is high in fetal and decreases with age due to phosphatases activation (Mawal-Dewan *et al.*, 1994; Rosner *et al.*, 1995). Thus, the most precise analysis on the expression of phosphorylation sites during development and adult life derives from immunohistochemical studies on the nervous tissue of animals that have been fixed by perfusion. This avoids postmortem delays that destroy phosphorylation sites (Mawal-Dewan *et al.*, 1994; Rosner *et al.*, 1995). Using AD2 antibody, we have observed that Tau immunostaining in the rat brain is intense in the fetal tissue. It decreases during development, becoming more diffuse and weak. Similar results are observed with a Western blot approach (Buée-Scherrer *et al.*, 1966a).

The phosphorylated Tau epitope located at Ser 396 and 404 and recognized by both mAb AD2 and PHF-1 (Fig. 5) is spontaneously expressed by nondegenerating human neuroblastoma cells (Dupont-Wallois *et al.*, 1995; Martin *et al.*, 1995). This epitope is also downregulated by cellular differentiation, induced by retinoic acid, and upregulated by the extracellular matrix components fibronectin and laminin (Martin *et al.*, 1995). One explanation of these data is that fibronectin maintains a population of SKNSH-SY5Y cells in a biochemical state of differentiation in which PHF-1 Tau is expressed. Therefore, molecules of the extracellular matrix can regulate the phosphorylation state of Tau (Martin *et al.*, 1995).

Tau phosphorylation can also be controlled by cell surface neurotransmitter receptors. Stimulation of m1 muscarinic acetylcholine receptor transfected in PC12 cells with two acetylcholine agonists, carbachol and AF102B, decreases Tau phosphorylation, as indicated by phosphorylation-dependent Tau monoclonal antibodies and by alkaline phosphatase treatment. The muscarinic effect is both time and dose dependent. In addition, a synergistic effect on Tau phosphorylation has been found between treatments with muscarinic agonists and nerve growth factor. Therefore, a link between the cholinergic signal transduction system and the neuronal cytoskeleton could be mediated by regulated phosphorylation of Tau MT-associated protein (Sadot *et al.*, 1996b).

Cold water stress induces an immediate (30–90 min) two- or three-fold increase in the phosphorylation of Tau proteins in rat brain, without direct involvement of the hypothalamic–pituitary–adrenal axis (Korneyev et al., 1995). Heat shock stress also induces modifications of Tau phosphorylation and the effect is different according to the sex (Papasozomenos, 1996).

Overall, these data show that phosphorylation seems to affect simultaneously several sites. However, this has to be clarified using a panel of monoclonal antibodies (mAbs) against different phosphorylation sites and following their fates during development. Furthermore, the state of phosphorylation is strongly modified during development because of the expression of several specific adult isoforms and because the ratio between kinases and phosphatases is also modified during development (Mawal-Dewan et al., 1994). Phosphorylation, in combination with the type of isoform, can modulate the properties of Tau proteins. In turn, Tau proteins give to the MT its own identity and physical character (rigidity, length, stability, and interactivity with other organelles). Therefore, indirectly, Tau proteins have a role in the functional organization of the neuron, i.e., axonal morphology, growth, and polarity.

### 4. Phosphorylation and Microtubule Assembly

Tau proteins effects on MT assembly depend partially on their phosphorylation state so that phosphorylated Tau proteins are less effective than nonphosphorylated Tau on MT polymerization (Lindwall and Cole, 1984; Drubin and Kirschner, 1986) (Fig. 6). Following phosphorylation, Tau proteins become longer and stiffer (Hagestedt et al., 1989).

Phosphorylation of Ser 262 alone dramatically reduces Tau affinity for MTs *in vitro* (Biernat et al., 1993). Nevertheless, this site alone is insufficient to eliminate Tau binding to MTs. It is present in fetal, adult Tau, as well as PHF-Tau (Seubert et al., 1995).

Studies of Tau phosphorylation on MT assembly must take into account the finding that native Tau proteins extracted from the brain are rapidly dephosphorylated by the strong phosphatase activity generated after cell death (see Section V,A).

## D. Other Regulations and Interactions

We have shown that Tau proteins interact with tubulin dimers (Fig. 6). However, associations of Tau proteins and MTs within axons may not be as stable as those described previously. For instance, the axonal transport rate for Tau proteins is distinct from that of tubulin and other MAPs, demonstrating the existence of dynamic changes in Tau–MT interactions in mature neurons *in vivo* (Mercken et al., 1995).

Tau proteins can also interact with some other proteins or be regulated with non-MT proteins. For example, laminin is capable of inducing axonal formation and MT stabilization in neurons arrested at stage II of neuritic development by Tau suppression (Caceres and Kosik, 1990; Caceres et al., 1991). Axonal formation in neurons growing in the presence of laminin is prevented by treatment of the cultures with a mixture of MAP-1B and Tau antisense oligonucleotides but not by the single suppression of any one of these MAPs. These data suggest, with regard to axonal elongation, that MAP-1B and Tau can be functionally substituted (Ditella et al., 1996).

Short-term exposure from neonatal rat brain to high concentrations of glutamate results in a substantial increase of both immunoreactivity to and mRNA levels of Tau proteins. Neurons preincubated with a specific Tau antisense oligonucleotide resisted glutamate treatment and did not show an increase in Tau immunoreactivity. These data indicate that neosynthesis of the cytoskeleton-associated Tau protein is a crucial step in the cascade of events promoted by glutamate leading to neurodegeneration (Pizzi et al., 1995).

There is evidence that Tau proteins interact with other components. It may allow interactions between MTs and mitochondria (Rendeon et al., 1990). Tau proteins also bind to spectrin (Carlier et al., 1984) and actin filaments (Henriquez et al., 1995; Griffith and Pollard, 1978; Correas et al., 1990). Tau cross-links F-actin *in vitro* in a phosphorylation-dependent manner (Griffith and Pollard, 1978; Sattilaro et al., 1981; Pollard, 1983). Through these interactions, Tau proteins may allow MTs to interconnect with other cytoskeletal components such as neurofilaments (Leterrier et al., 1982; Aamodt and Williams, 1984) and may restrict the flexibility of the MT (Matus, 1994).

Overall, Tau phosphorylation can be controlled or influenced, directly or indirectly, by many factors.

E. Localization

**1. Tau Sorting in Nerve Cells**

Tau is a phosphoprotein, as first revealed with a mAb named Tau-1 directed against a dephosphorylated site of phosphorylation located at positions 189–207 (Fig. 5). Because Tau-1 preferentially labels axons, Tau were tagged as "axonal proteins" (Binder et al., 1985). However, after the dephosphorylation of the tissue with alkaline phosphatase, Tau proteins are detected in neuronal cell bodies and dendrites. These results show that the state of phosphorylation of Tau proteins is different according to the cell compartments (Riederer and Binder, 1994), and that the labeling of cell bodies and dendrites with Tau-1, as well with other phosphorylation-independent antibodies such as Alz-50 (Fig. 5), demonstrates that these proteins are found in all compartments of the nerve cell and are not exclusively axonal proteins (Hamre et al., 1989).

In fact, Tau localization changes in developing hippocampal neurons. In early stages, Tau is abundant in soma, minor processes and their growth cones (stage 2), and in the newly formed axon (stage 3). Conversely, MAP2 is never found in growth cones (stages 2 and 3) and is restricted to the proximal part of the axon. In early stage 4, both MAP2 and Tau remain associated with the MTs of growing dendrites, but only Tau is associated with axonal MTs. By stage 5, Tau is found in axons (Mandell and Banker, 1995). Other observations indicate that Tau proteins remain unpolarized at all stages in culture (Dotti et al., 1987) or that they bind selectively to axonal MTs in cerebellar neurons at early stages of development (Ferreira and Caceres, 1989). These discrepancies may be related to differences in methodology or be due to posttranslational modifications such as acetylation of axonal MTs from hippocampal and cerebellar neurons (Ferreira and Caceres, 1989; Dotti and Banker, 1991).

Tau-1 antibody was also used in conjunction with alkaline phosphatase treatment to study the distribution and phosphorylation of Tau proteins in developing neurons. At early stages, Tau proteins are highly phosphorylated at the Tau-1 site (Fig. 5). In more matured neurons, 80% of phosphorylated Tau proteins at the Tau-1 site are located in the minor processes and proximal axon, whereas 20% are present in the distal axon and growth cone. Conversely, PHF-1 immunoreactivity is opposite to the gradient of phosphorylation at the Tau-1 site. This raises the possibility of independent regulation of multiple phosphorylation sites within subcellular domains (Mandell and Banker, 1995).

Finally, in cultured embryonic chicken and rat neurons, a clear differential distribution of immunostaining within growing axons is also observed. Tau proteins that are phosphorylated at Ser 202 (recognized by AT8; Fig. 5) appear to be concentrated in the axonal portion that is close to the cell body and they decrease in a proximal–distal direction. In contrast, Tau proteins carrying a Tau-1 epitope that contains dephospho-Ser 202 are enriched in the distal axon and the growth cone (Rabhan et al., 1995).

The use of biotinylated proteins allows the visualization of intracellular trafficking. Indeed, biotinylated Tau, MAP2C, or MAP2 microinjected into mature spinal cord neurons in culture had their fates analyzed by antibiotin immunocytochemistry. Initially, each was detected in axons and dendrites, although Tau persisted only in axons whereas MAP2C and MAP2 were restricted to cell bodies and dendrites. Thus, beyond contributions from mRNA localization and selective axonal transport, compartmentalization of each of the three major MAPs occurs through local differential turnover (Hirokawa et al., 1996). It has been shown that sequences located in the 3' untranslated *Tau* mRNA region include a *cis*-acting signal that is involved in *Tau* mRNA targeting. A minimal sequence of 91 nucleotides located in this region binds to two RNA-binding proteins. These data suggest that *cis*-acting signals, together with RNA-binding proteins, are involved in the targeting of *Tau* mRNA (Behar et al., 1995). Finally, transfection experiments in Sf9 cells with 3R and 4R *Tau* mRNA isoforms suggest that *Tau* is involved in axonal determination because transfected cells show long cell processes that resemble the axon-like structures observed in some neuron cultures (Knops et al., 1991). However, these changes are not observed in other cell systems (Kanai et al., 1989, 1992; Lee and Rook, 1992). Thus, Tau proteins may play a role in axonal stabilization. In mice lacking the *Tau* gene, axonal elongation is not affected in cultured neurons (Harada et al., 1994). However, in some small-caliber axons, MT stability is decreased and MT organization is significantly changed. An increase in MT-associated protein 1A, which may compensate for the functions of Tau in large-caliber axons, was observed. These results argue against the suggested role of Tau in axonal elongation but confirm that it is crucial in the stabilization and organization of axonal MTs in a certain type of axon.

Interestingly, it has been shown that Tau proteins are also found in the nucleus (Brady et al., 1995). The isoforms from nuclear Tau are similar to cytoplasmic Tau, but they show lower solubility. These data suggest that they have a specific modification (posttranslational, interaction with other proteins, etc.) that remains to be characterized. In neuroblastoma LAN5 cells, Tau proteins (more than 15%) are also found in the nucleus in the chromatin fraction containing DNA, chromatin, and associated proteins. Prior to their addressing in the nucleus, Tau proteins are phosphorylated in the cytoplasm (Greenwood and Johnson, 1995). The function of nuclear Tau and how it may be regulated by phosphorylation is still unknown.

Together, these observations show that Tau proteins are found in all cell compartments but with a different phosphorylation state. Inside the same compartment, variation of the degree of phosphorylation is observed during development. Both phosphorylation and transcriptional factors may be involved in Tau trafficking and cell sorting.

### 2. Tau Isoforms in Different Types of Neurons

MT stability could also be modulated by combinations of different Tau isoforms among the six that have been characterized in the adult human brain (Fig. 2). Indeed, it is likely that neurons can be distinguished by their content in *Tau* mRNAs (Goedert et al., 1989a). This is corroborated by

Tau studies on neurodegenerative disorders (see Section V,B,1). Big Tau, specifically found in CNS neurons projecting in the PNS also favors such an hypothesis (Georgieff et al., 1993).

### 3. Tau Proteins in Nonneuronal Cells

Tau proteins are MT-associated proteins found at low levels in neurons that contain by far many more MTs than other types of cells. Therefore, nonneuronal cells generally have trace amounts of Tau proteins, if any. Tau proteins can be expressed in glial cells, mainly in pathological conditions (see Section V,B,2,a). Tau proteins have been identified in oligodendrocytes *in situ* and *in vitro*. Tau localization, revealed by monoclonal antibody Tau-5, is confined to the cell soma (Lopresti et al., 1995). Tau-related MAPS are expressed in the pancreas (Michalik et al., 1995) and, more precisely, *Tau* mRNA is located in acinar cells in the exocrine pancreas (Neuville et al., 1995). Tau proteins are also found in calcitonin-producing cells in rat thyroid and in a calcitonin-producing cell line (Nishiyama et al., 1995).

Using Tau-1, it has been possible to detect Tau isoforms in numerous different tissues such as muscle, heart, lung, kidney, testis (Oyama et al., 1996), or fibroblasts (Ingelson and Lannfelt, 1996). Big Tau, small Tau, or both were detected in these nonneuronal tissues.

## IV. Pathological Tau Proteins

### A. Tau Accumulation in Neurodegenerative Disorders

Tau proteins aggregate in a number of degenerative disorders. Characterization of Tau proteins in these pathological conditions may allow a better understanding of their physiological role.

### 1. Neurofibrillary Tangles

One of the main features of many neurodegenerative disorders is the presence of NFTs in the cerebral cortex. NFTs are immunostained with antibodies against Tau proteins (Figs. 7A and 7B) (Joachim et al., 1987). They are commonly found in AD (Braak and Braak, 1995), amyotrophic lateral sclerosis/Parkinson–dementia complex of Guam (ALS/PDC) (Hof et al., 1991, 1994a), corticobasal degeneration (CBD) (Feany and Dickson, 1995; Feany et al., 1995, 1996), dementia pugilistica and head trauma (Hof et al., 1992a), Down's syndrome (Mann et al., 1989), a few patients with

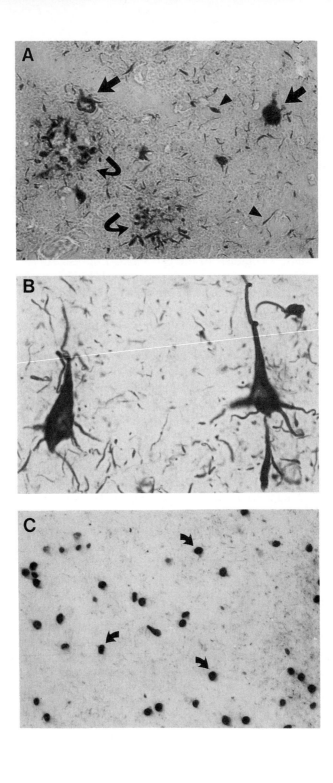

Gerstmann–Straussler–Scheinker (Ghetti *et al.*, 1995, 1996), Hallervorden–Spatz disease (Malandrini *et al.*, 1995), myotonic dystrophy (Kiuchi *et al.*, 1991), Niemann–Pick disease (Love *et al.*, 1995), Pick's disease (Hof *et al.*, 1994b), postencephalitic parkinsonism (Ishii and Nakamura, 1981), progressive supranuclear palsy (PSP) (Hof *et al.*, 1992b), subacute sclerosing panencephalitis (SSPE) (McQuaid *et al.*, 1994), and also in other rare conditions. They are also seen in normal aging (Bouras *et al.*, 1994). At the electron microscopic level it is possible to differentiate two categories: tangles made up of paired helical filaments, corresponding to most of the neurodegenerative disorders, and tangles composed of straight filaments, as in PSP and CBD (Kid, 1963; Tellez-Nagel and Wisniewski, 1973; Tomonaga, 1977; Montpetit *et al.*, 1985). Among these different conditions, both laminar and regional distributions of NFTs are different (Hof and Morrison, 1994).

## 2. Dystrophic Neurites

Dystrophic neurites are abnormally swollen neurites that are often Tau immunoreactive (Figs. 7A and 7B). They are generally observed in the vicinity of senile plaques in AD. They contain vesicles and subcellular organelles, accumulations of neuropeptides, and aggregated Tau proteins (Braak *et al.*, 1986; Praprotnik *et al.*, 1996). Neuropeptides seem to accumulate before the formation of PHF (Lenders *et al.*, 1989). Neurites with PHF are found in axons and dendrites (Braak and Braak, 1988). They generally constitute an extensive network that permeates almost the whole isocortex (Schmidt *et al.*, 1994). Extracellular amyloid deposits are sometimes surrounded by numerous dystrophic neurites, constituting the "neuritic plaques" (Fig. 7A). The number and distribution of neuritic plaques is extremely variable and always in high concentration in the brain of young AD patients.

## 3. Pick Bodies

Pick's disease, a form of progressive presenile dementia, is distinguished by the presence of neuronal lesions known as Pick bodies (PBs). PBs are

---

FIG. 7 NFT, dystrophic neurites, and Pick bodies. The different types of brain lesions immunolabeled with anti-Tau antibodies are presented. (A) Neuritic plaques (curved arrows), neurofibrillary tangles (arrows), and dystrophic neurites (arrowheads) scattered in the CA1 field of the hippocampus from an AD patient are labeled by an anti-tau polyclonal antibody. (B) AD2-immunoreactive neurofibrillary tangles at high magnification in superior frontal cortex. (C) Pick bodies (curved arrows) in the CA1 field of the hippocampus from a Pick patient are labeled by AD2 monoclonal antibody.

spherical argyrophilic inclusions. They are approximately the size of the neuronal nucleus and are often found in the apical region of the cell. They consist of randomly distributed straight fibrils, 10–20 nm wide, as well as larger (15–30 nm wide) constricted fibrils with a periodicity between constrictions of about 160 nm (Brion and Mikol, 1973; Takauchi *et al.*, 1984). They are immunostained by antibodies against different parts of the Tau molecule as well as against phosphorylated Tau epitopes (Hof *et al.*, 1994b; Delacourte *et al.*, 1996). Pick bodies are found in layers II and VI of the cortex and are extremely numerous in dentate gyrus granule cells in the hippocampus (Hof *et al.*, 1994b) (Fig. 7C).

## B. Pathological Tau Proteins as a Biochemical Marker for Alzheimer's Disease

### 1. Tau Phosphorylation in Alzheimer's Disease

***a. Tau Proteins Are Abnormally Phosphorylated during Alzheimer's Disease*** Alzheimer's disease is a common neurodegenerative disorder progressively leading to dementia, characterized by both clinical and neuropathological features. Memory loss is the first cognitive impairment, followed by aphasia, agnosia, apraxia, and behavior disturbances. Two types of brain lesions are observed: senile plaques, which are diffusely distributed throughout the cortical and subcortical brain areas, and neurofibrillary degeneration (NFD), which corresponds to the accumulation of abnormal fibrils within neurons of the hippocampal regions and the association isocortex. Senile plaques and NFD are not, *stricto sensu,* specific for AD, but the demonstration of both throughout the cerebral cortex is necessary to establish the diagnosis of definite AD, according to NINCDS-ADRDA criteria (McKhann *et al.*, 1984). Senile plaques result from the extracellular accumulation of a peptide referred to as A$\beta$ into amyloid deposits. A$\beta$ derives from a precursor, named amyloid precursor protein (APP). In familial AD, mutations were found on the APP gene, suggesting that it has a central role in etiopathogenesis (Hardy, 1992).

NFD results from the intraneuronal accumulation of PHF (see Section IV,A,1). The major antigenic components of PHF are Tau proteins (Brion *et al.*, 1985; Delacourte and Défossez, 1986; Grundke-Iqbal *et al.*, 1986) that share many epitopes with normal Tau proteins but also show several distinct properties such as a lower mobility on sodium dodecyl sulfate–polyacrylamide gel electrophoresis (Flament *et al.*, 1989; Delacourte *et al.*, 1990). Several groups have reported excessive or aberrant phosphorylation as the major modification in these proteins (Grundke-Iqbal *et al.*, 1986; Ihara *et al.*, 1986, Flament *et al.*, 1989, Greenberg *et al.*, 1992). Their biochemical characterization by immunoblotting revealed the presence of a triplet of Tau proteins (55, 64, and 69), also referred to as A68, or PHF-Tau, in all AD brains (Delacourte et al., 1990; Greenberg et al., 1992; Goedert et al., 1994). According to Goedert et al., (1992), Tau 55 results from the hyperphosphorylation of the two Tau variants 2−,3−,10− and 2−,3−,10+; Tau 64 from the variants 2+,3−,10− and 2+,3−,10+; and Tau 69 from the variants 2+,3+,10− and 2+,3+,10+ (Figs. 2 and 8). A 74-kDa component is also present in brain extracts from AD patients with a severe dementia. It corresponds to the longest Tau isoform (Delacourte et al., 1996). The term "pathological Tau" was coined in 1989 to designate abnormal biochemical differences between PHF-Tau and normal Tau that are due to a pathological degenerating process (Flament et al., 1989).

**b. Phosphorylated Sites in PHF-Tau and Normal Tau** Biochemical differences between PHF-Tau and normal Tau have been observed in postmortem tissue, after a comparison between control and AD cases. Generally, brain tissue is obtained after more than 5 h postmortem delay. However, a study using rapidly processed biopsies from normal human tissue has demonstrated that native Tau are more phosphorylated than previously thought (Matsuo et al., 1994). Most of the phosphorylation sites that were supposed to be specific for AD are also found on these native Tau.

**c. The Concept of Pathological Tau Proteins** Many phosphorylated sites on PHF-Tau are found on fetal and native adult Tau (Matsuo et al., 1994; Mawal-Dewan et al., 1994). In particular, this has been verified for Ser 202 and Thr 205 (Matsuo et al., 1994), Ser 262 (Seubert et al., 1995), Ser 396 and 404 (Matsuo et al., 1994; Buée-Scherrer et al., 1996a), and many other Ser-Pro sites (see Sections II,C,1 and III,B and Fig. 5). Despite the fact that many phosphorylation sites are common to PHF-Tau and native Tau, there are biochemical differences that differentiate them and support the concept of pathological Tau. First, two-dimensional immunoblot analysis reveals that PHF-Tau are more acidic than normal Tau from biopsy-derived samples (Sergeant et al., 1995). Second, these differences generate conformational differences that can be visualized by a few phosphorylation-dependent monoclonal antibodies such as AT100 (Matsuo et al., 1994) or AP422 (Hasegawa et al., 1996) (Figs. 5 and 8). Third, insoluble polymers of Tau are present exclusively in AD brain extracts, which are visualized as "smears" on Western blots. Therefore, the main difference between biopsy and postmortem tissues is that PHF-Tau are aggregated (Fig. 8), whereas Tau from biopsies are not (Fig. 9, lanes 1 and 4). This explains why PHF-Tau are not dephosphorylated during postmortem delays (because of the inaccessibility of phosphorylated sites to phosphatases), whereas native Tau are rapidly dephosphorylated (Matsuo et al., 1994). Using rat brain,

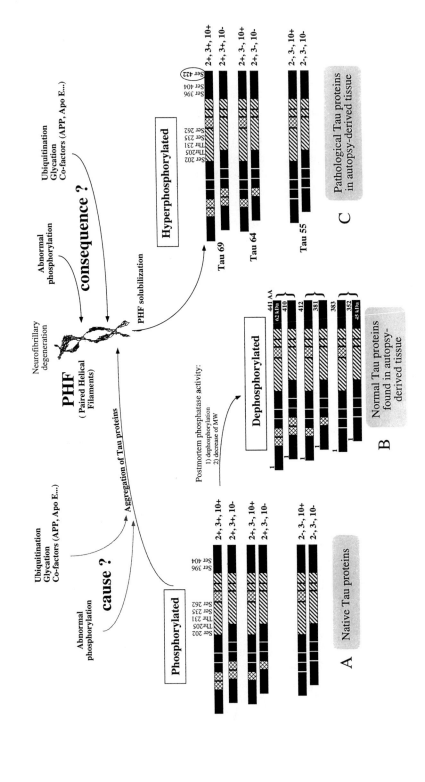

we observed that 80% of the AD2 immunoreactivity (against two phosphorylated sites at Ser 396 and 404 residues) disappeared after a postmortem delay of 2 h at room temperature (Buée-Scherrer *et al.*, 1996a).

Together, these results show that the main feature of pathological Tau is their aggregation into polymers that will constitute neurofibrillary lesions. In addition, specific phosphorylation sites are also present on PHF-Tau and may be in association with the aggregation process. Finally, the term "pathological Tau" is especially appropriate to show that NFD is found in numerous diseases, with or without PHF but with different profiles of pathological Tau proteins characteristic of different degenerating processes (see the following sections).

## 2. Tau Aggregation in Alzheimer's Disease

The most obvious pathological event of NFD is the aggregation of Tau proteins. This phenomenon could be linked, directly or indirectly, to other factors besides phosphorylation.

### a. Cofactors

*i. Apolipoprotein* Apolipoprotein E (ApoE) plays a critical role in lipid metabolism. It is a heterogeneous protein with three major isoforms in humans (E2, E3, and E4 corresponding to alleles ε2, ε3, and ε4). It is also a risk factor for AD in that the ε4 allele frequency is increased in AD patients when compared to the normal population. Following that finding, it was shown that ApoE is bound to senile plaques and NFT and could contribute to the formation of these lesions (Strittmatter *et al.*, 1994; Poirier *et al.*, 1995).

---

FIG. 8 Hypotheses on the aggregation of Tau proteins during neurodegenerative disorders. (A) Native Tau proteins are phosphorylated at numerous sites on the six Tau isoforms. During the neurodegenerative process, Tau proteins aggregate into filaments (for instance, PHF in AD) in different types of lesions (NFTs, dystrophic neurites, Pick bodies, etc.). Tau proteins modifications (glycation, phosphorylation, ubiquitination, etc.) and/or interactions (with APP, Apo E, etc.) can induce the aggregation (cause) or are generated after the aggregation (consequence). During the postmortem delay, there is no kinase activity. However, endogeneous phosphatases are still active and allow dephosphorylation of native Tau proteins into dephosphorylated Tau proteins with a lower apparent MW (B). (B) Normal Tau proteins found in autopsy-derived brain specimens result from the dephosphorylation of native Tau proteins. (C) Pathological Tau proteins are aggregated Tau proteins that are more phosphorylated than native Tau. They resist the postmortem phosphatase activity. After their solubilization (generally in SDS buffers), hyperphosphorylated Tau proteins are solubilized. Their analysis shows that they have abnormal phosphorylation sites including Ser422 (circle), AT100 epitope, leading to a character more acidic than native Tau proteins.

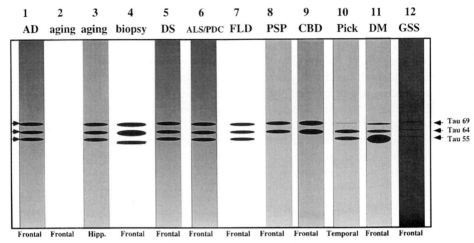

FIG. 9 Synopsis of the electrophoretic profiles of pathological Tau proteins. A phosphorylation-dependent monoclonal antibody referred to as AD2 and raised against phosphorylated Ser 396 and 404 (according to the longest human Tau isoform) was used to detect pathological Tau proteins in homogenates from controls and patients with different neurodegenerative disorders. A triplet of pathological Tau proteins (Tau 55, 64, and 69) is detected in almost all cortical areas from Alzheimer patients (lane 1) and in the hippocampus of nondemented elderly individuals (lane 3), but not in the neocortical areas (lane 2). In autopsy-derived samples from normal human brain, no labeling is found (lane 2). However, in biopsy-derived samples from normal human brain, a Tau triplet is immunodetected (lane 4). This Tau triplet immunoreactivity decreases with the time of postsurgical delay. After 4 h, AD2 immunoreactivity is completely abolished and similar to that found in lane 2. In Alzheimer's disease (lane 1) and all other conditions (lanes 5–12), AD2 immunoreactivity does not change with postmortem delay. A similar Tau triplet is found in cortical and subcortical structures of Down's syndrome (lane 5) and Guamanian ALS/PDC (lane 6). "Smears" corresponding to insoluble aggregated Tau proteins are also present. Intensity of smears is indicated by different shades of grays. The Tau triplet is exclusively found in frontotemporal regions of patients with frontal lobe degeneration (non-Alzheimer and non-Pick), with a characteristic absence of smears (lane 7). In "subcortical dementias," including PSP (lane 8) and CBD (lane 9), a Tau doublet (Tau 64 and 69) is detected. Pick disease is characterized by another Tau doublet (Tau 55 and 64) (lane 10) and in GSS, pathological Tau proteins that are likely to be a Tau triplet are hardly soluble and the intensity of smears AD2 immunoreactive is the highest one (lane 12). In myotonic dystrophy (DM), some NFTs are sometimes found in temporal brain areas. In these particular cases, a Tau triplet is also found with the predominance of Tau 55 compared to Tau 64 and 69 (lane 11). Pathological Tau proteins among these different conditions are correlated to the number of NFTs, dystrophic neurites, or Pick bodies. Their amounts are different according to the pathology, the area studied, and the severity of the disease. Thus, comparing one disorder to another, the relative AD2 immunoreactivity is not strictly respected.

It has been suggested that the deleterious effects of the ε4 allele of ApoE are related to its inability to interact with the MT-associated protein Tau and thereby to prevent its hyperphosphorylation. Synthetic peptides representing each of the four MT-binding domains of Tau bind more avidly to ApoE3 than to ApoE4. Phosphorylation of Ser 262 in domain I of Tau decreases Tau binding to MTs and also abolishes binding by ApoE3 (Huang *et al.*, 1995).

It was also suggested that ApoE plays a secondary role in NFT formation or accumulates within the neurons in response to reparative processes induced by NFT-associated neuronal damage (Benzing and Mufson, 1995).

*ii. APP and Amyloid* Extracellular deposition of amyloid fibrils and intraneuronal accumulation of PHFs are the neuropathological hallmarks of AD. However, there is no direct evidence of interconnection between these two pathological states. In the sequence of events leading to neurodegeneration during AD, APP dysfunction is likely to occur first, as demonstrated by genetic studies (Hardy, 1992). Following this demonstration, it would be interesting to determine whether A$\beta$ is able to trigger the formation of patholoical Tau proteins. In that perspective, Busciglio *et al.*, (1995) demonstrated that A$\beta$ fibrils induce Tau phosphorylation and loss of MT binding on human cortical neurons. In that model, all the basic components of AD seem to be present and to mimic perfectly well the etiopathogenesis of AD. However, it would be interesting to know if this model can produce pathological Tau epitopes specifically labeled by the mAb against Ser 422-P (Hasegawa *et al.*, 1996) or by AT10 (Matsuo *et al.*, 1994).

Conversely, other data suggest that hyperphosphorylated Tau proteins accumulating in dystrophic neurites after blockage of the fast axonal transport may interact with APP and act as a cofactor in A$\beta$ formation and aggregation into senile plaques (Praprotnik *et al.*, 1966). In fact, Tau directly interacts with a conformation-dependent domain of the APP encompassing residues APP714–723 (i.e., the transmembrane region of APP distal to A$\beta$). Antibodies to an epitope located between residues 713 and 723 of APP770 consistently labeled PHFs in the brain of Alzheimer patients. Furthermore, this APP peptide spontaneously formed fibrils *in vitro* and, in the presence of Tau, generated dense fibrillary assemblies containing both molecules. These data suggest that NFT and senile plaques could be synergistically interrelated and may play a role in each other's formation during the pathogenesis of AD (Smith *et al.*, 1995a; Giaccone *et al.*, 1996).

To determine whether amyloid fibril formation affects the phosphorylation state of Tau, primary cultures of fetal rat hippocampal and human cortical neurons were treated with A$\beta$ in a soluble, amorphous-aggregated, or fibrillar form. Fibrillar A$\beta$, but not soluble or amorphous-aggregated A$\beta$, markedly induces the phosphorylation of Tau at Ser 202 and Ser 396/Ser 404, resulting in a shift in the Tau MW in human cortical neurons. Hyperphosphorylated Tau accumulates in the somatodendritic compartment of fibrillar A$\beta$-treated neurons in a soluble form that is not associated with MTs and is incapable of binding to MTs *in vitro*. Dephosphorylation of Tau restores its capacity to bind to MTs. Thus, amyloid fibril formation alters the phosphorylation state of Tau, resulting in the loss of MT binding capacity and somatodendritic accumulation, properties also exhibited by Tau in the AD brain. Amyloid fibril formation may therefore be a cause of abnormal Tau phosphorylation in AD (Busciglio *et al.*, 1995). This Tau hyperphosphorylation may be mediated through the activation of GSK3-$\beta$. In fact, other results suggest that exposure of rat hippocampal neurons to A$\beta$ (25–35) leads to the activation of GSK3-$\beta$, the phosphorylation of Tau, and neuronal death (Takashima, 1996).

***b. Ubiquitination*** Ubiquitin is a stress protein implicated in the degradation of short-lived and abnormal proteins. It is a component of PHF in AD (Mori *et al.*, 1987). Tau proteins analyzed on immunoblots consist of the characteristic PHF-Tau triplet, proteolytic products, and smears that correspond to Tau polymers. These smears consist largely of the carboxyl-terminal portion of Tau and ubiquitin. The conjugation sites are localized to the MT-binding region. It is most likely that abnormally phosphorylated full-length Tau (PHF-Tau) accumulates as PHF, which is then gradually processed from its amino terminus and followed by ubiquitination (Morishima-Kawashima *et al.*, 1993).

***c. Glycation*** In AD, antibodies against glycation products label NFT (Ledesma *et al.*, 1994; Smith *et al.*, 1995b). Glycation is the reaction between the $NH_2$ of a side chain of an amino acid and the CHO of carbohydrate. The most suitable residues for glycation, lysines, present at the tubulin-binding motif of Tau protein, seem to be preferentially modified compared to lysines in other regions. Among the modified lysines, those located in the sequence comprising residues 318–336 (in the largest human Tau isoform) are found to be glycated, as determined by the reaction with an antibody that recognizes a glycated peptide containing this sequence. Because those lysines are present in a tubulin-binding motif of Tau protein, its modification could result in a decreased interaction of Tau with tubulin (Ledesma *et al.*, 1995). Furthermore, advanced glycation end products of Tau generate oxygen free radicals that could activate transcription via nuclear factor-$\kappa$ B, increase amyloid $\beta$ protein precursor, and release 4-kDa amyloid $\beta$ peptides. Therefore, glycated Tau could induce an oxidative stress that may contribute to the pathogenesis of AD (Yan *et al.*, 1995).

***d. Oxidation*** *In vitro* assembly of recombinant Tau proteins and constructs derived from it into PHFs depends on intermolecular disulfide bridges formed by the single Cys 322. Blocking the SH group, mutating Cys for Ala, or keeping Tau in a reducing environment all inhibited assembly. The data imply that the redox potential in the neuron is crucial for PHF assembly, independently or in addition to pathological phosphorylation reactions (Schweers et al., 1995).

It is clear that all these modifications may change Tau properties regarding their binding to MTs. However, it is unclear if these modifications follow or precede Tau aggregation into PHF (Fig. 9).

## C. Pathological Tau Proteins in Several Subtypes of Neurofibrillary Degeneration

Aggregated Tau proteins accumulate in several neurodegenerative disorders. Their biochemical signature can be different according to the degenerating process (Delacourte, 1994).

### 1. Disorders with PHF-Tau

***a. Down's Syndrome*** Due to the trisomy of chromosome 21, Down's syndrome (DS) patients have numerous dysfunctions that occur during development and lead to a cognitive impairment and in many cases to dementia. A$\beta$ from the APP gene on chromosome 21 is expressed at high levels from 21 gestational weeks to 61 years but it is undetectable in age-matched controls (Teller et al., 1996). Amyloid deposition is observed in DS after 15 years, then senile plaques are present, followed by neurofibrillary degeneration after 35 years with Tau accumulation (Mann et al., 1989; Hof et al., 1995).

Using an immunoblot approach, it has been demonstrated that an identical triplet of pathological Tau proteins is found in the isocortex of Down's syndrome patients over 35 years old (Fig. 9, lane 5) (Flament et al., 1990). This shows that biochemical dysfunctions linked to Tau pathology during Down's syndrome are very similar to those found in AD. These findings confirm that Down's syndrome can act as a model for the study of the molecular pathological events that occur during AD.

***b. Parkinsonism with Dementia*** The dementia in Parkinson's disease (PD) is generally considered to be of the subcortical type, as in other disorders with predominantly subcortical pathologies such as PSP or Huntingtons disease. PD is usually sporadic. Cognitive changes observed in PD evoke a frontal lobe dysfunction (Pillon et al., 1995) that may be a primary event in the disease. Approximately 15–20% of patients with PD develop dementia. PD dementia is different from that of AD, with a stronger involvement of frontal pathology. However, the nature and the neuropathological basis of the dementia remains controversial. This dementia may be related either to cell loss in subcortical nuclei (Albert, 1978) or to AD pathology because the incidence of neuropathologically proven AD in patients with PD appears to be higher than that in the general population (Hakim and Mathieson, 1979; Rewcastle, 1991).

Is there a Tau pathology in these neurological disorders? Vermersch *et al.*, (1993) analyzed samples from 24 nondemented Parkinson patients and Parkinson patients with various degrees of dementia. Patients with Lewy bodies in the isocortex were excluded from the study. In cases without dementia, Tau pathology was restricted to the hippocampal formation as described in nondemented elderly individuals (Vermersch *et al.*, 1992). The pathological Tau triplet referred to as Tau 55, 64, and 69 was found in all demented cases. It was in large amounts especially in the prefrontal area, temporal cortex, and entorhinal cortex of Parkinson patients with dementia. Therefore, Alzheimer-type proteins are found in PD with dementia, but the cortical distribution is different from the pattern seen in AD, with a significantly stronger involvement of prefrontal areas. The Tau pathology observed in the isocortex of PD patients may significantly contribute to the occurrence of cognitive changes (Vermersch *et al.*, 1993). Prevalence of PD with dementia is identical to AD and the major risk factor is age, as in AD. Therefore, pathological Tau proteins found in brains from PD patients may result from the development of the Alzheimer pathology in addition to Parkinson's disease. However, the degenerating process develops more extensively in the prefrontal regions that are already weakened by the Parkinson pathology.

*c. Guamanian ALS/PDC* Amyotrophic lateral sclerosis/parkinsonism dementia complex of Guam is a chronic neurodegenerative disorder highly prevalent in the native Chamorro population of Guam island in the Western Pacific (Hirano *et al.*, 1961; Garruto, 1991). The etiopathogenesis of this disorder is not yet elucidated, although it has been hypothesized that environmental factors such as aluminum or neurotoxins might be involved. Clinically, Guamanian ALS is indistinguishable from sporadic ALS and presents with fasciculations and lower and upper motor neuron signs. Parkinsonism dementia is characterized by an insidious progressive mental decline and extrapyramidal signs (Hirano *et al.*, 1961). Both aspects of the disease are frequently associated, but they are known to occur separately.

Neuropathologically, Guamanian ALS/PDC shows a severe cortical atrophy and neuronal loss. The neuropathological hallmark of ALS/PDC is the widespread NFT formation, especially in the isocortex and hippocampal formation (Hirano *et al.*, 1961, 1968). Although NFTs are present in large numbers in both AD and ALS/PDC, these two conditions are distinguished by differential NFT laminar distribution patterns and densities in isocortex. NFTs are preferentially distributed within layers II and III in the isocortical areas of Guamanian ALS/PDC cases and are relatively sparser in layers V and VI, whereas NFTs are denser in layers V and VI than in layers II and III in AD cases (Hof *et al.*, 1991, 1994b; Oyanagi *et al.*, 1994).

Immunohistochemical studies have also revealed that several cytoskeletal proteins (neurofilaments, MAP2, Tau, and ubiquitin) are present in NFTs of ALS/PDC patients (Shankar *et al.*, 1989). The ultrastructure of NFTs in Guamanian ALS/PDC consists of straight filaments and PHFs (Rewcastle, 1991), and PHFs have been shown to be essentially similar to those observed in AD (Hirano *et al.*, 1968).

In contrast to AD patients in which the Tau triplet is found mostly in cortical regions, a similar triplet was strongly detected in both cortical and subcortical areas in Guamanian patients (Fig. 9, lane 6). The Tau profile differed quantitatively from case to case, demonstrating that the AD-related Tau triplet had a heterogeneous regional distribution. These data suggest that the Tau triplet found in ALS/PDC is similar to that observed in AD, although the regional distribution of Tau proteins differs in these disorders (Buée-Scherrer *et al.*, 1995). Because the etiology of Gumanian ALS/PDC is different from that of AD, this also demonstrates that neurofibrillary degeneration with a Tau triplet 55, 64, and 69 is very likely a similar response to different types of neuronal insults.

*d. Niemann–Pick Disease Type C* Niemann–Pick type C disease (NPC) is a cholesterol storage disease with defects in the intracellular trafficking of exogenous cholesterol derived from low-density lipoproteins. NPC includes juvenile dystonic lipidosis, ophthalmoplegic lipidosis, neurovisceral storage disease with vertical supranuclear ophthalmoplegia, and juvenile Niemann–Pick disease. NPC onset may be in infancy, early childhood, adolescence, or, occasionally, in adulthood. Common neurological manifestations are clumsiness, ataxia, supranuclear gaze paresis, seizures, and psychomotor retardation (Turpin *et al.*, 1991). Neuropathologically, brains from NPC patients show neuronal distension in cortex and swollen axons in brain stem. In NPC cases with a chronic progressive course, NFTs with PHF are present (Love *et al.*, 1995). PHF-Tau in tangle-rich NPC brains is indistinguishable from PHF-Tau in AD brains (Auer *et al.*, 1995).

*e. Normal Aging* Numerous reports have shown that senile plaques and NFT can be found in various but generally low amounts in the central nervous tissue from normal individuals (Bouras *et al.*, 1994; Giannakopoulos *et al.*, 1995a). The presence of these lesions and amyloid deposits raise the problem of "normal" and "pathological aging" (Giannakopoulos *et al.*, 1995b). Nondemented individuals demonstrate a consistent pattern of progressive changes with increasing age characterized by the presence of NFT in layer II of the entorhinal cortex and in the CA1 field of the hippocampus (Braak and Braak, 1995; Bouras *et al.*, 1994; Giannakopoulos *et al.*, 1995a,b).

Using the immunoblot approach, PHF-Tau have been quantified in the different cortical areas of nondemented cases. PHF-Tau were consistently and exclusively found in the entorhinal cortex and the hippocampus of nondemented individuals over 75 years of age (Vermersch *et al.*, 1992; 1995a) (Fig. 9, lanes 2 and 3). PHF-Tau were similar to those found in Alzheimer brains, with the characteristic Tau triplet. However, lower amounts of PHF-Tau are found in aging. On the other hand, normal Tau proteins decrease in aging ($r = 0.32; p < 0.001$), with an average loss of 14% of soluble Tau per decade after the age of 20 years (Mukaeto-Valadinska *et al.*, 1996)

These results demonstrate that the entorhinal cortex is the most vulnerable neuronal region during aging and that the biochemical dysfunctions observed in this area are similar to Alzheimer pathology. Therefore, neuropathological and biochemical data converge to show that the entorhinal and hippocampal regions are affected during normal aging and AD. Accordingly, the presence of neurofibrillary lesions and PHF-Tau in this particular cortical region cannot really be used for the diagnosis of AD but can be useful for understanding the degenerating process (Braak and Braak, 1995). Because NFD is a general response to different types of insults (see previous and following sections), we cannot conclude that the affected hippocampal area during normal aging is the first step of AD. Brains of normal elderly individuals with tangles and PHF-Tau in the hippocampal region do not always exhibit amyloid deposits (A. Delacourte and L. Buée, manuscript in preparation). This suggests that amyloid deposition and NFD observed in aging are different processes occurring independently from each other. Thus, the degenerating process in the hippocampal area of aged controls could be due to a selective vulnerability of a subset of neurons and this "normal" condition is likely different from preclinical AD. Such data do not support the notion of normal and pathological aging.

*f. Others* Postencephalitic parkinsonism (PEP) cases display a Tau triplet in cortical and subcortical brain regions, in contrast to AD patients in which a similar triplet is predominantly found in the hippocampal formation and association neocortex. The regional distribution of the Tau triplet differs among PEP cases, suggesting some heterogeneity in the neurodegenerative process (Buée-Scherrer *et al.*, 1994). Because it is difficult to differentiate PSP from PEP from a neuropathological aspect (Hauw *et al.*, 1994; see the following section), the current biochemical approach is a useful tool to ascertain the diagnosis and may be complementary to the neuropathological analysis.

It is interesting to note that in ALS/PDC of Guam, the cases also exhibit the same electrophoretic Tau profile (Buée-Scherrer *et al.,* 1995). Thus, environmental toxins (ALS/PDC of Guam) and viruses (PEP; exposure to the pandemic of influenza in the period 1915–1930) may lead to a similar Tau neuropathology.

Finally, other disorders including familial presenile dementia with NFT are also characterized by the presence of a Tau triplet (Spillantini *et al.,* 1996).

### 2. Disorders with the Pathological Tau Triplet

***a. Frontal Lobe Degeneration (Non-Alzheimer, Non-Pick)*** Frontal lobe degeneration (FLD) is a common neurological disorder that has only recently been characterized, despite that fact that it is the most common dementing disorder of the presenium in Europe, after AD (Gustafson, 1993). Both diseases have a similar "frontal" pathology. However, whereas Pick's disease is easy to diagnose using the characteristic Pick bodies (see Section IV,A,3) immunostained with antibodies against phosphorylated Tau proteins such as AD2 and the doublet of pathological Tau proteins Tau 55 and 64 (Delacourte *et al.,* 1996), FLD has no specific neuropathological hallmarks. Morphological changes comprise a neuronal cell loss, spongiosis, and gliosis mainly in the superficial cortical laminae of the frontal and temporal cortex (Brun *et al.,* 1994). Interestingly, despite the fact that there is no NFD, and therefore an absence of NFT and dystrophic neurites with PHF, pathological Tau proteins are observed exclusively in the fronto-temporal region (Fig. 9, lane 7). The hippocampal region is not affected (Vermersch *et al.,* 1995b). The profile is a triplet, as in AD, but smears of Tau aggregates, always observed by immunoblotting in AD, are not present.

Thus, in FLD there is also a Tau pathology. Pathological Tau proteins are likely to form small aggregates that render them inaccessible to phosphatase activity generated during postmortem delays. However, these pathological Tau aggregates are different from those of AD, Pick's disease, and PSP because they are not detected at the optical level. Therefore, FLD is characterized by the presence of a specific Tau profile (Vermersch *et al.,* 1995b).

### 3. Tau 64 and 69

***a. Steele–Richardson–Olszewski Syndrome*** PSP, also named Steele–Richardson–Olszewski disease, is a neurodegenerative disorder characterized clinically by supranuclear ophthalmoplegia, pseudobulbar palsy, parkinsonism, and axial dystonia. Dementia is also a common feature at the end-stage of the disease. Neuropathologically, PSP is characterized by neuronal loss, gliosis, and NFTs. NFTs were first described in basal ganglia, brain stem, and cerebellum (Steele *et al.,* 1964). The subcortical localization of the neuropathological lesions has led to the definition of PSP as a model of "subcortical dementias" (Albert, 1978; Cummings and Benson, 1984). Recently, a cortical involvement of the degenerating process has been described, with the same features as subcortical NFTs (Hof *et al.,* 1992a; Hauw *et al.,* 1990). These studies demonstrated that, compared to AD, the primary motor cortex is more severely affected than isocortical association areas (Hauw *et al.,* 1990; Hof *et al.,* 1992a). Preliminary neuropathologic criteria for the diagnosis of PSP have been defined (Hauw *et al.,* 1994). Immunohistochemical studies have revealed that NFTs are denser in supragranular layers (III) than in infragranular layers (V) in PSP, whereas in AD, NFTs are denser in layer V than in layer III. Neurofibrillary tangles are found in many subcortical nuclei as well as in the isocortex, especially in the primary motor region (Brodmann area 4) (Hauw *et al.,* 1990); Hof *et al.,* 1992a). Ultrastructural analyses further support the difference between AD and PSP because PHFs are found in AD (Kidd, 1963), whereas straight filaments are observed in PSP (Tellez-Nagel and Wisniewski, 1973; Tomonaga, 1977).

Pathological Tau proteins in PSP are distributed in subcortical and cortical areas. The electrophoretic profile of pathological Tau proteins is significantly different from that of the AD because a characteristic doublet was found (Tau 64 and Tau 69) instead of the classical triplet found in AD (Flament *et al.,* 1991; Vermersch *et al.,* 1994) (Fig. 9, lane 8). Moreover, two-dimensional analysis reveals that the isoelectric properties of the doublet in PSP are different from those obtained in AD. Although Tau 64 is the most acidic component in AD, Tau 69 is the most acidic one in PSP (Flament *et al.,* 1991). However, most of the phosphorylation sites found in PHF-Tau are also encountered in pathological Tau proteins (Tau 64 and 69) from PSP patients (Schmidt *et al.,* 1996a). Frontal regions were the most affected, especially Brodmann area 4 (Vermersch *et al.,* 1994). The presence of Tau 64 and 69 in association areas suggests that the cortical pathology is likely to play a significant role in the cognitive changes observed in PSP.

***b. Corticobasal Degeneration*** Corticobasal degeneration is a slowly progressive neurodegenerative disorder characterized by an asymmetrical akinetic-rigid syndrome associated with cognitive (apraxia and aphasia) and extrapyramidal motor dysfunction (rigidity and dystonia). Moderate dementia sometimes emerges late in the course of the disease (Rinne *et al.,* 1994). Neuropathological examination shows gliosis, neuronal loss, the presence of achromatic ballooned neurons, neuritic changes, and NFTs (Paulus and Selim, 1990; Ksiezak-Reding *et al.,* 1994; Feany *et al.,* 1995; Buée-Scherrer *et al.,* 1996b).

There is an overlap between PSP and CBD (Feany *et al.*, 1996), and it would be most helpful to distinguish these two pathologies on a neuropathological or immunochemical basis. The profile of neurofibrillary tangles is similar to PSP, as well as the Tau electrophoretic profile (Ksiezak-Reding *et al.*, 1994; Feany and Dickson, 1995; Buée-Scherrer *et al.*, 1996b) (Fig. 9, lanes 8 and 9). The Tau 64 and 69 doublet immunoreactivity could be different in CBD and PSP. Indeed, the Tau doublet in CBD is not detected using antibodies against the region encoded by exon 3 (Ksiezak-Reding *et al.*, 1994). These data were confirmed by immunohistochemistry (Feany *et al.*, 1995b).

### 4. Tau 55 and 64

***a. Pick's Disease*** Pick's disease is a rare type of presenile dementia with a mainly "frontotemporal" pathology. It is characterized by a pan-laminar frontotemporal cortical atrophy, widespread degeneration of the white matter, chromatolytic neurons, and Pick bodies. The Pick body (see Section IV,A,3) is well immunostained with anti-Tau and anti-PHF antibodies (Hof *et al.*, 1994b). Its detection is at least twice as more sensitive than conventional silver staining using the AD2 mAb (Delacourte *et al.*, 1996).

Tau proteins were analyzed by one- and two-dimensional gel electrophoreses using a quantitative Western blot approach. In all specimens studied, a major 55- and 64-kDa Tau doublet was observed in the isocortex, in the limbic areas, and in subcortical nuclei. A very faint band is also observed at 69 kDa (Fig. 9, and lane 10). In the isocortex, all Brodmann areas of the frontal and temporal cortices were affected. Parietal cortex was frequently involved, whereas the occipital cortex was frequently spared. In the subcortical regions, the doublet of pathological Tau was found in the striatum, substantia nigra, locus coeruleus, and stem. The 55- and 64-kDa doublet is characteristic of Pick disease because it is different from the Alzheimer profile or the CBD profile (Buée-Scherrer *et al.*, 1996b; Delacourte *et al.*, 1996). Isoelectric points of Tau are less acidic than in AD (6.4–6.85 vs 5.9–6.7). The characteristic pattern of pathological Tau in Pick disease was well correlated with the presence of PB and a frontal-type syndrome (Delacourte *et al.*, 1996).

### 5. Tau 55

***a. Myotonic Dystrophy*** Myotonic dystrophy (DM), one of the most prevalent autosomal dominant defects, is a slowly progressive multisystemic disorder with myotonia, muscular atrophy, cataract, and endocrine dysfunction being the most prominent clinical features (Harper, 1989). Affected individuals present with a highly variable phenotype, ranging from an asymptomatic condition to a severe congenital form. The molecular basis of DM is an unstable CTG trinucleotide repeat in the 3' untranslated region of a gene encoding a putative serine/threonine protein kinase (Buxton *et al.*, 1992). Impairment of intellectual and cognitive function in DM is found in psychometric studies (Jaspert *et al.*, 1995). Typically, neuropathological observations of the brain in DM show reduced brain weight, minor abnormalities in gyral architecture and, microscopically, a disordered cortical cellular arrangement with neurons also present in subcortical white matter, and intracytoplasmic inclusion bodies in the thalamus, the striatum, the cerebral cortex, and the substantia nigra (Ono *et al.*, 1987). A few studies also reported the appearance of abnormally frequent NFTs in the temporal lobe, especially in the hippocampal region from patients with DM (Kiuchi *et al.*, 1991).

In DM patients, PHF-Tau are also immunodetected in the hippocampus, the entorhinal cortex, and in most of the temporal areas. Amounts of pathological Tau proteins are higher in the more severely affected cases but lower than those in AD brain homogenates. Their profile is different from the characteristic triplet of AD, with low amounts of the Tau 69 variant but higher amounts of the Tau 55 variant (Fig. 9, lane 11). PHF-Tau are less acidic in DM than in AD (Vermersch *et al.*, 1996). The interesting observation with this pathology is the possible link between the genetic dysfunction of the Ser/Thr protein kinase and the presence of pathological Tau proteins. These observations may provide valuable information about the cascade of events observed during AD. They demonstrate that changes in kinase expression can provoke numerous pathological events, including the formation of neurofibrillary tangles in specific brain areas (Vermersch *et al.*, 1996).

### 6. Smears

*a. Gerstmann–Straussler–Scheinker Disease* Exceptionally, NFTs with Tau proteins similar to Tau-PHF are also present during Gerstmann–Straussler–Scheinker disease (GSS), a neurodegenerative disorder that is an autosomal dominant disorder with a wide spectrum of clinical presentations including ataxia, spastic paraparesis, extrapyramidal signs, and dementia (Gerstmann *et al.*, 1936). Neuropathologically, prion plaques are observed in many brain areas, associated with a severe neuronal loss and spongiosis. A stop mutation at codon 145 of PRNP, the gene encoding the prion protein (PrP), is responsible for a PrP-cerebral amyloid angiopathy phenotype. In this disorder, NFTs are often associated with the amyloid angiopathy. Other genetic PrP-related disorders, with mutations at codons 198 and 217, also have NFTs (Ghetti *et al.*, 1996).

It has been reported that NFTs from GSS are composed of PHF-Tau similar to those of AD (Ghetti et al., 1995). One patient of a large French GSS family with the mutation in codon 117 developed NFD, especially around prion plaques, whereas other members did not. The Tau profile from the patient with GSS and NFTs displayed some common features with AD but also many differences. The three Tau-PHF (Tau 55, 64, and 69) were at extremely low levels, as well as catabolic products. Most of the Tau proteins were highly aggregated and immunodetected as "smears" intensely stained from the top to the bottom of the Western blot lane (see Fig. 9, lane 12; Section IV,B,2,b). Therefore, the GSS Tau profile has its own specificity characterized by an hyperstate of Tau aggregation. The Tau profile of the other members of the GSS family, who were without NFTs, is normal. The reason why some patients with GSS develop NFTs is still unknown (Tranchant et al., 1996).

## 7. Neuronal Injuries and the Phosphorylation State of Tau Proteins

Ischemia disrupts the neuronal cytoskeleton both by promoting proteolysis of its components and by affecting kinase and phosphatase activities that alter its assembly (Dewar et al., 1994; Buée et al., 1996). In a reversible model of spinal cord ischemia in rabbits, Tau is found to be dephosphorylated in response to ischemia with a time course that closely matches the production of permanent paraplegia. In a similar manner, $Ca^{2+}$/calmodulin-dependent kinase II activity is reduced only in the ischemic region. Thus, dephosphorylation of Tau is an early marker of ischemia as is the rapid loss of $Ca^{2+}$/calmodulin-dependent kinase II activity. Alterations in phosphorylation or degradation of Tau may affect MT stability, possibly contributing to disruption of axonal transport but also facilitating neurite plasticity in a regenerative response (Dewar and Dawson, 1995; Shackelford and Nelson, 1996; Buée et al., 1996).

## 8. Neurodegenerative Disorders without Pathological Tau Proteins

PHF-Tau are not systematically observed in neurodegenerative disorders. Using AD2 and other sensitive anti-PHF, we were unable to detect these proteins in prior diseases, except those rare cases reported previously. Other pathologies negative to PHF-Tau are Huntington's disease, ALS, PD, and primary gliosis.

In conclusion, all these studies demonstrate that the term "pathological Tau proteins" is relevant to denominate the Tau pathology observed in many different neurodegenerative disorders because the term PHF-Tau is restricted to NFTs with PHFs, observed in several neurodegenerative disorders but absent in PSP (straight filaments) or Pick disease (random filaments), and because of their close relationship with the pathology. Indeed, the most important point is that their presence in the association cortical areas is always strongly correlated with cognitive impairment and dementia, whatever the disease.

Together, these data show that Tau aggregation and hyperphosphorylation in all these disorders are likely to lead to MT assembly dysfunctions.

## V. Lessons Given by Pathological Tau Proteins

### A. Pathological Tau and Microtubule Assembly

Most of the data on MT assembly and pathological Tau has been obtained using PHF-Tau from AD patients. In particular, all studies converge to say that PHF-Tau proteins do not bind to MTs (Nieto *et al.*, 1991; Alonso *et al.*, 1994). After dephosphorylation, they recover a part of their ability to promote MT assembly. Phosphorylated Ser 396, Ser 404, and Ser 422 are much more resistant to alkaline phosphatase than are phosphorylated Ser 202 and Thr 205 (Hasegawa *et al.*, 1996). However, because most of the phosphorylation sites of PHF-Tau are also found in native Tau, MT assembly competence of human Tau proteins has to be reinvestigated using hyperphosphorylated PHF-Tau, dephosphorylated and phosphorylated Tau proteins derived from human brain biopsies in the absence of protein kinase and phosphatase activity. Currently, only one report is available (Garver *et al.*, 1996). Dephosphorylated Tau proteins are the most MT-assembly competent. PHF-Tau are virtually unable to direct MT assembly. Unmodified Tau proteins derived from human brain biopsies are intermediately MT-assembly competent (Garver *et al.*, 1996).

Because the abnormal phosphorylation of PHF-Tau has been clearly demonstrated (Matsuo *et al.*, 1994; Sergeant *et al.*, 1995; Hasegawa *et al.*, 1996), it is possible that kinases and phosphatases play a role in the etiopathogenic process, mainly as depolymerizing MT agents (Gustke *et al.*, 1992; Biernat *et al.*, 1993; Bramblett *et al.*, 1993; Merrick *et al.*, 1996). Kinases involved in the formation of PHF-Tau are likely those that induce more efficiently an aberrant phosphorylation and generate "PHF" epitopes. To induce a Tau hyperphosphorylation, phosphatase inactivation by okadaic acid is often used. Following this treatment, human neuroblastoma cell lines LAN5 (Vandenmeeren *et al.*, 1993) or SKNSH SY5Y (Sautiére *et al.*, 1994 ; Dupont-Wallois *et al.*, 1995) degenerate and express hyperphosphorylated Tau, with a simultaneous increase of Tau antigenicity, an increase of the apparent MW of Tau proteins, an acidification of the isoelectric point of Tau, the formation of PHF-Tau epitopes followed by an increase of immunoreactivity with AT8, PHF-1, and AD2, a simultaneous decrease of Tau-1 immunoreactivity, a collapse of the cytoskeleton, neurodegeneration, and cell death (Sautiere et al., 1994). In the SY5Y model, Ser 422 phosphorylation is induced (Caillet-Boudin and Delacourte, 1996). In that respect, mitogen-activated protein kinases were able to phosphorylate Ser 422 on recombinant Tau, whereas GSK3 failed (Hasegawa et al., 1996). These changes leading to a collapse of the cytoskeleton and cell death may be explained by the reduced capacity of hyperphosphorylated Tau formed after okadaic acid treatment to associate with MTs and support the hypothesis that Tau hyperphosphorylation may underlie MT breakdown in AD (Shea and Fischer, 1996).

This is also suggested in myotonic dystrophy, in which there is a genetic dysfunction of a putative kinase that may be responsible for many physiological disorders, including cognitive impairment and the presence of NFTs in a few brain areas (see Section IV,C,5) (Vermersch et al., 1996). In AD, modifications in the distribution or activities of kinases and phosphatases have already been reported. According to Arendt et al., (1995), the levels of immunochemically detected mitogen-activated protein kinase kinase and mitogen-activated protein kinase are both increased in AD by between 35 and 40% compared with age-matched controls. Elevation of mitogen-activated protein kinase kinase is most pronounced during early stages of AD and is inversely related to the tissue content of abnormally phosphorylated PHF-Tau. Also, in AD, a mitotic kinase activity is increased, as revealed with MPM-2, a marker antibody for mitotic phosphoepitopes. MPM-2 reacts strongly with neurofibrillary tangles, neuritic processes, and neurons in AD but has no staining in normal human brain. The accumulation of phosphoepitopes in AD may result from activation of mitotic post-translational mechanisms that do not normally operate in mature neurons of brain (Vincent et al., 1996). Finally, mitogen-activated protein kinase immunoreactivity is present in the same neurons as NFTs and in the same subcellular compartments as Tau, supporting a role for mitogen-activated protein kinases in Tau phosphorylation in AD (Hyman et al., 1994).

Abnormalities of phosphatase activities in the Alzheimer brain have also been reported. According to Gong et al. (1995), phosphatase activity is decreased approximately to 30% in brain of Alzheimer's disease patients compared with those of age-matched controls. Dysregulations in the balance of kinase/phosphatase may lead to axonal transport blockage by disruption of MTs (Iqbal and Grudke-Iqbal, 1995). Accumulation of perikaryal vesicles within dystrophic neurites also supports the view that fast axonal transport is blocked (Praprotnik et al., 1996). However, there is no evidence about sequence of events leading to neurodegeneration. It may be either a blockage of the axonal transport that leads to Tau aggregation or Tau hyperphosphorylation may induce MT disassembly.

## B. Pathological Tau Variants in Different Subtypes of Neurons

### 1. Distribution of Tau Isoforms in Each Population of Neurons

Among the different pathological disorders analyzed in this chapter, a large number of Tau electrophoretic profiles were found. In Alzheimer's disease, the Tau triplet is made of three Tau variants referred to as Tau 55, 64, and 69. According to Goedert *et al.* (1996), Tau 55 derives from the two shortest Tau isoforms, Tau 64 derives from the two isoforms with only one amino-terminus insert (exon 2), and Tau 69 derives from the two largest Tau isoforms. In PSP or CBD, a Tau doublet is present and a different one is present in Pick's disease. Each Tau profile is likely to reflect the Tau content of the affected neuronl population, which is specific to a disease. In fact, the degenerating neurons in AD are essentially the large pyramidal cells of layers III and V of the association cortex. In PSP, the smaller neurons of layer III are mainly involved, whereas Pick bodies of Pick's disease are found in layers II and VI of the isocortex. In Pick's disease, granular cells of the dentate gyrus contain Pick bodies. These neurons do not contain Tau proteins with four repeats (Goedert *et al.*, 1989a). These observations may explain why only Tau 55 and 64 are found in the hippocampus showing Pick bodies without NFTs (Buée-Scherrer *et al.*, 1996b; Delacourte *et al.*, 1996). In addition, we note that Tau isoforms from the CNS and PNS are different (Georgieff *et al.*, 1993) as well as is the Tau profile between the cerebellum and the isocortex (Parent *et al.*, 1988). Accordingly, it is possible that neuronal populations can be distinguished by their content of Tau isoforms.

Transgenic mice with the longest human Tau isoform were also studied. In brains of such animals, Tau immunoreactivity is found in both axonal and somatodendritic domains. These data suggest that the overproduction of Tau could override the mechanisms of Tau sorting and results in the accumulation of Tau proteins not only in axons but also in somatodendritic domain. These changes are similar to those encountered in the formation of NFTs. However, these mice do not develop NFTs (Gotz *et al.*, 1995).

In conclusion, different sets of Tau isoforms may be present in subtypes of neurons leading to various Tau cell sorting and distinct neuronal morphologies.

## 2. Tau Isoforms in Neurons

***a. Glial Cells*** Tau immunoreactivity is found in astrocytes and oligodendrocytes in several neurodegenerative disorders (Chin and Goldman, 1996). PHFs and straight tubules were present in astrocytes of three advanced cases of long-duration AD (>13 years). The PHFs and straight tubules were indistinguishable from those seen in neurons (Yamazaki *et al.*, 1995).

Both neurons and oligodendroglia are preferentially infected in subacute sclerosing panencephalitis. Massive argyrophilic and Tau-positive glial fibrillary tangles were found in oligodendroglia in autopsy cases of SSPE with NFTs. Glial fibrillary tangles shared common phosphorylated Tau epitopes with NFTs but were negative for ubiquitin. Electron microscopically, glial fibrillary tangles consisted of compact bundles of irregularly woven tubules. Thus, glial fibrillary tangles in SSPE differed from NFTs showing regular constriction of tubules and from GFT in some other cytoskeletal disorders in which glial fibrillary tangles reportedly consisted of straight tubules (Ikeda *et al.*, 1995; Nishimura *et al.*, 1995). Similarly, Tau-positive astrocytes are found in PSP (Yamada *et al.*, 1992; Abe *et al.*, 1994), CBD (Feany and Dickson, 1995; Feany, 1995; Buée-Scherrer *et al.*, 1996b), and postencephalitic parkinsonism (Ikeda *et al.*, 1993).

***b. Muscles*** Muscle fiber vacuoles containing inclusions from patients with sporadic inclusion body myositis and patients with autosomal recessive hereditary inclusion body myopathy are immunoreactive for ubiquitin, $\beta$-amyloid protein, as well as hyperphosphorylated Tau proteins. Tau-positive inclusions are composed of twisted tubulofilaments.

These studies demonstrate that abnormal Tau proteins also accumulate in a nonneural tissue and suggest that Tau aggregation may share comparable mechanisms (Askanas *et al.*, 1994; Murakami *et al.*, 1995; Chin and Goldman, 1996).

### C. Tau Pathology: Cause or Consequence?

The collapse of MTs is an important event of neurofibrillary degeneration, visualized by the aggregation of Tau proteins in nerve cells. Tau aggregates are found in numerous degenerative disorders. Because Tau proteins are factors of MT assembly, they could play a major role in the sequence of pathologic events leading to neurodegeneration. However, we do not know if there is first a depolymerization of MTs followed by Tau proteins release and aggregation or, alternatively, an abnormal and precise dysfunction of Tau proteins leading to MT disassembly. In other words, Tau pathology could be a cause or a consequence of NFD.

Tau pathology is a consequence. For example, we know that, in AD, a dysfunction of APP provokes a cascade of events leading to Tau pathology. At least it is true for familial AD with mutations on the APP gene and Down's syndrome (Hardy, 1992). In Guamanian ALS/PDC, amyloid deposits are not present but AD-like NFD is observed (Buée-Scherrer *et al.*, 1995). It is demonstrated that an environmental factor is likely to be involved. Thus, neurofibrillary degeneration is a rather unspecific event resulting from different types of neuronal injuries with distinct etiologic factors.

Tau pathology is a cause. A dysfunction of a putative myotonine kinase could be responsible for the presence of a neurofibrillary degeneration through the abnormal phosphorylation of Tau (Vermersch *et al.*, 1996). Also, we note the almost perfect correlation between Tau pathology and cognitive impairment in numerous degenerative diseases, and therefore it is tempting to speculate that Tau proteins play an early role in the cascade of events that lead to neuronal cell death (Delacourte, 1994). In that respect, Tau dysfunction could be a risk factor influencing the etiological factor. However, the difficulty for timing the dysfunctions is essentially because we are dealing with chronic diseases that take decades to develop, with a very long infraclinical stage, and that affect exclusively the "inaccessible" CNS.

## VI. Concluding Remarks

In conclusion, structure and posttranslational modifications of Tau proteins are key factors in Tau function. Both splicing and phosphorylation are developmentally regulated, providing additional evidence that these factors are important in the role of Tau proteins.

Cassette exons of Tau proteins modulate Tau function. Exons 2 and 3 belong to the projection domain of Tau proteins. Their alternative splicing is likely to play a role on the distance between MTs leading to changes in axonal diameter and on interactions with plasma membrane and subcellular organelles. Exon 10 contains an additional MT-binding domain and allows an anchoring of Tau proteins within MTs. Alternative splicing is different among neuronal cells and thus the Tau isoform content may lead to particular cell morphology and function.

Phosphorylation may be implicated in Tau trafficking. In fact, the degree of phosphorylation of Tau proteins is linked to its location: axon, dendrites, perikaryon, or nucleus. Phosphorylation of Tau proteins is also a key element in the regulation of MT assembly. A complex balance between kinases and phosphatases controls the phosphorylation of Tau proteins.

Tau proteins are also good markers of cell integrity. For instance, Tau proteins are completely dephosphorylated in ischemic conditions and phosphorylation is restored after reperfusion. In different neurodegenerative disorders, pathological Tau proteins are aggregated and are always found hyperphosphorylated. They inhibit MT assembly. They are characterized by specific electrophoretic profiles that are likely to reflect both different degrees of phosphorylation and Tau isoforms contents. Thus, they should allow the identification of which subtypes of neurons are degenerating and how. Tau proteins are essential keys in the understanding of neuronal pathophysiology.

Currently, the *Tau* gene is not fully explored and many regulations have yet to be discovered. The precision of the genetic and immunological tools at our disposal today, in association with a multidisciplinary approach, should give us in the near future a good understanding of the many roles of Tau.

For example, we mentioned that neuronal populations could be distinguished by their own content in Tau isoforms. This can be verified using phosphorylation-dependent monoclonal antibodies in combination with *in situ* hybridization techniques and probes specific to exons 2, 3, 4A, and 10. This approach should tell us which Tau isoforms are found in different subtypes of neurons, such as interneurons, pyramidal neurons, or motoneurons, and what is the relationship, if any, between a Tau electrophoretic pattern and a neuronal subtype.

In the same way, in myotonic dystrophy, which is a genetic disease, we have shown that there is likely a link between the dysfunction of a putative gene of a Ser/Thr kinase and a Tau pathology. Studies combining clinical data, genetic studies, neuropathology, immunochemistry, and cellular models should provide more specific information on the relationship between a kinase abnormality and the collapse of the cytoskeleton in specific brain areas. This pathology offers a real opportunity to connect a biochemical dysfunction to the production of NFTs.

Last, because the study of nerve cell physiology is complex and also relies on other cells (astrocytes, microglial cells, and endothelial cells), all approaches described previously should be developed in animal models. In that perspective, primate models of small size are more appropriate, such as *Microcebus murinus* (Bons *et al.,* 1995).

Overall, there is still important work to be done concerning the molecular dissection of Tau proteins in order to determine the role of each isoform and of each phosphorylation site. Also, it is important to analyze the sequence of molecular events leading to MT disruption and Tau aggregation during neurodegenerative disorders in order to define the targets for a pharmacological approach of neuroprotection.

## Acknowledgments

We thank A. Wattez for expert technical assistance; Professors Y. Agid, C. Bouras, R. Ravid, A. Destée, C. Di Menza, J-P Lejeune, D. Leys, D. M. A. Mann, C. W. Olanow, D. P. Perl, H. Petit, Y. Robitaille, D. Gauvreau, Dr. J.-P. David, and C. Tranchant for providing human biopsies and postmortem tissues; and Drs. V. Buée-Scherrer, M. L. Caillet-Boudin, P. R. Hof, and P. Vermersch for helpful discussion. We are also grateful to all members of INSERM U422-VCDN for their constant interest and support. This study was supported by the INSERM, the CNRS, the ADNA, and grants from Assistance Publique-Hôpitaux de Paris (Biologie du Vieillissement 94.00.05). AD2 was developed through a collaboration between UMR 9921 (Professor B. Paul and Dr. C. Mourton-Gilles, Montpellier University), Sanofi/Diagnostics Pasteur, and INSERM.

## References

Aamodt, E. J., and Williams, R. C., Jr. (1984). Microtubule-associated proteins connect microtubules and neurofilaments in vitro. *Biochemistry* **23,** 6023–6031.

Abe, H., Yagishita, S., and Amano, N. (1994). Unusual argyrophilic glial cells in progressive supranuclear palsy. *Neurodegeneration* **3,** 85–90.

Albert, M. S. (1978). Subcortical dementia. In "Alzheimer's Disease Senile, Dementia and Related Disorders, Aging" R. Katzman, R. D. Terry, and K. L. Bick, Eds., Vol. 7. Raven Press, New York.

Alonso, A. D., Zaidi, T., Grundke-Iqbal, I., and Iqbal, K. (1994). Role of abnormally phosphorylated tau in the breakdown of microtubules in Alzheimer's disease. *Proc. Natl. Acad. Sci. USA* **91,** 5562–5566.

Andreadis, A., Brown, W. M., and Kosik, K. S. (1992). Structure and novel exons of the human-Tau gene. *Biochemistry* **31,** 10626–10633.

Andreadis, A., Broderick, J. A., and Kosik, K. S. (1995). Relative exon affinities and suboptimal splice site signals lead to non-equivalence of two cassette exon. *Nucleic Acids Res.* **23,** 3585–3593.

Andreadis, A., Wagner, B. K., Broderick, J. A., and Kosik, K. S. (1996). A tau promoter region without neuronal specificity. *J. Neurochem.* **66,** 2257–2263.

Arendt, T., Holzer, M., Grossmann, A., Zedlick, D., and Bruckner, M. K. (1995). Increased expression and subcellular translocation of the mitogen activated protein kinase kinase and mitogen-activated protein kinase in Alzheimer's disease. *Acta Psychiatr. Scand.* **68,** 5–18.

Arnold, C. S., Cole, R. N., Johnson, G. V. W., and Hart, G. W. (1994). Tau is a glycoprotein. *Soc. Neurosci. Abstr.* **20,** 1035.

Askanas, V., Engel, W. K., Bilak, M., Alvarez, R. B., and Selkoe, D. J. (1994). Twisted tubulofilaments of inclusion body myositis muscle resemble paired helical filaments of Alzheimer brain and contain hyperphosphorylated tau. *Am. J. Pathol.* **144,** 177–187.

Auer, I. A., Schmidt, M. L., Lee, V. M. Y., Curry, B., Suzuki, K., Shin, R. W., Pentchev, P. G., Carstea, E. D., and Trojanowski, J. Q. (1995). Paired helical filament Tau (PHFtau) in Niemann-Pick type C disease is similar to PHFTau in Alzheimer's disease. *Acta Neuropathol.* **90,** 547–551.

Baudier, J., and Cole, R. D (1987). Phosphorylation of Tau proteins to a state like that in Alzheimer's brain is catalyzed by a calcium/calmodulin-dependent kinase and modulated by phospholipids. *J. Biol. Chem.* **262,** 17577–17583.

Baum, L., Seger, R., Woodgett, J. R., Kawabata, S., Maruyama, K., Koyama, M., Silver, J., and Saitoh, T. (1995). Overexpressed Tau protein in cultured cells is phosphorylated without formation of PHF: Implication of phosphoprotein phosphatase involvement. *Mol. Brain Res.* **34**, 1–17.

Baumann, K., Mandelkow, E. M., Biernat, J., *et al.* (1993). Abnormal Alzheimer-like phosphorylation of Tau-protein by cyclin-dependent kinases cdk2 and cdk5. *FEBS Lett.* **336**, 417–424.

Behar, L., Marx, R., Sadot, E., Barg, J., and Ginzburg, I. (1995). cis-acting signals and transacting proteins are involved in Tau mRNA targeting into neurites of differentiating neuronal cells. *Int. J. Dev. Neurosci.* **13**, 113–127.

Benzing, W. C., and Mufson, E. J. (1995). Apolipoprotein E immunoreactivity within neurofibrillary tangles: Relationship to Tau and PHF in Alzheimer's disease. *Exp. Neurol.* **132**, 162–171.

Biernat, J., Gustke, N., Drewes, G., Mandelkow, E. M., and Mandelkow, E. (1993). Phosphorylation of ser(262) strongly reduces binding of Tau-protein to microtubules—Distinction between PHF-like immunoreactivity and microtubule binding. *Neuron* 11, 153–163.

Binder, L. I., Frankfurter, A., and Rebhun, L. (1985). The distribution of Tau in the mammalian central nervous system. *J. Cell Biol.* **101**, 1371–1378.

Bons, N., Jallageas, V., Mestre-Frances, N., Silhol, S., Petter, A., and Delacourte, A. (1995). Microcebus murinus, a convenient laboratory animal model for the study of Alzheimer's disease. *Alzheimer's Res.* **1**, 83–87.

Boucher, D., Larcher, J. C., Gros, F., and Denoulet, P. (1994). Polyglutamylation of tubulin as a progressive regulator of in vitro interaction between the microtubule-associated protein Tau and tubulin. *Biochemistry* **33**, 12471–12477.

Bouras, C., Hof, P. R., Giannakopoulos, P., Michel, J. P., and Morrison, J. H. (1994). Regional distribution of neurofibrillary tangles and senile plaques in the cerebral cortex of elderly patients—A quantitative evaluation of a one-year-autopsy population from a geriatric hospital. *Cerebral Cortex* **4**, 138–150.

Boyne, L. J., Tessler, A., Murray, M., and Fischer, I. (1995). Distribution of Big Tau in the central nervous system of the adult and developing rat. *J. Comp. Neurol.* **358**, 279–293.

Braak, H., and Braak, E. (1988). Neuropil threads occur in dendrites of tangle-bearing nerve cells. *Neuropathol. Appl. Neurol.* **14**, 39–43.

Braak, H., and Braak, E., Grundke-Iqbal, I., and Iqbal, K. (1988). Occurrence of neuropils threads in the senile human brain and in Alzheimer's disease. A 3rd location of paired helical filaments outside of neurofibrillary tangles and neuritic plaques. *Neurosci. Lett.* **65**, 351–355.

Braak, H., and Braak, E. (1995). Staging of Alzheimer's disease-related neurofibrillary changes. *Neurobiol. Aging* **16**, 271–278.

Braak, H., Braak, E., and Strothjohann, M. (1994). Abnormally phosphorylated Tau protein related to the formation of neurofibrillary tangles and neuropil threads in the cerebral cortex of sheep and goat. *Neurosci. Lett.* **171**, 1–4.

Brady, R. M., Zinkowski, R. P., and Binder, L. I. (1995). Presence of Tau in isolated nuclei from human brain. *Neurobiol. Aging* **16**, 479–486.

Bramblett, G. T., Goedert, M., Jakes, R., Merrick, S. E., Trojanowski, J. Q., and Lee, V. M. Y. (1993). Abnormal Tau-phosphorylation at ser(396) in Alzheimer's disease recapitulates development and contributes to reduced microtubule binding. *Neuron* **10**, 1089–1099.

Brandt, R., Leger, J., and Lee, G. (1995). Interaction of Tau with the neural plasma membrane mediated by tau's amino-terminal projection domain. *J. Cell Biol.* **131**, 1327–1340.

Brion, J. P., Passareiro, H., Nunez, J., and Flament-Durand, J. (1985). Immunological detection of Tau protein in neurofibrillary tangles of Alzheimer's disease. *Arch. Biol.* **95**, 229–235.

Brion, S., and Mikol, J. (1973). Recent findings in Pick's disease. *Prog. Neuropathol.* **2**, 421.

Brun, A., Englund, B., Gustafson, L., Passant, U., Mann, D. M. A., Neary, D., and Snowden, J. S. (1994). Clinical and neuropathological criteria for frontotemporal dementia. *J. Neurol. Neurosurg. Ps.* **57**, 416–418.

Buée, L., Hof, P. R., Rosenthal, R. E., Delacourte, A., and Fiskum, G. (1996). Tau proteins phosphorylation and proteolysis in a canine model of cerebral ischemia/reperfusion. Submitted for publication.

Buée-Scherrer, V., Buée, L., Vermersch, P., et al. (1994). Tau pathology in neurodegenerative disorders: Biochemical analysis. Soc. Neurosci. Abstr. **20,** 1647.

Buée-Scherrer, V., Buée, L., Hof, P. R., Leveugle, B., Gilles, C., Loerzel, A. J., Perl, D. P., and Delacourte, A. (1995). Neurofibrillary degeneration in amyotrophic lateral sclerosis/parkinsonism–dementia complex of Guam—Immunochemical characterization of Tau proteins. Am. J. Pathol. **146,** 924–932.

Buée-Scherrer, V., Condamines, O., Mourton-Gilles, C., Jakes, R., Goedert, M., Pau, B., and Delacourte, A. (1996a). AD2, a phosphorylation-dependent monoclonal antibody directed against Tau proteins found in Alzheimer's disease. Mol. Brain Res. **39,** 79–88.

Buée-Scherrer, V., Hof, P. R., Buée, L., Leveugle, B., Vermersch, P., Perl, D. P., Olanow, C. W., and Delacourte, A. (1996b). Hyperphosphorylated Tau proteins differentiate corticobasal degeneration and Pick's disease. Acta Neuropathol. **91,** 351–359.

Buée-Scherrer, V., Buée, L., Perl, D. P., et al. (1996c). Pathological tau proteins in postencephalitic parkinsonism: Similarities and differences with Alzheimer's disease and other neurodegenerative disorders. Submitted for publication.

Burgoyne, R. D. (1991). "The Neuronal Cytoskeleton," pp. 1–323. Wiley-Liss, New York.

Busciglio, J., Lorenzo, A., Yeh, J., and Yanker, B. A. (1995). beta Amyloid fibrils induce Tau phosphorylation and loss of microtubule binding. Neuron **14,** 879–888.

Butner, K. A., and Kirschner, M. W. (1991). Tau-protein binds to microtubules through a flexible array of distributed weak sites. J. Cell. Biol. **115,** 717–730.

Buxton, J., Shelbourne, P., and Davies, J. (1992). Detection of an unstable fragment of DNA specific to individuals with myotonic dystrophy. Nature **355,** 547–548.

Caceres, A., and Kosik, K. S. (1990). Inhibition of neurite polarity by tau antisense oligonucleotides in primary cerebellar neurons. Nature **343,** 461–463.

Caceres, A., Potrebic, S., and Kosik, K. S. (1991). The effect of tau-antisense oligonucleotides on neurite formation of cultured cerebellar macroneurons. J. Neurosci. **11,** 1515–1523.

Caillet-Boudin, M. L., and Delacourte, A. (1996). Phosphorylated Ser 422, a specific Alzheimer epitope, is expressed by SKNSH-SY5Y cells after okadaic acid. Neuroreport **8,** in press.

Carlier, M. F., Simon, C., Cassoly, R., and Pradel, L. A. (1984). Interaction between microtubule-associated protein Tau and spectrin. Biochimie. **664,** 305–311.

Chapin, S. J., Lue, C. M., Yu, M. T., and Bulinski, J. C. (1995). Differential expression of alternatively spliced forms of MAP4: A repertoire of structurally different microtubule-binding domains. Biochemistry **34,** 2289–2301.

Chin, S. S. M., and Goldman, J. E. (1996). Glial inclusions in CNS degenerative diseases. J. Neuropathol. Exp. Neurol. **55,** 499–508.

Cohen, P., and Cohen, P. T. W. (1989). Protein phosphatases come of age. J. Biol. Chem. **264,** 21435–21438.

Correas, I., Padilla, R., and Avila, J. (1990). The tubulin-binding sequence of brain microtubule-associated proteins, Tau and MAP2, is also involved in actin binding. Biochem. J. **269,** 61–64.

Cummings, J. L., and Benson, D. F. (1984). Subcortical dementia: review of an emerging concept. Arch. Neurol. **41,** 874–879.

Delacourte, A. (1994). Pathological Tau proteins of Alzheimer's disease as a biochemical marker of neurofibrillary degeneration. Biomed. Pharmacother. **48,** 287–295.

Delacourte, A., and Défossez, A. (1986). Alzheimer's disease: Tau proteins, the promoting factors of microtubule assembly, are major components of paired helical filaments. J. Neurol. Sci. **76,** 173–186.

Delacourte, A., Flament, S., Dibe, E. M., Hublau, P., Sablonnière, B., Hemon, B., Scherrer, V., and Défossez, A. (1990). Pathological proteins Tau 64 and 69 are specifically expressed in the somatodendritic domain of the degenerating cortical neurons during Alzheimer's disease: Demonstration with a panel of antibodies against Tau proteins. *Acta Neuropathol.* **80,** 111–117.

Delacourte, A., Robitaille, Y., Sergeant, N., Buee, L., Hof, P. R., Wattez, A., Laroche-Cholette, A., Mathieu, J., Chagnon, P., and Gauvreau, D. (1996). Specific pathological Tau protein variants characterize Pick's disease. *J. Neuropathol. Exp. Neurol.* **55,** 159–168.

Dermott, J. B., Aamodt, S., and Aamodt, E. (1996) ptl-1, a gene whose products are homologous to the tau microtubule-associated proteins. *Biochemistry* **35,** 9415–9423.

Dewar, D., and Dawson, D. (1995). Tau protein is altered by focal cerebral ischaemia in the rat: An immunohistochemical and immunoblotting study. *Brain Res.* **684,** 70–78.

Dewar, D., Graham, D. I., Teasdale, G. M., and McCulloch, J. (1994). Cerebral ischemia induces alterations in Tau and ubiquitin proteins. *Dementia* **5,** 168–173.

Ditella, M. C., Feiguin, F., Carri, N., Kosik, K. S., and Caceres, A. (1996). Map-1B/Tau functional redundancy during laminin-enhanced axonal growth. *J. Cell Sci.* **109,** 467–477.

Dong, D. L. Y., Xu, Z. S., Chevier, M. R., Cotter, R. J., Cleveland, D. W., and Hart, G. W. (1993). Glycosylation of mammalian neurofilaments—Localization of multiple O-linked N-acetylglucosamine moieties on neurofilament polypeptides-L and polypeptides-M. *J. Biol. Chem.* **268,** 16679–16687.

Dotti, C. G., and Banker, G. (1991). Intracellular organization of hippocampal neurons during the development of neuronal polarity. *J. Cell. Sci. Suppl.* **15,** 75–84.

Dotti, C. G., Banker, G., and Binder, L. I. (1987). The expression and distribution of the microtubule-associated proteins Tau and microtubule-associated protein 2 in hippocampal neurons in the rat in situ and in cell culture. *Neuroscience* **23,** 121–130.

Drewes, G., Lichtenbergkraag, B., Doring, F., Mandelkow, E. M., Biernat, J., and Gori, X. (1992). Mitogen activated protein (MAP) kinase transforms Tau-protein into an Alzheimer-like state. *EMBO J.* **11,** 2131–2138.

Drewes, G., Trinczek, B., Illenberger, S., Biernat, J., Schmittulms, G., Meyer, H. E., Mandelkow, E. M., and Mandelkow, E. (1995). Microtubule-associated protein microtubule affinity-regulating kinase (p110(mark))—A novel protein kinase that regulates tau-microtubule interactions and dynamic instability by phosphorylation at the Alzheimer-specific site serine 262. *J. Biol. Chem.* **270,** 7679–7688.

Drubin, D. G., and Kirschner, M. W. (1986). Tau protein function in living cells. *J. Cell. Biol.* **103,** 2738–2746.

Drubin, D. G., Kobayashi, S., Kellogg, D., and Kirschner, M. W. (1988). Regulation of microtubule protein levels during cellular morphogenesis in nerve growth factor-treated PC12 cells. *J. Cell Biol.* **106,** 1583–1591.

Dudek, S. M., and Johnson, G. V. W. (1995). Postnatal changes in Ser/Thr protein phosphatases and their association with microtubules. *Dev. Brain Res.* **90,** 54–61.

Dupont-Wallois, L., Sautiere, P. E., Cocquerelle, C., Bailleul, B., Delacourte, A., and Caillet-Boudin, M. L. (1995). Shift from fetal-type to Alzheimer-type phosphorylated Tau proteins in SKNSH-SY 5Y cells treated with okadaic acid. *FEBS Lett.* **357,** 197–201.

Dustin, P. (1984). "Microtubules." Springer-Verlag, Berlin/New York.

Feany, M. B., and Dickson, D. W. (1995). Widespread cytoskeletal pathology characterizes corticobasal degeneration. *Am. J. Pathol.* **146,** 1388–1396.

Feany, M. B., Ksiezak-Reding, H., Liu, W. K., Vincent, I., Yen, S. H. C., and Dickson, D. W. (1995b). Epitope expression and hyperphosphorylation of Tau protein in corticobasal degeneration: Differentiation from progressive supranuclear palsy. *Acta Neuropathol.* **90,** 37–43.

Feany, M. B., Mattiace, L. A., and Dickson, D. W. (1996). Neuropathologic overlap of progressive supranuclear palsy, Pick's disease and corticobasal degeneration. *J. Neuropathol. Exp. Neurol.* **55,** 53–67.

Ferreira, A., and Caceres, A. (1989). The expression of acetylated microtubules during axonal and dendritic growth in cerebellar macroneurons which develop in vitro. *Dev. Brain Res.* **49,** 205–213.
Flament, S., Delacourte, A., Hemon, B., and Defossez, A. (1989). Characterization of two pathological tau-protein variants in Alzheimer brain cortices. *J. Neurol. Sci.* **92,** 133–141.
Flament, S., Delacourte, A., and Mann, D. M. A. (1990). Phosphorylation of Tau proteins: A major event during the process of neurofibrillary degeneration. A comparative study between Alzheimer's disease and Down's syndrome. *Brain Res.* **516,** 15–19.
Flament, S., Delacourte, A., Verny, M., Hauw, J. J., and Javoy-Agid F. (1991). Abnormal Tau proteins in progressive supranuclear palsy. Similarities and differences with the neurofibrillary degeneration of the Alzheimer type. *Acta Neuropathol.* **81,** 591–596.
Fleming, L. M., and Johnson, G. V. M. (1995). Modulation of the phosphorylation state of Tau in situ: The roles of calcium and cyclic AMP. *Biochem. J.* **309,** 41–47.
Garruto, R. M. (1991). Pacific paradigms of environmentally-induced neurological disorders: Clinical, epidemiological and molecular perspectives. *Neurotoxicology* **12,** 347–378.
Garver, T. D., Harris, K. A., Lehman, R. A. W., Lee, V. M. Y., Trojanowski, J. Q., and Billingsley, M. L. (1994). Tau phosphorylation in human, primate, and rat brain: Evidence that a pool of Tau is highly phosphorylated in vivo and is rapidly dephosphorylated in vitro. *J. Neurochem.* **63,** 2279–2287.
Garver, T. D., Lehman, R. A. W., and Billingsley, M. L. (1996). Microtubule assembly competence analysis of freshly-biopsied human tau, dephosphorylated tau, and Alzheimer tau. *J. Neurosci. Res.* **44,** 12–20.
Georgieff, I. S., Liem, R. K. H., Couchie, D., Mavilia, C., Nunez, J., and Shelanski, M. (1993). Expression of high molecular weight Tau in the central and peripheral nervous systems. *J. Cell Sci.* **105,** 729–737.
Gerstmann, J., Sträussler, E., and Schienker, I. (1936). Uber eigenartige hereditar-familiare Erkrankung des Zentralnervensystems. *Z. Neurol.* **154,** 736–762.
Ghetti, B., Dlouhy, S. R., Giaccone, G., Bugiani, O., Frangione, B., Farlow, M. R., and Tagliavini, F. (1995). Gerstmann–Straussler–Scheinker disease and the Indiana kindred. *Brain Pathol.* **5,** 61–75.
Ghetti, B., Piccardo, P., Spillantini, M. G., Ichimiya, Y., Porro, M., Perini, F., Kitamoto, T., Tateishi, J., Seiler, C., Frangione, B., Bugiani, O., Giaccone, G., Prelli, F., Goedert, M., Dlouhy, S. R., and Tagliavini, F. (1996). Vascular variant of prion protein cerebral amyloidosis with tau-positive neurofibrillary tangles: The phenotype of the stop codon 145 mutation in PRNP. *Proc. Natl. Acad. Sci. USA* **93,** 744–748.
Giaccone, G., Pedrotti, B., Migheli, A., Verga, L., Perez, J., Racagni, G., Smith, M. A., Perry, G., Degioia, L., Selvaggini, C., Salmona, M., Ghiso, J., Frangione, B., Islam, K., Bugiani, O., and Tagliavini, F. (1996). beta PP and Tau interaction: A possible link between amyloid and neurofibrillary tangles in Alzheimer's disease. *Am. J. Pathol.* **148,** 79–87.
Giannakopoulos, P., Hof, P. R., Giannakopoulos, A. S., Herrmann, F. R., Michel, J. P., and Bouras, C. (1995a). Regional distribution of neurofibrillary tangles and senile plaques in the cerebral cortex of very old patients. *Arch. Neurol.* **52,** 1150–1159.
Giannakopoulos, P., Hof, P. R., and Bouras, C. (1995b). Age versus ageing as a cause of dementia. *Lancet* **346,** 1486–1487.
Goedert, M., and Jakes, R. (1990). Expression of separate isoforms of human Tau protein: Correlation with the Tau pattern in brain and effects on tubulin polymerization. *EMBO J.* **9,** 4225–4230.
Goedert, M., Spillantini, M. G., Potier, M. C., Ulrich, J., and Crowther, R. A. (1989a). Cloning and sequencing of the cDNA encoding an isoform of microtubule-associated protein tau containing 4 tandem repeats—Differential expression of tau protein messenger RNAs in human brain. *EMBO J.* **8,** 393–399.

Goedert, M., Spillantini, M. G., Jakes, R., Rutherford, D., and Crowther, R. A. (1989b). Multiple isoforms of human microtubule-associated protein tau: Sequences and localization in neurofibrillary tangles of Alzheimer's disease. *Neuron* **3,** 519–526.

Goedert, M., Spillantini, M. G., Cairns, N. J., and Crowther, R. A. (1992). Tau-proteins of Alzheimer paired helical filaments—Abnormal phosphorylation of all six brain isoforms. *Neuron* **8,** 159–168.

Goedert, M., Jakes, R., Crowther, R. A., Cohen, P., Vanmechelen, E., Vandermeeren, M., and Cras, P. (1994). Epitope mapping of monoclonal antibodies to the paired helical filaments of Alzheimer's disease: Identification of phosphorylation sites in Tau protein. *Biochem. J.* **301,** 871–877.

Goedert, M., Jakes, R., and Vanmechelen, E. (1995). Monoclonal antibody AT8 recognizes Tau protein phosphorylated at both serine 202 and threonine 205. *Neurosci. Lett.* **189,** 167–170.

Gong, C. X., Shaikh, S., Wang, J. Z., Zaidi, T., Grundke-Iqbal, I., and Iqbal, K. (1995). Phosphatase activity toward abnormally phosphorylated tau: Decrease in Alzheimer disease brain. *J. Neurochem.* **65,** 732–738.

Goode, B. L., and Feinstein, S. C. (1994). Identification of a novel microtubule binding and assembly domain in the developmentally regulated Inter-Repeat region of tau. *J. Cell Biol.* **124,** 769–782.

Goto, S., Yamamoto, H., Fukunaga, K., Iwasa, T., Matsukado, Y., and Miyamoto, E. (1985). Dephosphorylation of microtubule-associated protein 2, Tau factor, and tubulin by calcineurin. *J. Neurochem.* **451,** 276–283.

Gotz, J., Probst, A., Spillantini, M. G., Schafer, T., Jakes, R., Burki, K., and Goedert, M. (1995). Somatodendritic localization and hyperphosphorylation of Tau protein in transgenic mice expressing the longest human brain Tau isoform. *EMBO J.* **14,** 1304–1313.

Greenberg, S. G., Davies, P. Schein, J. D., and Binder, L. I. (1992). Hydrofluoric acid-treated tau-PHF proteins display the same biochemical properties as normal tau. *J. Biol. Chem.* **267,** 564–569.

Greenwood, J. A., and Johnson, G. V. W. (1995). Localization and in situ phosphorylation state of nuclear tau. *Exp. Cell Res.* **220,** 332–337.

Greenwood, J. A., Scott, C. W., Spreen, R. C., Caputo, C. B., and Johnson, G. V. W. (1994). Casein kinase II preferentially phosphorylates human Tau isoforms containing an aminoterminal insert—Identification of threonine 39 as the primary phosphate acceptor. *J. Biol. Chem.* **269,** 4373–4380.

Griffith, L. M., and Pollard, T. D. (1978). Evidence for actin filament-microtubule interaction mediated by microtubule-associated proteins. *J. Cell Biol.* **78,** 958–965.

Grundke-Iqbal, I., Iqbal, K., Tung, Y. C., *et al.* (1986). Abnormal phosphorylation of the Map Tau in Alzheimer cytoskeletal pathology. *Proc. Natl. Acad. Sci. USA* **83,** 4913–4917.

Gustafson, L. (1993). Clinical picture of frontal lobe degeneration of non-Alzheimer type. *Dementia* **4,** 143–148.

Gustke, N., Steiner, B., Mandelkow, E. M., Biernat, J., Meyer, H. E., and Goedert, M. (1992). The Alzheimer-like phosphorylation of tau-protein reduces microtubule binding and involves Ser-Pro and Thr-Pro motifs. *FEBS Lett.* **307,** 199–205.

Gustke, N., Trinczek, B., Mandelkow, E. M., and Mandelkow, E. (1994). Domains of Tau protein and interactions with microtubules. *Biochemistry* **33,** 9511–9522.

Hagestedt, G., Lichtenberg, B., Wille, H., Mandelkow, E. M., and Mandelkow, M. (1989). Tau protein becomes long and stiff upon phosphorylation: Correlation between paracrystalline structure and degree of phosphorylation. *J. Cell Biol.* **109,** 1643–1651.

Hakim, A. M., and Mathieson, G. (1979). Dementia in Parkinson's disease: A neuropathologic study. *Neurology* **29,** 1204–1214.

Haltiwanger, R. S., Kelly, W. G., and Roquemore, E. P., Blomberg, M. A., Dong, D. L. Y., Kreppel, L., Chou, T. Y., and Hart, G. W. (1992). Glycosylation of nuclear and cytoplasmic proteins is ubiquitinous and dynamic. *Biochem. Soc. Trans.* **20,** 264–269.

Hamre, K. M., Hyman, B. T., Goodlett, C. R., West, J. R., and Vanhoesen, G. W. (1989). Alz-50 Immunoreactivity in the neonatal rat—Changes in development and co-distribution with Map-2 immunoreactivity. *Neurosci. Lett.* **98,** 264–271.

Hanger, D. P., Hughes, K., Woodgett, J. R., Brion, J. P., and Anderton, B. H. (1992). Glycogen synthase kinase-3 induces alzheimers disease-like phosphorylation of Tau-generation of paired helical filament epitopes and neuronal localisation of the kinase. *Neurosci. Lett.* **147,** 58–62.

Harada, A., Oguchi, K., Okabe, S., Kuno, J., Terada, S., Ohshima, T., Satoyoshitake, R., Takei, Y., Noda, T., and Hirokawa, N. (1994). Altered microtubule organization in small-calibre axons of mice lacking Tau protein. *Nature* **369,** 488–491.

Hardy, J. (1992). An anatomical cascade hypothesis for Alzheimer's disease. *Trends Neurosci.* **15,** 200–201.

Harper, P. S. (1989). "Myotonic Dystrophy," 2nd ed. Saunders, London.

Hasegawa, M., Morishima-Kawashima, M., Takio, K., Suzuki, M., Titani, K., and Ihara, Y. (1992). Protein sequence and mass spectrometric analyses of Tau in the Alzheimer's disease brain. *J. Biol. Chem.* **267,** 17047–17054.

Hasegawa, M., Watanabe, A., Takio, K., Suzuki, M., Arai, T., Titani, K., and Ihara, Y. (1993). Characterization of 2 distinct monoclonal antibodies to paired helical filaments—Further evidence for fetal-type phosphorylation of the Tau in paired helical filaments. *J. Neurochem.* **60,** 2068–2077.

Hasegawa, M., Jakes, R., Crowther, R. A., Lee, V. M. Y., Ihara, Y., and Goedert, M. (1996). Characterization of mAb AP422, a novel phosphorylation-dependent monoclonal antibody against Tau protein. *FEBS Lett.* **384,** 25–30.

Hauw, J. J., Verny, M., Delaere, P., Cervera, P., He, Y., and Duyckaerts, C. (1990). Constant neurofibrillary changes in the neocortex in progressive supranuclear palsy—Basic differences with Alzheimer's disease and aging. *Neurosci. Lett.* **119,** 182–186.

Hauw, J. J., Daniel, S. E., Dickson, D., Horoupian, D. S., Jellinger, K., Lantos, P. L., Mckee, A., Tabaton, M., and Litvan, I. (1994). Preliminary NINDS neuropathologic criteria for Steele–Richardson–Olszewski syndrome (progressive supranuclear palsy). *Neurology* **44,** 2015–2019.

Henriquez, J. P., Cross, D., Vial, C., and Maccioni, R. B. (1995). Subpopulations of Tau interact with microtubules and actin filaments in various cell types. *Cell Biochem. Funct.* **13,** 239–250.

Higgins, L. S. Rodems, J. M., Catalano, R., Quon, D., and Cordell, B. (1995). Early Alzheimer disease-like histopathology increases in frequency with age in mice transgenic for beta-APP751. *Proc. Natl. Acad. Sci. USA* **92,** 4402–4406.

Himmler, A. (1989). Structure of the bovine tau-gene—Alternatively spliced transcripts generate a protein family. *Mol. Cell Biol.* **9,** 1389–1396.

Hirano, A., Kurland, L. T., Krooth, R. S., and Lessel, S. (1961). Parkinsonism–dementia complex, and endemic disease on the island of Guam: I. Clinical features. *Brain* **84,** 642–661.

Hirano, A., Dembitzer, H. M., and Kurland, L. Y. (1968). The fine structure of some intraganglionic alterations. Neurofibrillary tangles, granulovacuolar bodies and "rod-like" structures as seen in Guam amyotrophic lateral sclerosis and Parkinson's dementia complex. *J. Neuropathol. Exp. Neurol.* **67,** 167–182.

Hirokawa, N., Shiomura, Y., and Okabe, S. (1988). Tau proteins: The molecular structure and mode of binding on microtubules. *J. Cell Biol.* **107,** 1449–1459.

Hirokawa, N., Funakoshi, T., Satoharada, R., and Kanai, Y. (1996). Selective stabilization of Tau in axons and microtubule-associated protein 2C in cell bodies and dendrites contributes to polarized localization of cytoskeletal proteins in mature neurons. *J. Cell Biol.* **132,** 667–679.

Hof, P. R., and Morrison, J. H. (1994). The cellular basis of cortical disconnection in Alzheimer's disease and related dementing conditions. *In* "Alzheimer's Disease" (R. D. Terry, R. Katzman, and K. L. Bick, Eds.), pp. 197–229. Raven Press, New York.

Hof, P. R., Perl, D. P., Loerzel, A. J., and Morrison, J. H. (1991). Neurofibrillary tangle distribution in the cerebral cortex of Parkinsonism-dementia cases from Guam—Differences with Alzheimer's disease. *Brain Res.* **564**, 306-313.

Hof, P. R., Bouras, C., Buée, L., Delacourte, A., Perl, D. P., and Morrison, J. H. (1992a). Differential distribution of neurofibrillary tangles in the cerebral cortex of Dementia Pugilistica and Alzheimer's disease cases. *Acta Neuropathol.* **85**, 23-30.

Hof, P. R., Delacourte, A., and Bouras, C. (1992b). Distribution of cortical neurofibrillary tangles in progressive supranuclear palsy. A quantitative analysis of 6 cases. *Acta Neuropathol.* **84**, 45-51.

Hof, P. R., Nimchinsky, E. A., Buée-Scherrer, V., Buée L., Nasrallah, J., Hottinger, A. F., Purohit, D. P., Loerzel, A. J., Steele, J. C., Delacourte, A., Bouras, C., Morrison, J. H., and Perl, D. P. (1994a). Amyotrophic lateral sclerosis/parkinsonism–dementia complex of Guam: Quantitative neuropathology, immunohistochemical analysis of neuronal vulnerability, and comparison with related neurodegenerative disorders. *Acta Neuropathol.* **88**, 397-404.

Hof, P. R., Bouras, C., Perl, D. P., and Morrison, J. H. (1994b). Quantitative neuropathologic analysis of picks disease cases—Cortical distribution of pick bodies and coexistence with alzheimers disease. *Acta Neuropathol.* **87**, 115-124.

Hof, P. R., Bouras, C., Perl, D. P., Sparks, D. L., Mehta, N., and Morrison, J. H. (1995). Age-related distribution of neuropathologic changes in the cerebral cortex of patients with Down's syndrome: Quantitative regional analysis and comparison with Alzheimer's disease. *Arch. Neurol.* **52**, 379-391.

Huang, D. Y., Weisgraber, K. H., Goedert, M., Saunders, A. M., Roses, A. D., and Strittmatter, W. J. (1995). ApoE3 binding to Tau tandem repeat I is abolished by Tau serine(262) phosphorylation. *Neurosci. Lett.* **192**, 209-212.

Hyman, B. T., Elvhage, T. E., and Reiter, J. (1994). Extracellular signal regulated kinases—Localization of protein and mRNA in the human hippocampal formation in Alzheimer's disease. *Am. J. Pathol.* **144**, 565-572.

Ihara, Y., Nukina, N., Miura, R., and Ogawara, M. (1986). Phosphorylated tau protein is integrated into paired helical filaments in Alzheimer's disease. *J. Biochem. (Tokyo)* **99**, 1807-1810.

Ikeda, K., Akiyama, H., Kondo, H., and Ikeda, K. (1993). Anti-tau-positive glial fibrillary tangles in the brain of postencephalitic parkinsonism of economo type. *Neurosci. Lett.* **162**, 176-178.

Ikeda, K., Akiyama, H., Kondo, H., Arai, T., Arai, N., and Yagishita, S. (1995). Numerous glial fibrillary tangles in oligondendroglia in cases of subacute sclerosing panencephalitis with neurofibrillary tangles. *Neurosci. Lett.* **194**, 133-135.

Ingebristen, T. S., and Cohen, P. (1983). Protein phosphatases: Properties and role in cellular regulation. *Science* **221**, 331-338.

Ingelson, M., and Lannfelt, L. (1996). Tau in fibroblasts with and without the Swedish APP 670/671 mutation. *Neurobiol. Aging* **17** (Suppl. 4), S101.

Iqbal, K., and Grundke-Iqbal, I. (1995). Alzheimer abnormally phosphorylated tau is more hyperphosphorylated than the fetal tau abd causes the disruption of microtubules. *Neurobiol. Aging* **16**, 375-379.

Ishii, T., and Nakamura, Y. (1981). Distribution and ultrastructure of Alzheimer's neurofibrillary tangles in postencephalitic Parkinsonism of Economo type. *Acta Neuropathol.* **55**, 59-62.

Jaspert, A., Fahsold, R., Grehl, H., and Claus, D. (1995). Myotonic dystrophy: Correlation of clinical symptoms with the size of the CTG trinucleotide repeat. *J. Neurol.* **242**, 99-104.

Joachim, C. L., Morris, J. H., Kosik, K. S., and Selkoe, D. J. (1987). Tau antisera recognize neurofibrillary tangles in a range of neurodegenerative disorders. *Ann. Neurol.* **22**, 514-520.

Johnson, G. V. W. (1992). Differential phosphorylation of Tau by cyclic AMP-dependent protein kinase and Ca-2+/calmodulin-dependent protein kinase-II—Metabolic and functional consequences. *J. Neurochem.* **59**, 2056-2062.

Kanai, Y., Takemura, R., Oshima, T., Mori, H., Ihara, Y., Yanagisawa, M., Masaki, T., and Hirokawa, N. (1989). Expression of multiple tau isoforms and microtubule bundle formation in fibroblasts transfected with a single tau cDNA. *J. Cell Biol.* **109**, 1173–1184.

Kanai, J., Chen, J., and Hirokawa, N. (1992). Microtubule bundling by Tau proteins in vivo: Analysis of functional domains. *EMBO J.* **11**, 3953–3961.

Kidd, M. (1963). Paired helical filaments in electron microscopy of Alzheimer's disease. *Nature* **197**, 192–193.

Kiuchi, A., Otsuka, N., Namba, Y., Nakano, I., and Tomonaga, M. (1991). Presenile appearance of abundant Alzheimer's neurofibrillary tangles without senile plaques in the brain in myotonic dystrophy. *Acta Neuropathol.* **82**, 1–5.

Knops, J., Kosik, K. S., Lee, G., Pardee, J. D., Cohengould, L., and Mcconlogue, L. (1991). Overexpression of tau in a nonneuronal cell induces long cellular processes. *J. Cell Biol.* **114**, 725–733.

Korneyev, A., Binder, L., and Bernardis, J. (1995). Rapid reversible phosphorylation of rat brain Tau proteins in response to cold water stress. *Neurosci. Lett.* **191**, 19–22.

Kosik, K. S., Orecchio, L. D., Bakalis, S., and Neve, R. L. (1989). Developmentally regulated expression of specific tau sequences. *Neuron* **2**, 1389–1397.

Ksiezak-Reding, H., Morgan, K., Mattiace, L. A., Davies, P., Liu, W. K., Yen, S. H., Weidenheim, K., and Dickson, D. W. (1994). Ultrastructure and biochemical composition of paired helical filaments in corticobasal degeneration. *Am. J. Pathol.* **145**, 1496–1508.

Ledesma, M. D., Bonay, P., Colaco, C., and Avila, J. (1994). Analysis of microtubule-associated protein Tau glycation in paired helical filaments. *J. Biol. Chem.* **269**, 21614–21619.

Ledesma, M. D., Bonay, P., and Avila, J. (1995). Tau protein from Alzheimer's disease patients is glycated at its tubulin-binding domain. *J. Neurochem.* **65**, 1658–1664.

Lee, G., and Rook, S. L. (1992). Expression of tau protein in non-neuronal cells—Microtubule binding and stabilization. *J. Cell Sci.* **102**, 227–237.

Lee, G., Cowan, N., and Kirschner, M. (1988). The primary structure and heterogeneity of Tau protein from mouse brain. *Science* **239**, 285–289.

Lee, G., Neve, R. L., and Kosik, K. S. (1989). The microtubule binding domain of tau protein. *Neuron* **2**, 1615–1624.

Lenders, M. B., Peers, M. C., Tramu, G., Delacourte, A., Defossez, A., Petit, H., and Mazzuca, M. (1989). Dystrophic peptidergic neurites in senile plaques of Alzheimers disease hippocampus precede formation of paired helical filaments. *Brain Res.* **481**, 344–349.

Leterrier, J. F., Liem, R. K., and Shelanski, M. L. (1982). Interactions between neurofilaments and microtubule-associated proteins: A possible mechanism for intraorganellar bridging. *J. Cell Biol.* **95**, 982–986.

Lindwall, G., and Cole, R. D. (1984). Phosphorylation affects the ability of Tau protein to promote microtubule assembly. *J. Biol. Chem.* **255**, 5301–5305.

Litersky, J. M., and Johnson, G. V. W. (1992). Phosphorylation by cAMP-dependent protein kinase inhibits the degradation of tau by calpain. *J. Biol. Chem.* **267**, 1563–1568.

Liu, W. K., Williams, R. T., Hall, F. L., Dickson, D. W., and Yen, S. H. (1995). Detection of a Cdc2-related kinase associated with Alzheimer paired helical filaments. *Am. J. Pathol.* **146**, 228–238.

Lobert, S., Isern, N., Hennington, B. S., and Correia, J. J. (1994). Interaction of tubulin and microtubule proteins with vanadate oligomers. *Biochemistry* **33**, 6244–6252.

Lopresti, P. Szuchet, S., Papasozomenos, S. C., Zinkowski, R. P., and Binder, L. I. (1995). Functional implications for the microtubule-associated protein tau: Localization in oligodendrocytes. *Proc. Natl. Acad. Sci. USA* **92**, 10369–10373.

Love, S., Bridges, L. R., and Case, C. P. (1995). Neurofibrillary tangles in Niemann–Pick disease type C. *Brain* **118**, 119–129.

Malandrini, A., Cavallaro, T., Fabrizi, G. M., Berti, G., Salvestroni, R., Salvadori, C., and Guazzi, G. C. (1995). Ultrastructure and immunoreactivity of dystrophic axons indicate a different pathogenesis of Hallervorden-Spatz disease and infantile neuroaxonal dystrophy. *Virchows Archiv. A* **427,** 415-421.

Mandelkow, E., and Mandelkow, E. M. (1994). Microtubule structure. *Curr. Opin. Struct. Biol.* **4,** 171-179.

Mandell, J. W., and Banker, G. A. (1995). The microtubule cytoskeleton and the development of neuronal polarity. *Neurobiol. Aging* **16,** 229-237.

Mann, D. M. A., Prinja, D., Davies, C. A., Ihara, Y., Delacourte, A., Defossez, A., Mayer, R. J., and Landon, M. (1989). Immunocytochemical profile of neurofibrillary tangles in Downs syndrome patients of different ages. *J. Neurol. Sci.* **92,** 247-260.

Martin, H., Lambert, M. P., Barber, K., Hinton, S., and Klein, W. L. (1995). Alzheimer's-associated phospho-Tau epitope in human neuroblastoma cell cultures: Up-regulation by fibronectin and laminin. *Neuroscience* **66,** 769-779.

Marya, P. K., Syed, Z., Fraylich, P. E., and Eagles, P. A. M. (1994). Kinesin and Tau bind to distinct sites on microtubules. *J. Cell Sci.* **107,** 339-344.

Matsuo, E. S., Shin, R. W., Billingsley, M. L., Vandevoorde, A., O'connor, M., Trojanowski, J. Q., and Lee, V. M. Y. (1994). Biopsy-derived adult human brain Tau is phosphorylated at many of the same sites as Alzheimer's disease paired helical filament tau. *Neuron* **13,** 989-1002.

Matus, A. (1994). Stiff microtubules and neuronal morphology. *Trends Neurosci.* **17,** 19-22.

Mawal-Dewan, M., Henley, J., Vandevoorde, A., Trojanowski, J. Q., and Lee, V. M. Y. (1994). The phosphorylation state of Tau in the developing rat brain is regulated by phosphoprotein phosphatases. *J. Biol. Chem.* **269,** 30981-30987.

McKhann, G. M., Drachman, D., Folstein, M., Katzman, R., Price, D., and Stadlan, E. M. (1984). Clinical diagnosis of Alzheimer's disease. *Neurology* **34,** 939-944.

McQuaid, S., Allen, I. V., Mcmahon, J., and Kirk, J. (1994). Association of measles virus with neurofibrillary tangles in subacute sclerosing panencephalitis—A combined in situ hybridization and immunocytochemical investigation. *Neuropathol. Appl. Neurol.* **20,** 103-110.

Mercer, J. A., Albanesi, J. P., and Brady, S. T. (1994). Molecular motors and cell motility in the brain. *Brain Pathol.* **4,** 167-179.

Mercken, M., Vandermeeren, M., Lubke, U., Six, J., Boons, J., and Vandevoorde, A. (1992). Monoclonal antibodies with selective specificity for Alzheimer Tau are directed against phosphatase-sensitive epitopes. *Acta Neuropathol.* **84,** 265-272.

Mercken, M. Fischer, I., Kosik, K. S., and Nixon, R. A. (1995). Three distinct axonal transport rates for tau, tubulin, and other microtubule-associated proteins: Evidence for dynamic interactions of Tau with microtubules in vivo. *J. Neurosci.* **15,** 8259-8267.

Merrick, S. E., Demoise, D. C., and Lee, V. M. Y. (1996). Site-specific dephosphorylation of tau protein at Ser(202)/Thr(205) in response to microtubule depolymerization in cultured human neurons involves protein phosphatase 2A. *J. Biol. Chem.* **271,** 5589-5594.

Michalik, L., Neuville, P., Vanier, M. T., and Launay, J. F. (1995). Pancreatic Tau related maps: Biochemical and immunofluorescence analysis in a tumoral cell line. *Mol. Cell Biochem.* **143,** 107-112.

Mitake, S., Inagaki, T., Niimi, T., Shirai, T., Yamamoto, M. (1989). Development of Alzheimer neurofibrillary changes in two autopsy cases of myotonic dystrophy. *Clin. Neurol. (Tokyo)* **29,** 488-492.

Montpetit, V., Clapin, D. F., and Guberman, A. (1985). Substructure of 20-nm filaments of progressive supranuclear palsy. *Acta Neuropathol.* **68,** 311-318.

Mori, H., Kondo, J., and Ihara, Y. (1987). Ubiquitin is a component of paired helical filaments in Alzheimer's disease. *Science* **235,** 1641-1644.

Morishima-Kawashima, M., Hasegawa, M., Takio, K., Suzuki, M., Titani, K., and Ihara, Y. (1993). Ubiquitin is conjugated with amino-terminally processed Tau in paired helical filaments. *Neuron* **10,** 1151-1160.

Morishima-Kawashima, M., Hasegawa, M., Takio, K., Suzuki, M., Yoshida, H., Titani, K., and Ihara, Y. (1995). Proline-directed and non-proline-directed phosphorylation of PHF-tau. *J. Biol. Chem.* **270,** 823–829.

Mukaeto-Valadinska, E., Harrington, C. R., Roth, M., and Wischik, C. M. (1996). Alterations in Tau protein metabolism during normal aging. *Dementia* **7,** 95–103.

Murakami, N., Ishiguro, K., Ihara, Y., Nonaka, I., Sugita, H., and Imahori, K. (1995). Tau protein immunoreactivity in muscle fibers with rimmed vacuoles differs from that in regenerating muscle fibers. *Acta Neuropathol.* **90,** 467–471.

Nelson, P. T., and Safer, C. B. (1995). Ultrastructure of neurofibrillary tangles in the cerebral cortex of sheep. *Neurobiol. Aging.* **16,** 315–323.

Neuville, P., Vanier, M. T., Michalik, L., and Launay, J. F. (1995). In situ localization with digoxigenin-labelled probes of tau-related mRNAs in the rat pancreas. *Histochem. J.* **27,** 565–574.

Neve, R. L., Harris, P., Kosik, K., Kurnit, D. M., and Donlon, A. (1986). Identification of cDNA clones for the human microtubule-associated protein Tau and chromsmal location of genes for Tau and microtubule-associated protein 2. *Mol. Brain Res* **1,** 271–280.

Nieto, A., Correas, I., Lopezotin, C., and Avila, J. (1991). Tau-related protein present in paired helical filaments has a decreased tubulin binding capacity as compared with microtubule-associated protein tau. *Biochim. Biophys. Acta* **1096,** 197–204.

Nishimura, M., Tomimoto, H., Suenaga, T., Namba, Y., Ikeda, K., Akiguchi, I., and Kimura, J. (1995). Immunocytochemical characterization of glial fibrillary tangles in Alzheimer's disease brain. *Am. J. Pathol.* **146,** 1052–1058.

Nishiyama, I., Oota, T., and Ogiso, M. (1995). Expression of tau-like microtubule-associated proteins in calcitonin-producing cells. *Biomed. Res.* **16,** 59–62.

Ono, S., Inoue, K., Mannen, T., Kanda, F., Jinnai, K., and Takahashi, K. (1987). Neuropathological changes of the brain in myotonic dystrophy. Some new observations. *J. Neurol. Sci.* **81,** 301–320.

Ono, T., Yamamoto, H., Tashima, K., Nakashima, H., Okumura, E., Yamada, K., Hisanaga, S. I., Kishimoto, T., Miyakawa, T., and Miyamoto, E. (1995). Dephosphorylation of abnormal sites of Tau factor by protein phosphatases and its implication for Alzheimer's disease. *Neurochem. Int.* **26,** 205–215.

Otvos, L., Feiner, L., Lang, E., Szendrei, G. I., Goedert, M., and Lee, V. M. Y. (1994). Monoclonal antibody PHF-1 recognizes Tau protein phosphorylated at serine residues 396 and 404. *J. Neurosci. Res.* **39,** 669–673.

Oyama, F., Gu, Y., Murakami, N., Nonaka, I., and Ihara, Y. (1996). Nonneuronal tau, transient upregulation and subsequent accumulation of big Tau and small Tau in chloroquine neuropathy. *Neurobiol. Aging* **17**(Suppl. 4), S188. [Abstract 759]

Oyanagi, K., Makifushi, T., Ohtoh, T., Ikuta, F., Chen, K. M., Chase, T. N., and Gajdusek, D. C. (1994). Topographic investigation of brain atrophy in Parkinsonism dementia complex of Guam: A comparison with Alzheimer's disease and progressive supranuclear palsy. *Neurodegeneration* **3,** 301–304.

Panda, D., Goode, B. L., Feinstein, S. C., and Wilson, L. (1995). Kinetic stabilization of microtubule dynamics at steady state by Tau and microtubule-binding domains of tau. *Biochemistry* **34,** 11117–11127.

Papasozomenos, S. C. (1996). Heat shock induces rapid dephosphorylation of Tau in both female and male rats followed by hyperphosphorylation only in female rats: Implications for Alzheimer's disease. *J. Neurochem.* **66,** 1140–1149.

Parent, M., Delacourte, A., Défossez, A., Hémon, B., Han, K-K., and Petit, H. (1988). Alzheimer's disease: Study of the distribution of paired helical filaments tau proteins in the human central nervous system. *C. R. Acad. Sci.* **306,** 391–397.

Paulus, W., and Selim, M. (1990). Corticonigral degeneration with neuronal achromasia and basal neurofibrillary tangles. *Acta Neuropathol.* **81,** 89–94.

Pillon, B., Gouiderkhouja, N., Deweer, B., Vidailhet, M., Malapani, C., Dubois, B., and Agid, Y. (1995). Neuropsychological pattern of striatonigral degeneration: Comparison with Parkinson's disease and progressive supranuclear palsy. *J. Neurol. Neurosurg. Ps.* **58,** 174–179.

Pizzi, M., Valerio, A., Arrighi, V., Galli, P., Belloni, M., Ribola, M., Alberici, A., Spano, P., and Memo, M. (1995). Inhibition of glutamate-induced neurotoxicity by a Tau antisense oligonucleotide in primary culture of rat cerebellar granule cells. *Eur. Neurosci.* **7,** 1603–1613.

Poirier, J., Minnich, A., and Davignon, J. (1995). Apolipoprotein E, synaptic plasticity and Alzheimer's disease. *Ann. Med.* **27,** 663–670.

Pollard, T. D. (1983). Phosphorylation of microtubule-associated proteins regulates their interaction with actin filaments. *J. Biol. Chem.* **258,** 7064–7067.

Pope, W., Enam, S. A., Bawa, N., Miller, B. E., Ghanbari, H. A., and Klein, W. L. (1993). Phosphorylated Tau epitope of Alzheimer's disease is coupled to axon development in the avian central nervous system. *Exp. Neurol.* **120,** 106–113.

Praprotnik, D., Smith, M. A., Richey, P. L., Vinters, H. V., and Perry, G. (1996). Filament heterogeneity within the dystrophic neurites of senile plaques suggests blockage of fast axonal transport in Alzheimer's disease. *Acta Neuropathol.* **91,** 226–235.

Rebhan, M., Vacun, G., and Rosner, H. (1995). Complementary distribution of Tau proteins in different phosphorylation states within growing axons. *Neuroreport* **6,** 429–432.

Rendeon, A. D., Jung, D., and Jancsik, V. (1990). Interaction of microtubules and microtubule-associated proteins (MAPs) with rat brain mitochondria. *Biochem. J.* **269,** 555–556.

Reszka, A. A., Seger, R., Diltz, C. D., Krebs, E. G., and Fischer, E. H. (1995). Association of mitogen-activated protein kinase with the microtubule cytoskeleton. *Proc. Natl. Acad. Sci. USA* **92,** 8881–8885.

Rewcastle, N. B. (1991). Degenerative diseases of the central nervous system. *In* "Textbook of Neuropathology, second edition" (R. L. Davis and D. M. Robertson, Eds.), pp. 903–961. Williams & Wilkins, Baltimore.

Riederer, B. M., and Binder, L. I. (1994). Differential distribution of Tau proteins in developing cat cerebellum. *Brain Res. Bull.* **33,** 155–161.

Rinne, J. O., Lee, M. S., Thompson, P. D., and Marsden, C. D. (1994). Corticobasal degeneration: A clinical study of 36 cases. *Brain* **117,** 1183–1196.

Rosner, H., Rebhan, M., Vacun, G., and Vanmechelen, E. (1995). Developmental expression of Tau proteins in the chicken and rat brain: Rapid down-regulation of a paired helical filament epitope in the rat cerebral cortex coincides with the transition from immature to adult Tau isoforms. *Int. J. Dev. Neurosci.* **13,** 607–617.

Sadot, E., Heicklenklein, A., Barg, J., Behar, L., Ginzburg, I., and Fisher, A. (1996a). Identification of a Tau promoter region mediating tissue-specific-regulated expression in PC12 cells. *J. Mol. Biol.* **256,** 805–812.

Sadot, E., Gurwitz, D. B., Lazarovici, P., and Ginzburg, I. (1996b). Activation of m(1) muscarinic acetylcholine receptor regulates Tau phosphorylation in transfected PC12 cells. *J. Neurochem.* **66,** 877–880.

Saito, T., Ishiguro, K., Uchida, T., Miyamoto, E., Kishimoto, T., and Hisanaga, S. (1995). In situ dephosphorylation of Tau by protein phosphatase 2A and 2B in fetal rat primary cultured neurons. *FEBS Lett.* **376,** 238–242.

Saoudi, Y., Paintrand, I., Multigner, L., and Job, D. (1995). Stabilization and bundling of subtilisin-treated microtubules induced by microtubule associated proteins. *J. Cell Sci.* **108,** 357–367.

Sattilaro, R. F., Dentler, W. L.; and LeChuyse, E. L. (1981). Microtubule-associated proteins (MAPs) and the organization of actin filaments in vitro. *J. Cell Biol.* **90,** 467–473.

Sautière, P. E., Caillet-Boudin, M. L., Wattez, A., and Delacourte, A. (1994). Detection of Alzheimer-type Tau proteins in okadaic acid-treated Sknsh-Sy 5Y neuroblastoma cells. *Neurodegeneration* **3,** 53–60.

Sawa, A., Oyama, F., Matsushita, M., *et al.* (1994). Molecular diversity at the carboxyl terminus of human and rat Tau. *Mol. Brain Res.* **27**, 111–117.

Schmidt, M. L., Didario, A. G., Lee, V. M. Y., and Trojanowski, J. Q. (1994). Extensive network of PHF tau-rich dystrophic neurites permeates neocortex and nearly all neuritic and diffuse amyloid plaques in Alzheimer disease. *FEBS Lett.* **344**, 69–73.

Schmidt, M. L. Huang, R., Martin, J. A., Henley, J., Mawal-Dewan, M., Hurtig, H. I., Lee, V. M. Y., and Trojanowski, J. Q. (1996a). Neurofibrillary tangles in progressive supranuclear palsy contain the same tau epitopes identified in Alzheimer's disease PHFtau. *J. Neuropathol. Exp. Neurol.* **55**, 534–539.

Schmidt, M. L., Huang, R., Martin, J. A., Henley, J., Mawal-Dewan, M., Hurtig, H. I., Lee, V. M. Y., and Trojanowski, J. Q. (1996b). Neurofibrillary tangles in progressive supranuclear palsy contain the same tau epitopes identified in Alzheimer's disease PHFtau. *J. Neuropathol. Exp. Neurol.* **55**, 534–539.

Schoenfeld, T. A., and Obar, R. A. (1994). Diverse distribution and function of fibrous microtubule-associated proteins in the nervous system. *Int. Rev. Cytol.* **151**, 67–137.

Schweers, O., Mandelkow, E. M., Biernat, J., and Mandelkow, E. (1995). Oxidation of cysteine-322 in the repeat domain of microtubule-associated protein Tau controls the in vitro assembly of paired helical filaments. *Proc. Natl. Acad. Sci. USA* **92**, 8463–8467.

Sergeant, N., Bussiere, T., Vermersch, P., Lejeune, J. P., and Delacourte, A. (1995). Isoelectric point differentiates PHF-Tau from biopsy-derived human brain Tau proteins. *Neuroreport* **6**, 2217–2220.

Seubert, P., Mawaldewan, M., Barbour, R., Jakes, R., Goedert, M., Johnson, G. V. W., Litersky, J. M., Schenk, D., Lieberburg, I., Trojanowski, J. Q., and Lee, V. M. Y. (1995). Detection of phosphorylated Ser(262) in fetal tau, adult tau, and paired helical filament tau. *J. Biol. Chem.* **270**, 18917–18922.

Shackelford, D. A., and Nelson, K. E. (1996). Changes in phosphorylation of Tau during ischemia and reperfusion in the rabbit spinal cord. *J. Neurochem.* **66**, 286–295.

Shankar, S. K., Yanagihara, R. M., Garruto, I., Grundke-Iqbal, K. S., Kosik, D. C., and Gajdusek, C. (1989). Immunocytochemical characterization of neurofibrillary tangles in amyotrophic lateral sclerosis and Parkinsonism–Dementia of Guam. *Ann. Neurol.* **25**, 146–151.

Shea, T. B., and Fischer, I. (1996). Phosphatase inhibition in human neuroblastoma cells alters Tau antigenicity and renders it incompetent to associate with exogenous microtubules. *FEBS Lett.* **380**, 63–67.

Shea, T. B., Beermann, M. L., Nixon, R. A., and Fischer, I. (1992). Microtubule-associated protein tau is required for axonal neurite elaboration by neuroblastoma cells. *J. Neurosci. Res.* **32**, 363–374.

Shinozaki, T., Watanabe, H., Arita, S., and Chigira, M. (1995). Amino acid phosphatase activity of alkaline phosphatase—A possible role of protein phosphatase. *Eur. J. Biochem.* **227**, 367–371.

Singh, T. J., Zaidi, T., Grundke-Iqbal., and Iqbal, K. (1996). Non-proline-dependent protein kinases phosphorylate several sites found in Tau from Alzheimer disease brain. *Mol. Cell. Biochem.* **154**, 143–151.

Smith, M. A., Siedlak, S. L., Richey, P. L., Mulvihill, P., Ghiso, J., Frangione, B., Tagliavini, F., Giaccone, G., Bugiani, O., Praprotnik, D., Kalaria, R. N., and Perry, G. (1995a). Tau protein directly interacts with the amyloid beta-protein precursor: Implications for Alzheimer's disease. *Nature Med.* **1**, 365–369.

Smith, M. A., Sayre, L. M., Monnier, V. M., and Perry, G. (1995b). Radical AGEing in Alzheimer's disease. *Trends Neurosci.* **18**, 172–176.

Song, J. S., and Yang, S. D. (1995). Tau protein kinase I/GSK-3 beta/kinase F-A in heparin phosphorylates Tau on Ser(199), Thr(231), Ser(235), Ser(262) Ser(369), and Ser(400) sites phosphorylated in Alzheimer disease brain. *J. Protein Chem.* **14**, 95–105.

Sontag, E., Nunbhakdicraig, V., Bloom, G. S., and Mumby, M. C. (1995). A novel pool of protein phosphatase 2A is associated with microtubules and is regulated during the cell cycle. *J. Cell Biol.* **128,** 1131–1144.

Soulié, C., Lepagnol, J., Delacourte, A., and Caillet-Boudin M. L. (1996). Dephosphorylation studies of SKNSH-Sy 5Y cell Tau proteins by endogenous phosphatase activity. *Neurosci. Lett.* **206,** 189–192.

Sperber, B. R., Leight, S., Goedert, M., and Lee, V. M. Y. (1995). Glycogen synthase kinase-3 beta phosphorylates Tau protein at multiple sites in intact cells. *Neurosci. Lett.* **197,** 149–153.

Spillantini, M. G., Crowther, R. A., and Goedert, M. (1996). Comparison of the neurofibrillary pathology in Alzheimer's disease and familial presenile dementia with tangles. *Acta Neuropathol.* **92,** 42–48.

Steele, J. C., Richardson, J. C., and Olzewski, J. (1964). Progressive supranuclear palsy. A heterogeneous degeneration involving brain stem, basal ganglia and cerebellum with vertical gaze and pseudobulbar palsy, nuchal dystonia and dementia. *Arch. Neurol.* **10,** 333–359.

Strittmatter, W. J., Saunders, A. M., Geodert, M., Weisgraber, K. H., Dong, L. M., Jakes, R., Huang, D. Y., Pericakvance, M., Schmechel, D., and Roses, A. D. (1994). Isoform-specific interactions of apolipoprotein E with microtubule-associated protein tau: Implications for Alzheimer disease. *Proc. Natl. Acad. Sci. USA* **91,** 11183–11186.

Takashima, A., Noguchi, K., Michel, G., Mercken, M., Hoshi, M., Ishiguro, K., and Imahori, K. (1996). Exposure of rat hippocampal neurons to amyloid beta peptide (25–35) induces the inactivation of phosphatidyl inositol-3 kinase and the activation of Tau protein kinase I glycogen synthase kinase-3 beta. *Neurosci. Lett.* **203,** 33–36.

Takauchi, S., Hosomi, M., Marasigan, S., Sato, M., Hayashi, S., and Miyoshi, K. (1984). An ultrastructural study of Pick bodies. *Acta Neuropathol.* **644,** 344–348.

Teller, J. K., Russo, C., Debusk, L. M., Angelini, G., Zaccheo, D., Dagnabricarelli, F., Scartezzini, P., Bertolini, S., Mann, D. M. A., Tabaton, M., and Gambetti, P. (1996). Presence of soluble amyloid beta-peptide precedes amyloid plaque formation in Down's syndrome. *Nature Med.* **2,** 93–95.

Tellez-Nagel, I., and Wisniewski, H. M. (1973). Ultrastructure of neurofibrillary tangles in Steele–Richardson–Olszewski syndrome. *Arch. Neurol.* **29,** 324–327.

Tomonaga, M. (1977). Ultrastructure of neurofibrillary tangles in progressive supranuclear palsy. *Acta Neuropathol.* **37,** 177–181.

Tranchant, C., Sergeant, N., Wattez, A., Mohr, M., Warter, J. M., and Delacourte, A. (1996). Biochemical study of the Tau protein in Gerstmann–Straussler–Scheinker syndrome with neurofibrillary tangles. Submitted for publication.

Turpin, J. C., Masson, M., and Baumann, N. (1991). Clinical aspects of Niemann–Pick type C disease in the adult. *Dev. Neurosci.* **13,** 304–306.

Vandermeeren, M., Lubke, U., Six, J., and Cras, P. (1993). The phosphatase inhibitor okadaic acid induces a phosphorylated paired helical filament Tau-Epitope in human LA-N-5 neuroblastoma cells. *Neurosci. Lett.* **153,** 57–60.

Vermersch, P., Frigard, B., David, J. P., Fallet-Bianco, C., and Delacourte, A. (1992). Presence of abnormally phosphorylated Tau proteins in the entorhinal cortex of aged Non-Demented subjects. *Neurosci. Lett.* **144,** 143–146.

Vermersch, P., Delacourte, A., Javoy-Agid, F., Hauw, J. J., and Agid, Y. (1993). Dementia in Parkinson's disease—Biochemical evidence for cortical involvement using the immunodetection of abnormal Tau-Proteins. *Ann. Neurol.* **33,** 445–450.

Vermersch, P., Robitaille, Y., Bernier, L., Wattez, A., Gauvreau, D., and Delacourte, A. (1994). Biochemical mapping of neurofibrillary degeneration in a case of progressive supranuclear palsy: Evidence for general cortical involvement. *Acta Neuropathol.* **87,** 572–577.

Vermersch, P., David, J. P., Frigard, B., Fallet-Bianco, C., Wattez, A., Petit, H., and Delacourte, A. (1995a). Cortical mapping of Alzheimer pathology in brains of aged nondemented subjects. *Prog. Neuro-Psychopharmacol. Biol. Psychiatr.* **19,** 1035–1047.

Vermersch, P., Bordet, R., Ledoze, F., Ruchoux, M. M., Chapon, F., Thomas, P., Destée, A., and Lechevallier, B. (1995b). Demonstration of a specific profile of pathological Tau proteins in frontotemporal dementia cases. *C. R. Acad. Sci.* **318,** 439–445.

Vermersch, P., Sergeant, N., Ruchoux, M. M., Hofmann-Radvanyi, H., Petit, H., Dewailly, P., and Delacourte, A. (1996). Specific Tau variants in the brain from patients with myotonic dystrophy. *Neurology* **47,** 711–717.

Vincent, I., Rosado, M., and Davies, P. (1996). Mitotic mechanisms in Alzheimer's disease? *J. Cell Biol.* **132,** 413–425.

Vulliet, R., Halloran, S. M., Braun, R. K., Smith, A. J., and Lee, G. (1992). Proline-directed phosphorylation of human Tau-protein. *J. Biol. Chem.* **267,** 22570–22574.

Weingarten, M. D., Lockwood, A. H., Hwo, S-Y., and Kirschner, M. W. (1975). A protein factor essential for microtubule assembly. *Proc. Natl. Acad. Sci. USA* **72,** 1858–1862.

Wisniewski, T., Lalowski, M., Bobik, M., Russell, M., Strosznajder, J., and Frangione, B. (1996). Amyloid beta 1-42 deposits do not lead to Alzheimer's neuritic plaques aged dogs. *Biochem. J.* **313,** 575–580.

Yamada, T., McGeer, P. L., and McGeer, E. G. (1992). Appearance of paired nucleated, tau-positive glia in progressive supranuclear palsy brain tissue. *Neurosci. Lett.* **135,** 99–102.

Yamamoto, H., Saitoh, Y., Fukunaga, K., Nishimura, H., and Miyamoto, E. (1988). Dephosphorylation of microtubule proteins by brain protein phosphatases 1 and 2A, and its effect on microtubule assembly. *J. Neurochem.* **50,** 1614–1623.

Yamamoto, H., Hasegawa, M., Ono, T., Tashima, K., Ihara, Y., and Miyamoto, E. (1995). Dephosphorylation of fetal-Tau and paired helical filaments-Tau by protein phosphatases 1 and 2A and calcineurin. *J. Biochem.* **118,** 1224–1231.

Yamazaki, M., Nakano, I., Imazu, O., Kaieda, R., and Terashi, A. (1995). Paired helical filaments and straight tubules in astrocytes: An electron microscopic study in dementia of the Alzheimer type. *Acta Neuropathol.* **90,** 31–36.

Yan, S. D., Yan, S. F., Chen, X., Fu, J., Chen, M., Kuppusamy, P., Smith, M. A., Perry, G., Godman, G. C., Nawroth, P., Zweiter, J. L., and Stern, D. (1995). Nonenzymatically glycated Tau in Alzheimer's disease induces neuronal ozidant stress resulting in cytokine gene expression and release of amyloid beta-peptide. *Nature Med.* **1,** 693–699.

# Differentiation and Transdifferentiation of the Retinal Pigment Epithelium

Shulei Zhao,* Lawrence J. Rizzolo,† and Colin J. Barnstable*
*Department of Ophthalmology and Visual Science and †Section of Anatomy, Department of Surgery, Yale University School of Medicine, New Haven, Connecticut 06520

---

The retinal pigment epithelium (RPE) lies between the retina and the choroid of the eye and plays a vital role in ocular metabolism. The RPE develops from the same sheet of neuroepithelium as the retina and the two derivatives become distinguished by different expression patterns of a number of transcription factors during embryonic development. As the RPE layer differentiates it expresses a set of unique molecules, many of which are restricted to certain regions of the cell. RPE cells undergo both a loss of polarity and a loss of expression of many of these cell type-specific molecules when placed in monolayer culture. The RPE of many species, including mammals, can be induced to transdifferentiate by growth factors such as basic fibroblast growth factor. Under the influence of such factors the RPE is triggered to alter expression of a wide array of molecules and to take on a retinal epithelium fate, from which differentiated retinal cell types including rod photoreceptors can be produced.

**KEY WORDS:** Development, Retina, Cell polarity, Transdifferentiation, Gene expression, Cell culture, Organ culture, Retinal pigment epithelium.

---

## I. Introduction

The mature retinal pigment epithelium (RPE) is a mosaic of polygonal cells interposed between the choroid and the neural retina. The apical side of RPE cells closely associates with the outer segments of cones and rods, whereas the base of the cell attaches to Bruch's membrane (Fig. 1). Like all epithelial and endothelial cells, mature RPE cells show a morphological and functional polarity along their apical-to-basal axis. The basal plasma membrane is markedly infolded and contains fibrils, some of which extend

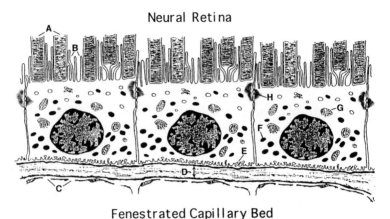

FIG. 1 A schematic drawing of the cross section of the RPE cell and outer segments of photoreceptors (adapted from Clark, 1986). The apex of the RPE are modified into villi that envelop the impinging tips of photoreceptor outer segments. These processes are involved in the phagocytosis of shed outer segments of photoreceptors. A, photoreceptor outer segments; B, microvilli; C, fenestrations in capillary wall; D, Bruch's membrane; E, infoldings of basal plasma membrane; F, melanin; G, phagosome; H, apical junctional complex with intermixed gap, adherens, and tight junctions.

into the underlying Bruch's membrane (Zinn and Benjamin-Henkind, 1979). These fine connections may account for the greater cohesion between the RPE and Bruch's membrane than between RPE and retina. The lateral cell walls of RPE have few or no desmosomes and are fairly smooth, with only occasional interdigitations compared to the highly convoluted basal surface. The apical junctional complexes are also unusual for an epithelium. They are located at the apical end of the lateral wall and consist of intermixed tight, adherens, and gap junctions (Hudspeth and Yee, 1973). The tight junctions form a permeability barrier characterized by fusion of adjacent RPE cell membranes (Fig. 1). The apical microvilli are unusual because they interact with a cellular tissue, retinal photoreceptors. This interaction modulates the polarity that is induced by basal and lateral membrane interactions (Rodriguez-Boulan and Nelson, 1989).

RPE cells are involved in many activities that are essential for function and survival of retinal photoreceptors. RPE cells form tight intercellular junctions with one another and serve as part of the blood–retina barrier regulating the flow of nutrients to the outer retina and preventing toxic levels of many serum components. The outer segments of photoreceptors interdigitate with processes from the RPE which provides physical protection for these sensory neurons (Steinberg and Wood, 1979). Melanin granules in RPE cells absorb light and, therefore, reduce scattered background, block light coming through the sclera, and improve image sharpness. RPE cells phagocytoze shed outer segments of rods and cones and synthesize numerous enzymes to digest these debris from photoreceptors (Clark, 1986). RPE cells also synthesize and transport many substances including glycosaminoglycans and vitamin A metabolites to and from retinal cells and the choroidal circulation (Zinn and Benjamin-Henkind, 1979).

RPE cells have long been an object of study because of their ability to transdifferentiate and allow the regeneration of a neural retina in many species. As well as being of intrinsic biological interest for understanding the regulation and maintenance of the differentiated state, this finding has important medical implications in that it may provide a way of treating a variety of retinal degenerative diseases.

The availability of modern molecular biology techniques has provided powerful tools to examine the development and physiological functions of RPE and significant progress has been made in the past decade. In this chapter we review (i) current understanding of normal RPE development including phenotype determination in the optic vesicle, cell polarization, pigmentation, and some molecular markers useful for studying RPE differentiation; (ii) stability of the RPE phenotype in altered environments such as retinal injuries, transplantation, and tissue cultures; and (iii) *in vitro* RPE transdifferentiation to other tissue types, particularly the neural retina.

## II. Development of the RPE

During vertebrate embryogenesis, the optic vesicle grows out from the ventrolateral forebrain neuroepithelium (Morse and McCann, 1984; Barnstable, 1987, 1991). The neural retina develops from the anterior of the optic vesicle under the influence of head ectoderm (Coulombre and Coulombre, 1965). The optic vesicle, in a complex series of interactions with the developing lens vesicle, invaginates to form the optic cup. The inner and outer layers of the optic cup give rise to neural retina and RPE, respectively, and the margins of these two layers form the ciliary epithelium. The cells of the optic stalk connect the optic cup to the forebrain and eventually form some of the glia of the optic nerve (Barnstable, 1991). In rats, formation of the optic vesicle begins at Day 10 of gestation (E10, where E0 is defined as the day of conception), is clearly visible at E11, and has become an optic cup by E12 (Morse and McCann, 1984). In mice the process occurs at approximately 1 day earlier, reflecting the shorter gestation period of this species. In chick the optic cup forms by E2 and pigmentation first appears in the RPE on E3.

At the same time as the optic vesicle is growing out from the forebrain, changes occur in surrounding tissues that lead to the formation of other components of the eye. Contact of the optic vesicle with the surface ectoderm induces the formation of the lens vesicle and the corneal epithelium. The nature of interaction of the developing lens with the optic vesicle as it becomes the optic cup is not fully understood. It has been suggested that the invagination of the optic vesicle into the optic cup involves calcium/calmodulin-mediated changes in cytoskeleton organization or activity (Hilfer et al., 1981). The invagination may also be aided by a close apposition or fusion of extracellular matrices of the two structures. In addition to the neuroepithelium and the surface ectoderm, the developing eye receives contributions from the cranial neural crest. Crest cells migrate into the developing eye and form the melanocytes of the choroid, part of the iris, and the corneal endothelium. The signals integrating the movements and differentiation of these many cell types are not yet known.

## A. Patterns of Gene Expression and Tissue Type Specification

Genes encoding putative transcription factors containing homeobox and/or paired box motifs have often been implicated in regional specification during development of invertebrates and vertebrates. Some of these genes are expressed in early eye rudiments, and their distribution often defines future ocular structures. Many of these genes were first defined in the larger structures of the brain and so the majority of them are expressed primarily in the more closely related retinal layer of the optic cup. To date, there has not been a systematic search for the expression of these classes of genes using the developing RPE layer as starting material.

Two Msh-like homeobox genes, *Hox-7.1* and *Hox-8.1* (or *Msx-1* and *Msx-2* according to Scott, 1992) are expressed in a complementary manner, suggesting that these genes are involved in pattern formation in the developing mammalian eye (Monaghan et al., 1991). The expression patterns of these genes define domains of the neural retina, cornea, and ciliary body in the mouse eye several days before morphological differentiation is easily detected. The prospective lens epithelium, expressing *Hox-8.1*, and the corneal epithelium, expressing both *Hox-7.1* and *Hox-8.1*, have differentiated from their adjacent surface epithelium at E9.5. At this stage, the anterior portion of the optic vesicle, the prospective neural retina, has contacted the surface epithelium and been distinguished from the adjacent optic vesicle tissue by expressing *Hox-8.1*. At E11, *Hox-8.1* expression marks the prospective lens and the inner layer of the optic cup, the neural retina. At E12.5, the expression of *Hox-7.1* has been localized in the presumptive ciliary epithelium, suggesting that the ciliary epithelium is distinct from the neural retina and RPE approximately 2 days before the morphogenesis of the ciliary body begins and also well before the contact of lens with the rim of folded neuroepithelium.

The expression pattern of the paired-box-containing gene, *Pax-2*, suggests that this gene has at least two distinct roles during vertebrate eye development: (i) morphogenesis of the optic cup and stalk prior to axon growth into the stalk, and (ii) axonogenesis (Nornes *et al.*, 1990). At E9, *Pax-2* is expressed in the anterior region of the optic vesicle but not in the optic stalk. As the optic vesicle invaginates, the pattern changes to expression in the ventral half of the optic cup and optic stalk. *Pax-2* transcripts eventually disappear from the neural retina and become restricted to the optic stalk as axons of retinal ganglion cells start to grow out.

Another paired-box gene, *pax-6*, is expressed in the forebrain region that begins to evaginate to form the optic vesicle (Walther and Gruss, 1991; Li *et al.*, 1994). At later stages it is found expressed at high levels in the retinal epithelium, the lens, and the presumptive corneal epithelium, but it is expressed at much lower levels in the RPE. It has been suggested that the level of expression is related to the rate of cell division, which is already much lower in the RPE layer by the optic cup stage. Mutations in *pax-6* are lethal in the homozygous state but lead to the *small eye* phenotype in the mouse and the *aniridia* phenotype in humans. The homologous gene in *Drosophila*, eyeless, is not only essential for eye formation but when expressed ectopically can lead to eye formation in place of other structures (Quiring *et al.*, 1994). This has led to the suggestion that *pax-6* may be a master gene for eye formation. In vertebrates this would probably include formation of the RPE.

Two vertebrate homeobox genes, *Otx-1* and *Otx-2*, related to the *Drosophila* gene *orthodenticle*, were expressed in restricted regions of the developing mouse brain (Simeone *et al.*, 1993). At the optic vesicle stage (E10), *Otx-1* is strongly expressed in the margin of anterior and the posterior layer of the vesicle which will become ciliary epithelium and RPE, respectively. Expression of *Otx-1* in the presumptive neural retina in the optic vesicle is much weaker. Expression of *Otx-2* is found in almost the entire optic vesicle at E10. Neither of these two genes are expressed in the optic stalk at this stage. At the time the optic cup is formed (E12), expression of *Otx-1* is restricted to the ventricular zone of the neural retina, the RPE, and the optic stalk through which axons of retinal ganglion cells grow whereas *Otx-2* is only expressed in the RPE layer of the optic cup. At a later stage (E17), *Otx-1* is only found in the ciliary body/iris region, whereas *Otx-2* is expressed in the RPE and outer layer of neural retina.

Liu *et al.* (1994) identified a murine homeobox gene, *Chx10*, which appears to be important in both determination of neural retina in the optic vesicle and the formation of specific retinal cell types. *Chx10* transcripts can be detected by *in situ* hybridization in the optic vesicle at E9.5 but not in the overlying head ectoderm. Expression of *Chx10* is restricted to the anterior region of the optic vesicle that will become the neural retina. No *Chx10* expression is seen in the optic stalk or the posterior regions of the optic vesicle from which the RPE is derived. When the optic cup has formed at E11.5, *Chx10* is strongly expressed in all the cells of the actively proliferating and thickening neural retina layer. No expression is observed in any other parts of the developing eye. During murine embryogenesis, *Chx10* is also expressed in developing rhombencephalon, spinal cord, thalamus, and lips. In the adult mouse, *Chx10* mRNA is expressed exclusively in the central nervous system, with a high level in retina and low levels in pons, medulla, and spinal cord. Immunocytochemistry with Chx10 antibodies showed that the protein is present in the inner nuclear layer of the mature retina and appears to be restricted to bipolar cells. The temporal and spatial expression patterns of *Chx10* suggest that it may have two distinct unctions in retinal development: (i) involvement in specification of retinal fate and retinal cell proliferation, and (ii) determination and maintenance of bipolar cell phenotype. Mutations in Chx10 cause the ocular retardation phenotype (Burmeister *et al.*, 1996). Loss of Chx10 led to both reduced proliferation of retinal progenitors and the specific absence of bipolar cells. The relationship between these two effects has yet to be fully determined.

A novel homeodomain gene *Rx* was isolated from early *Xenopus* embryos (Mathers and Jamrich, 1995). This gene is expressed at the beginning of neurulation in a large area of the anterior neural plate and becomes restricted to the anterior of the optic vesicle. *Rx* mRNA is present throughout embryogenesis in the proliferating cells of the retina. As development proceeds further, *Rx* expression remains in the ciliary margin. The same expression pattern of *Rx* was also observed in zebrafish and mouse. Two additional new homeobox genes were discovered in the chick embryo and their expression patterns suggest their roles in establishment of the eye primordia during organogenesis (Sundin *et al.*, 1995).

It is not clear whether any of the gene products listed previously function in a linked cascade or whether they act in multiple independent and parallel pathways. For example, some of the transcription factors may be involved in cell proliferation and have nothing to do with the fate or differentiation of cell types. Similarly, it is not known whether any single gene can specify the fate of a cell layer or region, or whether the genes act in a combinatorial manner such that the definition of a retinal epithelium or of an RPE layer is the coincident expression of a specific group of transcription factors. On the other hand, there are transcription factors that are expressed by specific regions of the optic cup that are clearly involved in the regulation of expression of genes that define cellular phenotype.

Perhaps the best characterized of these, microphthalmia gene (*Mi*), plays a critical role in differentiation of pigmented cells including the RPE. Since the first mouse *Mi* mutation was discovered over 50 years ago (Hertwig, 1942), at least 17 mutant alleles have been identified and genetically characterized (Silvers, 1979; Green, 1989; Tachibana et al., 1992; Krakowsky et al., 1993). Common to all *Mi* alleles are defects in neural crest-derived melanocytes, which lead to coat color dilution, white spotting, or complete loss of pigmentation and to deafness. In their most severe manifestations, these cellular defects also lead to microphthalmia and osteopetrosis. Various mouse mutant phenotypes seem to be associated with different mutations in *Mi* gene (Steingrimsson et al., 1994). It was found that mutations in human *Mi* gene (*MITF*) caused Waardenburg syndrome type 2, a dominantly inherited syndrome associated hearing loss and pigmentary disturbances (Tachibana et al., 1994; Hughes et al., 1994; Tassabehji et al., 1994).

*Mi* encodes a basic helix–loop–helix leucine zipper transcription factor (Hodgkinson et al., 1993; Hughes et al., 1993). During mouse development, expression of this gene is limited to the RPE and other melanocytes in the skin and surrounding the otic vesicle (Hodgkinson et al., 1993). Mi protein binds DNA as a homodimer or heterodimer with another closely related transcription factor such as TFEB, TFE3, or TFEC (Hemesath et al., 1994).

During development of the mi/mi mouse, rather than becoming a thin monolayer of epithelium the outer layer of the optic cup (the RPE) increased significantly in thickness to form a pseudostratified structure, resembling that of neural retina of the early stage (Packer, 1967; Hero et al., 1991). This observation raises the possibility that *Mi* may be an important factor that determines the RPE phenotype as the optic vesicle invaginates to form the two layers of the optic cup.

Summarized in Fig. 2 are expression patterns of a few genes selected to show their possible roles in defining regions of different tissue types during early stages of mouse eye development. At the optic vesicle stage (Fig. 2A), *Chx10* and *Rx* are expressed in the presumptive neural retina (the anterior of the optic vesicle), whereas *Otx-1* strongly expressed in presumptive RPE (the posterior). When the optic cup is formed (Fig. 2B), regions of the neural retina, RPE, and ciliary body/iris are defined, respectively, by (i) *Chx10* and *Rx*, (ii) *Otx-2* and *Mi*, and (iii) *Otx-1* and *Msx-1* expression patterns. This figure is not meant to suggest that these genes are sufficient to specify the various tissue types in the developing eye. There are temporal differences in the patterns of expression of some of these genes and there are undoubtedly many others remaining to be discovered. Nevertheless, understanding how the expression of some of these genes becomes restricted to one or another layer of the optic cup will provide important

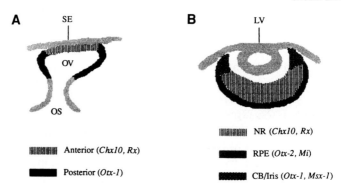

FIG. 2 Patterns of gene expression in the optic vesicle and cup. (A) At the optic vesicle stage, homeobox genes *Rx* and *Chx10* are expressed exclusively in the anterior region, the presumptive neural retina, whereas *Otx-1* is strongly expressed in the posterior region, the presumptive RPE. (B) In the optic cup, the neural retina (NR) is defined by the expression domain of *Rx* and *Chx10*, the RPE by the *Otx-2* and *Mi*. In a later stage, the ciliary body (CB) and iris are defined by restricted expression of *Otx-1* and *Msx-1*. SE surface ectoderm; OV, the optic vesicle; OS, the optic stalk; LV, lens vesicle.

clues as to the mechanisms governing the induction of the differentiated state in these cells.

### B. Pigmentation and the Tyrosinase Gene

#### 1. Tyrosinase and Melanin Synthesis

One of the functions of the RPE is to absorb light and reduce scattering and, therefore, improve the quality of images that are received by the retina. The light-absorbing molecule in the RPE is melanin, which binds to other proteins and aggregates to form pigment granules, also called melanosomes. Melanin absorbs radiant energy in the visible as well as ultraviolet spectrum and dissipates the absorbed energy in the form of heat (Zinn and Benjamin-Henkind, 1979). Melanin in the RPE is also capable of binding with metal ions and it has also been suggested that sequestration of heavy metal ions by melanin is a protective mechanism against oxidative damage to RPE cells (Sarna, 1992). In addition to these roles of pigment granules in visual reception and RPE physiology, pigmentation is a naturally visible indicator for RPE development and change of phenotype *in vivo* and *in vitro*.

The synthesis of the melanin begins with the accumulation of an enzyme, tyrosinase, in premelanosomes that are derived from smooth endoplasmic reticulum through an intermediary vesicular stage and contain filaments.

In the rat, tyrosinase mRNA is detectable at E12, 1 day before the protein can be detected by immunohistochemistry, and reaches approximately adult levels at E14 (Zhao et al., 1995). During this process there is a partial obscuration of the fine filaments within the premelanosome by the newly synthesized electron-dense melanin. When the melanosome is fully melanized, tyrosinase activity within the melanosome disappears (Zinn and Benjamin-Henkind, 1979). Tyrosinase is an unusual enzyme that can catalyze three different reactions in the biosynthetic pathway of melanin (Prota, 1992). The melanin produced can be of two basic types—eumelanins (brown or black) and phaeomelanins (red or yellow) (Hearing and Tsukamoto, 1991; Ito, 1993). Melanins usually found in humans are mixtures of both types in different proportions. Tyrosinase is the rate-limiting enzyme in melanin synthesis and is specifically expressed in differentiated pigmented cells.

In addition to the tyrosinase gene that is mapped to the albino locus, two genes, *trp-1* and *trp-2*, encoding tyrosinase-related proteins (TRP) have been cloned and mapped to the *brown* and *slaty* loci, respectively, in mice (Hearing and King, 1993; Kwon, 1993). TRP-1 and TRP-2 have structures and features similar to tyrosinase but with distinct catalytic capabilities in the synthetic pathway of melanin biopolymer. They catalyze the steps following the action of tyrosinase and modulate the quantity and quality of the melanin produced. (Tsukamoto et al., 1992; Kobayashi et al., 1994a,b). Mutations in *trp-1* result in the formation of brown rather than black melanin, whereas the coat color of the *trp-2* mutant mice is dark gray rather than the dark black color of congenic black mice.

## 2. Regulation of Tyrosinase Gene Transcription

The restricted expression of the tyrosinase gene in cells producing pigment suggests the presence of *cis*-regulatory elements and *trans*-acting tissue-specific factors. A tyrosinase minigene under control of 5.5 kb of upstream sequences has been shown to direct expression to pigmented cell types and restore pigmentation in albino mice (Beermann et al., 1990; Yokoyama et al., 1990; Tanaka et al., 1990). In addition, the simian virus 40 early region under control of tyrosinase upstream sequence led to tumor formation of pigmented cell types (Klein-Szanto et al., 1991; Bradl et al., 1991). It has been further demonstrated that 270 bp of 5′-flanking sequence are sufficient for specific expression of the transgene in the pigmented cells (Klüppel et al., 1991). The timing of this transgene expression also mimicked precisely the pattern of the endogenous tyrosinase gene, suggesting that sequences in the immediate vicinity of the mouse tyrosinase gene are sufficient to provide cell-type specificity and developmental regulation in retinal pigment epithelium and melanocytes (Beermann et al., 1992a,b).

In vitro assays demonstrated that there exist one negative and two positive elements within the 270-bp sequence (Ganss et al., 1994b,c). One of the positive elements includes the M-box, a sequence motif shared with the promoter of two other melanocyte-specific genes, trp-1 and trp-2. An M-box contains an E-box (or AP-1 element). Cotransfection experiments have provided evidence that the basic helix–loop–helix zipper protein encoded by the (Mi) transactivated the tyrosinase promoter, probably by binding to the M-box. Another strong melanoma cell-specific enhancer element at 12 kb upstream of the mouse tyrosinase gene was identified (Ganss et al., 1994a). Transgenic mice and transient transfection assays confirmed its specific activity in the melanocyte. Full activity of this enhancer element was obtained with a minimal sequence of 200 bp, which was bound by a melanoma cell-specific complex.

More extensively investigated is the activity of the human tyrosinase gene promoter. Ponnazhagan et al. (1994) identified many consensus cis-acting elements including five E-boxes in a −2020-bp promoter that yielded high activity. However, when deletion reached −1739 bp, the promoter activity was dramatically reduced, indicating that important enhancer elements are present between −2020 and −1739. Further deletions also resulted in significant decreases in the promoter activity. When the deletion reached −550 bp, there was only 26% promoter activity compared to that of −2020 promoter. This study demonstrated that human tyrosinase gene is controlled by multiple regulatory elements. Bentley et al. (1994) showed that as small as 115 bp of the upstream sequence was sufficient to direct tissue-specific expression. This 115-bp stretch contains three positive elements: an M-box, an Sp1 site, and a highly conserved element located between −14 and +1 comprising an E-box motif and an overlapping octamer element. Beyond this 115-bp stretch, there exist one positive and one negative element between positions −185 and −150 and positions −150 and −115, respectively. It was also found that the transcription factor encoded by the microphthalmia gene can transactivate the tyrosinase promoter via the M-box and the conserved E-box. Yasumoto et al. (1994) identified a distal enhancer element (−1861 to −1842) in the 5′-flanking region of the human tyrosinase gene. The product of Mi specifically transactivated promoter activity of the tyrosinase gene probably through the E-box contained in this enhancer element. It has been shown that in melanocytes transformed by adenovirus E1A, expression of Mi was significantly inhibited (Yavuzer et al., 1995). As a result, melanocyte-specific transcription was repressed. However, this transcription activity was restored by ectopic expression of Mi (Yavuzer et al., 1995). It has been described in a previous section that mutations in the Mi gene result in dilution or complete loss of pigmentation. All the evidence strongly suggests that Mi protein is a positive and specific factor regulating expression of tyrosinase and related genes in pigmented cells in both mouse and human. It may act in cooperation with other factors on different enhancer elements in promoters of tyrosinase and related genes in different pigment cell types or during different developmental stages.

## C. Molecular Markers of RPE during Differentiation

An approach widely adopted in the study of cell differentiation is to monitor expression of cell type-specific markers. These markers may provide information such as cell identities, timing for onset of differentiation or maturation, and change of cell phenotype. A marker is usually detected either by using antibody to the protein or by *in situ* hybridization of a DNA or RNA probe to the mRNA for the protein. Here, we summarize some of the molecular markers that have be used to study RPE differentiation *in vivo* and *in vitro*.

Change of expression of two membrane antigens, N-CAM and RPE-specific cell surface antigen RET-PE2 (50–55 kDa), was studied during normal rat optic cup development and during culture of RPE (Neill and Barnstable, 1990). RET-PE2 antibody clearly recognizes both layers of the optic cup at E13, but it stains only the RPE in the adult. Loss of this antigen from the retinal epithelium follows a center to periphery time course. It is of interest that both layers of the optic cup at the ciliary margin are RET-PE2 positive at E17 because this is about the time in rat that ciliary epithelium begins to form. The pattern of RET-PE2 antigen expression suggests that the marginal cells are different at this stage and it might be the portion of the nonpigmented layer that is responsive to signals stimulating ciliary epithelium differentiation. The distribution of the cell-adhesion molecule N-CAM (180- and 140-kDa forms) during optic cup development is in contrast to that of RET-PE2. This molecule is expressed in neuroepithelia at high concentrations from the earliest times after neural induction. In the optic cup, N-CAM is initially expressed on cells of both layers but then lost from the RPE. The loss of antigen occurs after differentiation of the RPE has begun. By E19, N-CAM is strongly expressed on all cells of the inner retina and weakly expressed in the outer retina. The distribution of the carbohydrate epitope recognized by antibody HNK-1 shows a similar pattern to that of the N-CAM polypeptide early in development but differs in later stages and adult. The HNK-1 epitope is gradually lost from retinal cells such that in the adult only horizontal cells and amacrine cells are positive. This suggests that biosynthesis of the HNK-1 epitope is regulated separately from the expression of N-CAM polypeptide. The loss of expression of RET-PE2 by retina during development seems to be permanent. When retina and RPE cells are grown in culture, RPE cells can be labeled by antibody RET-PE2 but retinal cells are not stained. Loss of N-CAM from RPE cells is under more dynamic control because it is reexpressed when RPE cells are placed in tissue culture.

A 65-kDa RPE-specific microsomal protein RPE65, recognized by antibody RPE9, was found in vertebrate species (Hamel et al., 1993a,b, 1994). This same protein was identified by Båvik et al. (1993) as a component of the receptor for retinol binding protein. In the rat, expression of this protein is not detectable until Postnatal Day 4 (PN4), just before photoreceptors develop their outer segments. RPE cells cultured for 7 weeks contained RPE65 mRNA in amounts equivalent to fresh RPE. However, RPE65 protein detectable by immunocytochemistry was lost from these cells within 2 weeks in culture, while they maintained confluency and characteristic epithelial morphology. Disappearance of this protein in culture was likely due to the cessation of translation (Hammel et al., 1993b). Neill et al. (1993) identified a RPE-specific cytoplasmic polypeptide (61 kDa), RET-PE10, with an expression pattern similar to that of RPE65. In the rat, RET-PE10 is expressed late in eye development, with a faint initial labeling of the RPE in the central region at PN9. Its expression increases to adult level and extent by PN14. RET-PE10 became undetectable by immunocytochemistry in cultured cells within a week. It is unknown whether mRNA for this protein was still expressed in these cultured cells. RET-PE10 may be associated with part of the intermediate filament network and required for maintaining the fully mature RPE state. The maintained expression of RET-PE10 depends on extrinsic factors other than contact with Bruch's membrane or light-induced retinal activity (Neill et al., 1993).

Anti-melanoma monoclonal antibody HMB45 is widely used in diagnostic pathology owing to its great specificity and sensitivity in identifying pigmented tumors such as malignant melanoma. This antibody was also found to be able to specifically label human RPE but only during prenatal and infantile periods (Kapur et al., 1992). The antigen recognized by this antibody is a sialated glycoconjugate associated with immature melanosomes. Changes in immunoreactivity with maturation may be a reflection of posttranslational modification. Another anti-melanoma monoclonal antibody, HMB-50, recognizes RPE specifically in human and bovine eyes (Kim and Wistow, 1992). The antigen is a membrane-anchored glycoprotein. Expression of this RPE-specific protein during development was not reported.

A tetrameric protein transthyretin (TTR; 55 kDA) is involved in transport of thyroxine and retinol through its interaction with the retinol-binding protein. Major sites of TTR synthesis include liver, choroid plexus (CP) epithelium, the visceral yolk sac, and the eye. The RPE was found to be the source of TTR in the rat eye (Cavallaro et al., 1990). The mRNA was present in abundance in the primordial CP before organogenesis (E10–12).

In the eye, TTR mRNA was first detected at considerably lower levels after organogenesis (E16) and only in a subset of RPE cells in the equatorial region. The relative abundance of the mRNA in RPE rose gradually until E21, but a surge was seen on the first day of life. The expression reached the adult level by PN7 (Mizuno, et al., 1992). It was hypothesized that the requirement for TTR and its function in CP and RPE may differ during development. The postnatal surge of TTR mRNA expression in RPE raises the possibility that transcription of the TTR gene may be responsive to environmental stimuli such as light (Mizuno et al., 1992).

Many antibodies against RPE-specific antigens have been raised in recent years. Some of these antigens are known proteins (Philp et al., 1990; Okami et al., 1990; Chu and Grunwald, 1991; Suzuki et al., 1993; Kitaoka et al., 1994), but most of them are unknown proteins and/or their expression patterns during development are uncharacterized (Donoso et al., 1988; Hooks et al., 1989; Klein et al., 1990; Chu and Grunwald, 1990a,b; Sagara and Hirosawa, 1991; Kobayashi et al., 1991; Elner et al., 1992; Vinores et al., 1993; Janssen et al., 1994; Zhou et al., 1994). Some of these antibodies recognized RPE cells from various vertebrates, whereas others were only tested for a particular species such as human, rat, chicken, or newt. Many of these antibodies need to be characterized more carefully before they are used for studying RPE differentiation.

D. Polarization of RPE Cells

Although the studies of gene expression described previously suggest that the fate of RPE begins to be determined early, at the optic vesicle stage, morphology and function mature gradually during later development. This maturation is marked by the appearance or disappearance of various antigens at different developmental stages (Hooks et al., 1989; Klein et al., 1990; Chu and Grunwald, 1990b; Neill and Barnstable, 1990; Neill et al., 1993; Hamel et al., 1993a). Paralleling these changes are shifts in the distribution of membrane proteins between the apical (facing the neural retina), basal (facing Bruch's membrane), and lateral plasma membranes. There is a corresponding maturation in the junctional complexes that separate the apical and lateral membranes. Studies are just beginning to reveal how changes in the environment regulate the maturation of RPE.

RPE cells exhibit morphological and functional polarity. They distribute various organelles and membrane proteins asymmetrically along the apical–basal axis (Zinn and Benjamin-Henkind, 1979; Kortc et al., 1994). The distribution of many membrane proteins is typical of most epithelia. Several isoforms of integrin are confined to the basal membrane (Philp and Nachmias, 1987; Philp et al., 1990; Rizzolo et al., 1994). Aminopeptidase N is in the apical membrane (Gundersen et al., 1991). Furthermore, enveloped viruses bud from the appropriate membranes (Bok et al., 1992). This commonality extends to various aspects of the cytoskeleton. The centrosome and microtubule organizing center are subjacent to the apical membrane (Rizzolo and Joshi, 1993). Spectrin and ankyrin are components of the cortical cytoskeleton that localize to the basal and lateral membranes, and to a region at the base of the microvilli (Huotari et al., 1995; Rizzolo and Zhou, 1995). A similar distribution is observed in the intestine and proximal kidney tubule.

Although the basolateral membranes of most epithelia share many functions, the apical membranes are specialized. Consequently, the structure, composition, and function of the microvilli vary. For example, RPE microvilli contain ezrin and bundled actin but lack villin (Höfer and Drenckhahn, 1993). Unlike most other epithelia, they contain myosin VIIa (Hasson et al., 1995). The microvillar localization of several proteins is consistent with their presumed functions. Mannose receptors and CD 36 are thought to participate in the phagocytosis of outer segment membranes (Ryeom et al., 1996; Sparrow et al., 1996; Tarnowski et al., 1988). N-CAM (in rat) and cadherin-like peptides (in chicken) have been proposed to interact with the neural retina (Gundersen et al., 1993; Huotari et al., 1995). In contrast to most epithelia, the $Na^+K^+$-ATPase localizes to the apical membrane (Bok, 1982; Gundersen et al., 1991; Rizzolo, 1990; Okami et al., 1990). In most epithelia, the $Na^+K^+$-ATPase is linked via ankyrin to spectrin along the lateral or basolateral membranes. This led to the hypothesis that the cortical cytoskeleton helps maintain the polarized distribution of membrane proteins (Hammerton et al., 1991; Rodriguez-Boulan and Nelson, 1989). However, in chick RPE, several laboratories demonstrated that the $Na^+K^+$-ATPase is segregated from spectrin. The $Na^+K^+$-ATPase is found in the microvilli, but the apical pool of spectrin is confined to a terminal web-like region at the base of the microvilli (Fig. 3; Huotari et al., 1995; Rizzolo and Zhou, 1995). This observation suggests that RPE microvilli sequester the $Na^+K^+$-ATPase by a specialized mechanism that supersedes the housekeeping mechanism common to all epithelia. This specialized mechanism likely includes ankyrin or an ankyrin-like peptide. Besides its colocalization with spectrin, ankyrin immunoreactivity is also in microvilli (Rizzolo and Zhou, 1995). This is consistent with data obtained in rat, in which the $Na^+K^+$-ATPase was readily cross-linked to ankyrin but poorly cross-linked to a complex with spectrin (Gundersen et al., 1991). The localization of ankyrin to microvilli is unusual and may be due to a specialized isoform.

The fine structure and composition of the apical junctional complexes gradually change throughout the development of chick RPE. These complexes lie at the apical end of the lateral membranes and form the boundary with the apical membrane (Fig. 1). In vertebrates, the complexes are unusual

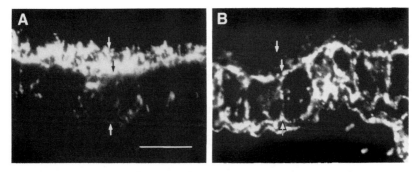

FIG. 3 The Na$^+$K$^+$-ATPase, but not spectrin, localizes to the entire length of RPE microvilli. Sections from E16 eyecups were prepared for immunofluoresence using biotinylated secondary antibodies and Texas Red conjugated avidin. The immunolabel for the Na$^+$K$^+$-ATPase (A) was evident in the apical microvilli. The immunolabel for $\alpha$-spectrin (B) was evident only along the basal and lateral membranes and near the base of the microvilli. No label was observed in the microvilli above the level of background. This distribution of spectrin is typical of epithelia in the intestine or proximal kidney tubule. Downward arrows emphasize the difference between the base and the tips of the microvilli. Upward arrows, basal membrane of RPE; bar = 5 $\mu$m. Modified from Rizzolo and Zhou (1995).

because they contain intermixed, gap, adherens, and tight junctions (Hudspeth and Yee, 1973). In chick, the mature adherens junction contains intramembranous disc-like structures (Sandig and Kalnins, 1988). During development, these structural subunits appear between E4 and E7 (Sandig and Kalnins, 1990). During this time, an amorphous, electron-dense material coalesces into distinct intramembranous discs. The discs appear immature because although they are readily observed in the adult, they require tannic acid for visualization before hatching. The cytoplasmic plaque associated with the junction also develops gradually and is not fully defined until after hatching. These morphological changes are accompanied by molecular changes (Williams and Rizzolo, 1995). Two isoforms of ZO-1 have been described in mammals (Balda and Anderson, 1993). Two additional isoforms were observed in chick that we have termed ZO-1' and ZO-1''. ZO-1 is a component of the cytoplasmic plaque and preferentially associates with tight junctions. However, it is also found in some cells that express only the adherens junction. Occludin is a transmembrane protein that is specific for tight junctions (Furuse et al., 1993). On E3, the entire neuroepithelium of the optic cup expresses ZO-1'' and occludin. In the presumptive neural retina, occludin expression decreases between E7 and E8 when the photoreceptors begin to differentiate morphologically. By contrast, ZO-1'' incorporates into the outer limiting membrane (an adherens junction between the apical microvilli of Müller cells and photoreceptors). In the

RPE, ZO-1' gradually replaces ZO-1'' between E7 and E16, and the ratio of ZO-1/occludin decreases 9 or 10 times. These changes likely relate to the formation of the outer blood–retinal barrier (see below). The presence of occludin suggests the presence of tight junctions in the undifferentiated neuroepithelium. If so, they are rudimentary structures. Fujisawa et al. (1976) detected adherens junctions in the inner and outer layers of the optic cup before E7. Although tight junction-like structures were also observed by transmission electron microscopy, tight junctions were not detected by freeze-fracture in the neural retina. Because the polarized functions of the RPE depend on the integrity of the junctional complex, it is likely that these functions also develop gradually.

Like the junctional complex, the polarization of membrane proteins and the cytoskeleton matures gradually. However, there is no obvious correlation between the two phenomena. In chick, different components polarize at different times, which indicates that multiple mechanisms are involved. By E3, $\gamma$-tubulin, a marker of the microtubule organizing center, is subjacent to the apical membrane (Rizzolo and Joshi, 1993). This study suggests the epithelioid organization of the microtubular cytoskeleton is established at the earliest stage of RPE differentiation. Integrins are a family of cell surface and extracellular matrix receptors. Like $\gamma$-tubulin, the integrin $\alpha 3\beta 1$ is polarized early, in this case to the basal plasma membrane. By contrast, other members of the integrin family, including $\alpha 6\beta 1$, gradually become basally polarized between E9 and E14 (Rizzolo and Heiges, 1991; Rizzolo et al., 1994). Another integral membrane protein that shifts its distribution from nonpolarized to basally polarized is related to the monocarboxylate transporters (Philp et al., 1995). This RPE-specific protein, REMP, shifts its distribution by E14. A pump protein, the $Na^+K^+$-ATPase shifts its distribution in the opposite direction (Rizzolo and Heiges, 1991). A gradient of polarization was observed between the central region (near the optic nerve head) and the periphery. On E7, the $Na^+K^+$-ATPase was apically polarized centrally but nonpolarized peripherally. By E11, the $Na^+K^+$-ATPase was apically polarized throughout the epithelium. This observation is consistent with the fact that the differentiation of the neural retina proceeds from center to periphery.

The cortical cytoskeleton changed its distribution beginning on E12. This cytoskeleton lies beneath the plasma membrane and is thought to help maintain the polarized distribution of membrane proteins. Initially, two components of this cytoskeleton, spectrin and ankyrin, localize to the base of the microvilli and along lateral membrane (Huotari et al., 1995; Rizzolo and Zhou, 1995). On E12, patches of spectrin and ankyrin appear on the basal membrane. These patches grow in size and number until a continuous distribution along the basal membrane is observed on E16. This shift in distribution might correspond to the formation of basal infoldings during this time. Notably, one membrane protein changed its distribution in parallel with the appearance of ankyrin and spectrin in the basal membrane. The 5A11 antigen is a member of the immunoglobulin superfamily of surface receptors and mediates interactions between RPE or Müller cells and neurons (Fadool and Linser, 1993). It appears identical to a protein found in capillaries of the central nervous system after the blood-brain barrier develops (Seulberger *et al.*, 1990, 1992). Early in RPE development, 5A11 is in the apical and lateral membranes. At about the time that the blood-brain barrier develops in chick, 5A11 also appears in the basal membrane (Fadool and Linser, 1994; Rizzolo and Zhou, 1995). Between E12 and E16, 5A11 colocalizes with spectrin and ankyrin in patches along the basal membrane (Rizzolo and Zhou, 1995). It is unknown if 5A11 complexes with spectrin and ankyrin or if it colocalizes by another mechanism.

There are interesting similarities between the polarization of RPE during normal development and the polarization of regenerating RPE during wound healing *in vivo* (Korte and Wanderman, 1993; Korte *et al.*, 1994). Both processes are likely influenced by their environment. In general, epithelial polarity is not an inherent property but is induced by interactions along the basal and lateral plasma membranes (Rodriguez-Boulan and Nelson, 1989). For RPE, this induction is modulated by interactions at the apical membrane. The basement (Bruch's) membrane induces the basal polarity of integrins in cultured RPE, even if RPE is isolated from E7 chick embryos when the $\beta 1$ subunit of integrin is nonpolarized (Rizzolo, 1991). Polarity is also induced by laminin and collagen IV but not by fibronectin. These molecules are present in the basement membrane *in vivo* as early as E4 (Turksen *et al.*, 1985; Lin and Essner, 1994). Early in development, the apical distribution of certain integrins depended on interactions with the neural retina (Rizzolo *et al.*, 1994). In organ culture, $\alpha 6\beta 1$ is maintained in the apical membrane of E7 RPE only if the outer (apical) aspect of the primordial neural retina is apposed to the apical membrane of RPE (Fig. 4). If the primordial retina is removed, or the inner (ganglion cell) aspect is apposed to the RPE, $\alpha 6\beta 1$ is observed only in the basal membrane. This suggests that some integrins are maintained in the apical membrane by a ligand on the primordial neural retina or in the subretinal extracellular matrix. Because integrins transmit signals across the plasma membrane, this might provide the RPE one mechanism to sense changes in the developing neural retina and coordinate the differentiation of the two tissues.

The basal membrane interactions described previously failed to polarize the $Na^+K^+$-ATPase (Rizzolo, 1991). Apical interactions might affect the structure and composition of microvilli that are required to polarize the $Na^+K^+$-ATPase appropriately. Several studies suggest possible mechanisms. In developing rat RPE, N-CAM was observed in the apical mem-

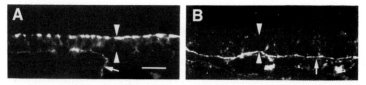

FIG. 4  On E7, the α6-containing integrins were maintained in the apical membrane by the neural retina. Eyecups were prepared from E7 chicken embyros and cultured for 18 h with the neural retina in its normal orientation (A) or with the ganglion cell aspect apposed to the RPE (B). Prior to fixation, the neural retina was removed. Using monoclonal antibodies to α6, the tissues were processed for immunofluorescence and photographed. When cultured with the neural retina in its normal orientation, α6 was retained on the apical membrane (downward arrowheads). The neural retina was ineffective when the opposite (ganglion) side of the retina was apposed to the RPE. Without the influence of the neural retina, Bruch's membrane induced α6 to relocated to the basal membrane (upward arrowheads). Arrows, endothelium or mesenchyme; bar = 5 μm. Modified from Rizzolo et al. (1994).

brane *in vivo* and in explant cultures (Gundersen *et al.*, 1993). Interestingly, the N-CAM was only basolateral in cells that grew out of the explant. In other words, cells that dedifferentiated, migrated, and redifferentiated in the absence of the neural retina failed to reestablish an apical pool of N-CAM. Even though the neural retina was absent, cells in the center of the explant maintained polarity. N-CAM is not expressed in chick RPE or more mature rat RPE (Neill and Barnstable, 1990; Rizzolo *et al.*, 1994). However, a cadherin-immunoreactive protein was detected in the RPE microvilli and photoreceptor outer segments (Huotari *et al.*, 1995). In principle, this protein could substitute for N-CAM in mediating RPE–photoreceptor interactions. In several culture systems that lack apical interactions with the neural retina, N-CAM and cadherin-like proteins were basolateral, the microvilli were sparse, and the $Na^+K^+$-ATPase was basolateral or nonpolarized (Nabi *et al.*, 1993; Rizzolo, 1991; Rizzolo and Li, 1993; Huotari *et al.*, 1995). These, or similar, interactions between the microvilli and outer segments may account for the changes observed following retinal detachment (Anderson *et al.*, 1983). Within 1 day of an experimentally induced detachment, RPE microvilli become shorter and undifferentiated.

The maturation of the RPE is also regulated by diffusible factors produced by the neural retina. Evidence for this concept was obtained in studies of barrier function in culture. Early in development, the endothelia and epithelia of the central nervous system are leaky and the tight junctions are permeable to small proteins. At different times, in different regions the blood–brain barrier forms as tight junctions are converted from leaky to tight forms (reviewed in Dermietzel and Krause, 1991). Astrocytes produce diffusible factors that regulate this conversion (see Rubin *et al.*, 1991, and references therein). An analogous process was observed for RPE (Rizzolo and Li, 1993). RPEs isolated from E7 chick embryos are leaky in primary cell culture. The permeability of the cultures decreases when they are maintained in medium that was conditioned by E14 neural retinas (Fig. 5). The effectiveness of the conditioned medium depends on when it is presented to the RPE. After several days in culture, the RPE loses the ability to respond to conditioned medium. Despite the decrease in permeability, there is minimal effect on the distribution of the $Na^+K^+$-ATPase.

### III. Instability of the RPE Differentiated State

A. Effects of Environmental Changes *in Vivo*

RPE cells in the adult normally form a quiescent monolayer, but they retain the ability to divide and do so in response to environmental changes. RPE

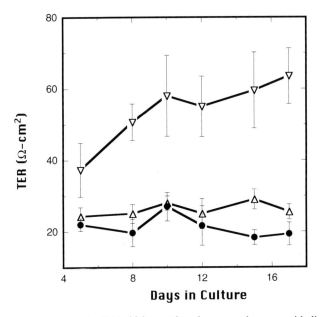

FIG. 5 Medium conditioned by E14 chicken retinas increases the transepithelial electrical resistance of RPE that was isolated from E7 embryos. The RPE was cultured on permeable filters and conditioned medium was added to the apical side of the monolayer. When the conditioned medium was size fractionated using Centricon filters (Amicon, Beverly, MA) with a nominal 10-kDa cutoff, the bulk of the activity was obtained in the filtrate ($\triangledown$) with little or no activity in the retentate ($\triangle$). The apical chamber of control cultures contained growth medium ($\bullet$). Modified from Rizzolo and Li (1993).

proliferates readily *in vivo* in response to a variety of stimuli such as photocoagulation and physical injury. Histologic patterns of reactive proliferation of RPE vary greatly and depend on the nature and severity of the stimulus (Tso, 1979). When the RPE is rendered necrotic after acute insults, macrophages quickly arrive in the subretinal space to remove the cellular debris. RPE cells at the edge of the lesion flatten and proliferate. In mild injuries involving only RPE and photoreceptor cells, proliferating RPE cells quickly reestablish a continuous monolayer over Bruch's membrane. Regeneration of RPE after selective damage by sodium iodate in the animal model has been carefully studied and reviewed (Korte *at al.,* 1994). A gradient of morphology and polarity was observed from the flattened, nonpolarized, proliferating cells at the edge of the lesion to cuboidal, polarized cells distal to the edge.

Retinal detachments lead to a number of changes. In animal models, RPE microvilli shortened within 1 day (Anderson *et al.,* 1983; Guérin *et al.,* 1990). In several days there are cytoskeletal changes as vimentin expression is induced. With prolonged detachments, the RPE proliferates to form multilayers. These changes only partially reverse with reattachment of the retina. Similar changes are observed in human patients, in which RPE cells proliferate in finger-like projections that invade the retina. These invading cells eventually form chorioretinal adhesions. Also typically observed in patients with longstanding retinal detachment is tubular proliferation. The histologic pattern of this type of proliferation is that cuboidal RPE cells are arranged in an oval fashion around a central lumen and surrounded by a thin layer of basement membrane. In some pathologic conditions, the RPE proliferates to form three or four layers of cells on Bruch's membrane. This type of proliferation, called diffuse hyperplasia, is observed in patients with choroidal melanomas or diffuse uveitis. There are many other proliferation patterns associated with various eye diseases (Tso, 1979). Proliferating RPE cells *in vivo* contained melanin granules, but the level of pigmentation varies from one location to another. In some areas the level of pigmentation is significantly lower than that of quiescent RPE cells. The morphology of individual proliferating cells also varies, depending on the pattern of growth. In most cases, cells are oval or cuboid like, whereas in some cases the cells become spindle like or slightly flattened.

Detached RPE cells can be found in the vitreous following ocular trauma or after cryosurgery to treat detached retina. These cells, alone or in combination with other cell types, proliferate and attach to the inner surface of the retina where they form contractile strands that cause further retinal detachments and tears (Machemer, 1988). Although there have been some tissue culture studies of RPE proliferation and migration, the factors that cause or present abnormal RPE migration and proliferation *in vivo* remain unknown. Some of the changes in protein expression observed in certain culture models mimic changes observed in RPE within epiretinal membranes (Grisanti and Guidry, 1995; Vinores *et al.*, 1995).

Studies of RPE following retinal detachment in animals have identified two types of RPE cells—macrophage- and fibroblast-like (or spindle-like) cells (Machemer and Laqua, 1975; Machemer *et al.*, 1978; Grierson *et al.*, 1994). These cells were actively proliferating and migratory. When cultured RPE cells were injected into the rabbit vitreous they also became macrophage- and fibroblast-like cells (Mandelcorn *et al.*, 1975). These cells can revert to the epithelial phenotype under certain conditions. For example, when placed in culture, many of these cells spread out and packed tightly to form mosaics resembling the original RPE (Mazure and Grierson, 1992).

## B. RPE Cells in Culture

The plasticity of RPE in culture is well documented. Proliferating cultures grown on plastic, in gel foam, or vitreous have been used to model wound healing or the dedifferentiated RPE in proliferative retinopathies (e.g., Grisanti and Guidry, 1995; Vinores *et al.*, 1995). In certain circumstances, RPE will transdifferentiate into lens or other retinal cells (discussed under Section IV). This section focuses on attempts to re-create a normal phenotype *in vitro*. Many cultures yield hexagonal arrays of cuboidal cells with circumferential bands of actin and morphologically distinct junctional complexes, but these are minimal criteria. Fully polarized cells with elongated, fully differentiated microvilli and low permeability tight junctions have been more difficult to achieve. Adult human RPE appears to be heterogeneous (McKay and Burke, 1994; Burke *et al.*, 1996). Using selective trypsinization, fusiform and epitheliod cells can be isolated separately. These phenotypes are stable and cannot be interconverted. Although the epitheliod cells can recover a fair degree of polarity, this requires months in culture (McKay *et al.*, 1992). After weeks in culture, multiple epithelial phenotypes become apparent (Burke *et al.*, 1996). Primary cultures rapidly dedifferentiate at the molecular level and markers, such as RET-PE10 (Neill *et al.*, 1993) and RPE65 (Hammel *et al.*, 1993b), rapidly become undetectable. NCAM is expressed by the neuroepithelium but in chick and rat is turned off early in RPE development. In culture, some forms of NCAM are reexpressed (Neill and Barnstable, 1990). The expression of intermediate filaments is also altered. Nonavian species lack vimentin but express cytokeratins 8 and 18 *in vivo*. In culture, human RPEs express vimentin and cytokeratins 7, 8, 18, and 19 (Hunt and Davis, 1990). Keratin 19 was expressed only by proliferating cells. These intermediate filament proteins were also expressed in a stable RPE cell line with epithelial properties (Davis *et al.*, 1995). Although this cell line expressed some enzymes of the visual cycle, it lacked others. A tubulin isoform that is normally expressed only in neurons ($\beta$-tubulin isotype III) is also expressed by human RPE in culture (Vinores et al., 1995). RPE cells with morphological features resembling the original RPE cell are usually negative for $\beta$-tubulin isotype III, but cells appearing in the early stages of dedifferentiation express the marker. Notably, keratin 19 and $\beta$-tubulin isotype III were also described in the RPE of epiretinal membranes that were isolated from patients with proliferating vitreoretinopathy.

In addition to changes in protein expression, there is a partial loss of polarity. The centrosomes and $\gamma$-tubulin are perinuclear, as in nonepitheliod cells, which implies a reorganization of the tubulin cytoskeleton (Rizzolo and Joshi, 1993; Turksen et al., 1983). The distribution of talin, a cytoskeletal-associated protein, is disorganized (Philp et al., 1990). As noted earlier, the distribution of the $Na^+K^+$-ATPase becomes nonpolar or basolaterally polarized (Huotari et al., 1995; Rizzolo, 1990, 1991; Rizzolo and Li, 1993). As cells are passaged, even morphological polarity is lost (Grisanti and Guidry, 1995).

Several culture systems have retained a significant degree of polarity (Chang et al., 1991; Hu et al., 1994; Mircheff et al., 1990). These systems use fully defined or low serum medium. No systematic study has been reported to explain why these systems are so successful. Among these, the cultures with the lowest permeability have used human fetal RPE (Hu et al., 1994; Mircheff et al., 1990). Notably, this phenotype is only achieved after months in culture. These cultures express all the enzymes that participate in the visual cycle and exhibit a remarkable degree of polarity. Even here, the $Na^+K^+$-ATPase was present on both the apical and basolateral membranes of long-term cultures, but it was more abundant on the apical membrane. It is unclear whether basolateral $Na^+K^+$-ATPase is present *in vivo* or whether the mechanisms that maintain polarity are inefficient in the cultured cells. Immunofluorescence data suggest the bulk of the $Na^+K^+$-ATPase is in the apical membrane *in vivo*. Notably, this culture system uses a retinal extract. The importance of the retinal extract was highlighted by a spontaneously arising human RPE cell line, ARPE-19 (Dunn et al., 1996). This cell line retained some differentiated characteristics in a low serum medium. However, the permeability of the cultures decreased when cultures were maintained in serum-free medium that contained retinal extract. Previously, we described how basement membrane interactions could induce a limited amount of polarity, but that interactions with the neural retina were also important. One attempt to uncover physiologic, retinal-derived factors used embryonic chick RPE (Rizzolo and Li, 1993). As described under Section II,D, there is evidence that retinal factors decrease junction permeability (Fig. 5). The active factors had a molecular mass of less than 10 kDa and were protease resistant. These studies indicate that cultures can be used to identify physiologic factors that regulate various aspects of the differentiated phenotype.

## C. RPE Transplantation

Retinal photoreceptor degeneration in the RCS rat is a consequence of a defect in the ability of the RPE to phagocytose photoreceptor outer segments. This has been defined as a tissue-specific defect in the expression of the scavenger receptor CD36 in the RPE (Sparrow *et al.*, 1996). Similar defects in human RPE are thought to underly some forms of macular degeneration. This has generated interest in transplantation of healthy RPE into subretinal space to rescue photoreceptor cells (Lopez *et al.*, 1987; Sheedlo *et al.*, 1992, 1993; Wongpichedchai *et al.*, 1992). The RPE to be transplanted is first maintained in culture where it partially dedifferentiates. This process is apparently reversed when the cells are transplanted and transplants can show a normal relationship with Bruch's membrane and photoreceptors. Functionality of the transplanted RPE has been demonstrated in several ways. For example, the retinal degeneration of the RCS rat was retarded or prevented. Transplanted RPE cells resumed outer segment phagocytosis, maintained photoreceptor transduction, and inhibited degeneration of the retinal vascular bed and the subsequent neovascularization (LaVail *et al.*, 1992; Seaton *et al.*, 1994; Yamamoto *et al.*, 1993). RPE transplantation has also been shown to delay age-related photoreceptor death in the rat (Yamaguchi *et al.*, 1993). Despite these beneficial effects, it is unclear whether a blood–retinal barrier was reestablished.

However, not all transplanted RPEs establish normal relationships. When suspensions of RPE cells are transplanted, cells can form clumps that invade the vitreous or retina. This led several investigators to transplant the RPE as sheets with or without an underlying support (Bhatt *et al.*, 1994; Sheng *et al.*, 1995). These patch transplants were probably functional because adjacent outer segments of the photoreceptors of the host appeared healthy and underwent phagocytosis by the transplanted RPE. RPE cells transplanted into subretinal space remain stable and do not undergo massive cell division. When unsupported sheets were transplanted on top of host RPE, the host RPE became vacuolated and transplanted RPE insinuated itself onto Bruch's membrane. This indicates that either factors present in that area inhibit RPE proliferation or the mitogenic factor(s) for RPE cells is absent. Melatonin and retinoic acid were shown to have an inhibitory effect on the proliferation of RPE cells *in vitro* (Yu *et al.*, 1993; Vinores *et al.*, 1995). Whether these are the factors inhibiting adult RPEs proliferation *in vivo* is not known.

## IV. Transdifferentiation of RPE

Studies of transdifferentiation can be used to address the questions raised in the preceding section regarding why RPE behaves differently in different environments, which environmental factors regulate different pathways of differentiation, and how these signals are translated into changes in gene expression. Altering the environment of the RPE of many species of vertebrates can induce it to transdifferentiate into other ocular tissues such as neural retina and lens (Stone and Steinitz, 1957; Stroeva and Mitashov, 1983; Park and Hollenberg, 1993; Eguchi, 1987; Eguchi and Kodama, 1993). However, the capacity for transdifferentiation varies among different species.

The RPE from *Triturus* and some other amphibian species such as frogs is capable of transdifferentiation throughout life, whereas the RPE from other amphibians can transdifferentiate into retina only during larval life (Coulombre and Coulombre, 1965; Lopashov and Sologub, 1972; Reh and Nagy, 1987). In fishes and birds, the RPE can transdifferentiate only during early embryonic stages. The RPE in *Acipenserid* fishes cannot change after pigmentation has appeared (Stroeva, 1960), whereas the already pigmented RPE in teleosts can form retina (Sologub, 1975). Chick RPE can regenerate retina at least as late as E4, Hamburger and Hamilton stage 22–25, (Hamburger and Hamilton, 1951), by which time it has become pigmented (Coulombre and Coulombre, 1965).

### A. Transdifferentiation of Amphibian RPE

The first systematic study of RPE transdifferentiation was carried out in newt by Stone (1950), although the basic observations go back to the eighteenth and nineteenth centuries (see references in Stone, 1950). In the experiments of Stone it was observed that, after complete removal of the retina, cells in the RPE layer entered mitosis. It was suggested that one daughter cell migrated away from Bruch's membrane and formed the pool of stem cells from which a new retina was formed. The other daughter cell remained attached to Bruch's membrane and regained the properties of the mature RPE. In cases in which only a small portion of retina was removed, the RPE was still capable of generating cells to fill in and replace the lost tissue; no repair came from the adjacent retinal tissue. Two important conclusions can be drawn from these studies. The first is that the signals inducing RPE transdifferentiation are very local because removal of a small piece of retina induced a local reaction. The second is that transdifferentiation correlates with detachment from Bruch's membrane. From the earliest experiments it was realized that RPE cells could transdifferentiate into lens as well as into retina. In a systematic study in adult *Triturus,* it was shown that the frequency of lens formation was much lower that than of retina and it only occurred when large pieces of RPE were transplanted into host eyes (Stone and Stenitz, 1957). It was suggested that influences from cells underlying the RPE in the choroid might inhibit lens formation and only at the margins were these influences diminished. The possible role of choroidal cells in regulating RPE cells has still not been adequately explored.

### B. *In Vitro* Transdifferentiation of Chick RPE

It was shown that retinal regeneration from embryonic chick RPE *in vivo* required the presence of a piece of neural retina or some other neural tissues (Coulombre and Coulombre, 1965). Work has shown that this neural tissue can be replaced by a plastic implant slowly releasing basic fibroblast growth factor (bFGF), suggesting that the role of those tissues may be to release a growth factor, most probably bFGF (Park and Hollenberg, 1989). The induction of RPE transdifferentiation by bFGF has been confirmed by *in vitro* experiments (Pittack *et al.,* 1991; Guillemot and Cepko, 1992).

Pittack *et al.* (1991) found that bFGF stimulated proliferation of dissociated embryonic chick RPE cells in culture and caused morphological changes in these cells, including loss of pigmentation. However, no transdifferentiation to neuronal phenotypes was observed. In contrast, when small sheets of RPE were cultured in the presence of bFGF, a large number of retinal progenitor cells were generated. An independent study by another group showed that both aFGF and bFGF were able to induce transdifferentiation of embryonic chick RPE to neuronal cells, with aFGF more potent than bFGF (Guillemot and Cepko, 1992). This study also demonstrated that cell division was not required for the conversion of a cell from RPE to neuronal phenotype.

### C. *In Vitro* Transdifferentiation of Rat RPE

In contrast to the extensive data on transdifferentiation of RPE in amphibians, fishes, and birds, very few studies have been carried out with mammalian RPE. Stroeva (1960) transplanted embryonic rat eyes into the anterior chamber of adult rat eyes and found that the presumptive RPE of very early stages transformed into retina-like tissue if it remained free floating, but that the RPE became pigmented if it remained attached to mesenchyme or the RPE became attached to the iris. However, this study was solely based on morphological observations and the identities of cell types in implanted tissues remained unknown. To further investigate the transdifferentiation potential of mammalian RPE and molecular factors that influence this process, rat RPE of early embryonic stages has been studied in defined culture media supplemented with or without various growth factors. The differentiation pathway of embryonic rat RPE could be changed in culture by bFGF to produce cells expressing neuronal and retina-specific markers, but this could occur only during a narrow period of early eye development (Zhao et al., 1995).

As described in previous sections, differentiation of neural retina and RPE lineages starts at least as early as the optic vesicle stage (E9 or E10 in mouse and E10 or E11 in rat) and is associated with the expression of a number of transcription factors. The presumptive RPE can be distinguished from the neural retina morphologically by E12 in rat embryos. At this stage the outer layer of the optic cup (neural retina) is much thicker than the inner layer (presumptive RPE) and the border between these two layers becomes very clear. Pigment granules start to appear in the RPE layer at E13 and can be clearly seen microscopically in tissue sections. Pigmentation becomes visible to the naked eye at E14. If placed in serum-free medium UC, E12 or E13 RPE differentiation proceeded in a similar way to that *in vivo* to become a pigmented monolayer as judged by pigmentation (Fig. 6A) and expression of the RPE-specific marker, RET-PE2 (Zhao et al., 1995).

When cultured in medium containing bFGF, E12 or E13 rat RPE started to grow and form a pseudostratified layer (Fig. 6B). Expression of tyrosinase gene and pigmentation were inhibited. However, RPE from later rat embryos (E14 or E15) did not respond to bFGF and its appearance in a serum-free medium with and without bFGF was essentially the same, except that the bFGF-treated tissue was slightly less pigmented after 4 or 5 days in culture. These results suggested that even though differentiation of RPE has started earlier than E12, the phenotype of the cell can still be changed but with diminishing plasticity; and that the commitment to the RPE phenotype became irreversible at about E14.

In bFGF-treated E12 or E13 RPE tissues, a number of neuron-specific markers became detectable by immunohistochemistry after a few days in culture, including neurofilament and $\beta$-tubulin isotype III. These two markers are expressed only in ganglion cells in embryonic retina. The strong labeling of anti-$\beta$-tubulin isotype III at the outer margin of tissue was due to axons of ganglion cells growing out of the tissue and encircling it (Figs. 6C, and 6D). A large bundle of axons was clearly seen in the whole mount labeling with an anti-neurofilament antibody. Other molecular markers for specific retinal cell types, such as HPC-1 and Thy-1, were also expressed in these tissues. HPC-1 and Thy-1 are expressed in amacrine and ganglion cells, respectively, in the retina (Barnstable et al., 1985; Barnstable and Drager, 1984). When the RPE was cultured for about 2 weeks, rod opsin

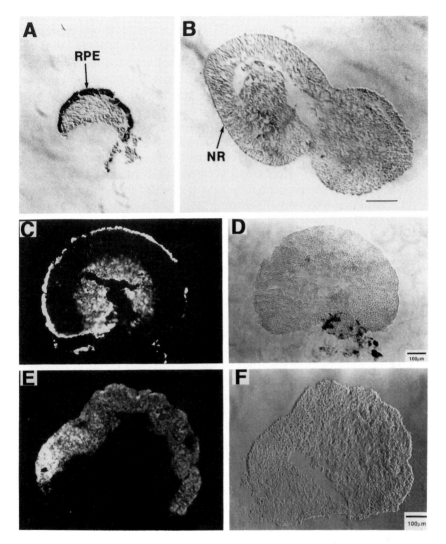

FIG. 6 (A, B) Sections of E12 rat RPE tissues cultured for 4 days. Tissue became pigmented in serum-free medium UC (A). In UC supplemented with bFGF, E12 RPE grew to form a neural retina-like pseudostratified layer (NR) (B). The tissue underneath the RPE layer was the attached mesenchyme. (C, D) E12 rat RPE tissues cultured for 8 days in serum-free medium supplemented with bFGF. Section of the transdifferentiated RPE tissue labeled with an antibody to $\beta$-tubulin isotype III, a retinal ganglion cell marker. Cells in the inner layer of the tissue in C strongly expressed this marker. The bright labeling at the outer margin of the tissue was axons of ganglion cells growing out of the tissue and wrapping around it. Some radially elongated cells were also $\beta$-tubulin isotype III positive. They are believed to be differentiating ganglion cells migrating from the outer to the linner layer. (E, F) Section of E13 rat RPE treated with bFGF and maintained in culture for 24 days. Immunocytochemistry with anti-rod opsin antibody RET-P1 (E) showed that nearly all the cells in the outer layer of the tissue differentiated to become rod photoreceptors. Bars = 100 $\mu$m. (A and B adapted from Zhao et al., 1995.)

was detected in isolated cells using antibody RET-P1 (Zhao et al., 1995). When the culture was maintained for 3 weeks or longer, strong expression rod opsin was observed in nearly all cells in the outer layer of the tissue (Figs. 6E, and 6F). However, these cells did not have the morphology of the mature photoreceptor because inner and outer segments did not develop. This arrested development was characteristic of explant cultures of neonatal retina (Sparrow et al., 1990).

The action of bFGF to change the phenotype of embryonic RPE cells could be either direct or indirect. The former would imply a transduction pathway from activation of FGF receptors to changes in transcription of key genes. An indirect effect might occur if bFGF was simply a mitogen for RPE and then some postmitotic cells responded to their environment by differentiating into retinal neurons. Several results suggest that bFGF has a direct effect on RPE phenotype.

First, E12 RPE was cultured in medium containing bFGF for 10–12 h and then the growth factor was removed from the medium. The RPE stopped growing and appeared to return to its original differentiation pathway and became pigmented within 2 days. If removal of bFGF took place at least 24 h after culture, the RPE tissue continued to proliferate and form stratified layers that had the same morphological and immunocytochemical properties as those of tissues maintained in bFGF-containing medium all the time. The length of time needed for the irreversible change was dependent on the stage of embryos from which the RPE was taken. If E13 RPE was used in the initial culture, 48–60 h was required to make the bFGF effect permanent. This result indicated that bFGF was required in the culture for a certain period of time to irreversibly change the differentiation pathway of RPE, and thereafter bFGF was not required to sustain cell growth and differentiation. A similar finding has been reported for FGF-induced transdifferentiation of chick RPE cells (Opas and Dziak, 1994).

Second, E12 and E13 RPE tissues were treated with mitomycin C, which prevents DNA replication (Tomasz et al., 1987), before being cultured in either UC or UC supplemented with bFGF. Mitomycin C-treated RPE tissues cultured in UC became pigmented within 2 days, whereas those cultured in UC supplemented with bFGF showed no pigmentation when observed under dissecting microscope. Thus, the effects of bFGF on RPE transdifferentiation were separate from any effects on cell proliferation. This result confirms those of studies in the chick that had led to the same conclusion (Guillemot and Cepko, 1992).

What is unclear is whether FGF acts on the fully differentiated RPE cell or on a cell that has already responded to the experimental treatment. Evidence in favor of the latter has been gained by studying the FGF effects on chick RPE cells (Zhou and Opas, 1994). These authors found that, as described previously, when RPE cells were placed in culture they began to lose pigment and to express N-CAM. These cells had a lower resting pH than the fully pigmented cells. FGF was found to cause a response (measured as a cytoplasmic alkalinization) only in those cells that had already begun to express N-CAM. The observed need for a "priming" step may be very similar to the observations of Stone (1950) that RPE daughter cells that move away from Bruch's membrane are the precursors of a new retina.

The molecules detected in the transdifferentiating RPE were all consistent with both the earliest cell types formed during normal retinal development—ganglion and amacrine cells—and the last cells to be formed—rod photoreceptors. Although not studied systematically, it also appears that these molecules appear in approximately the same sequence as in normal retinal development. This suggests that the multilayered tissue formed following bFGF treatment was a complete retinal epithelium and that formation of retinal cell types is a property intrinsic to a retinal epithelium rather than needing instructive influences from other tissues. This does not exclude influences from extrinisc factors. For example, we have shown that bFGF can influence the timing of retinal ganglion cell formation and the organization of the the outer retina in retinal explant cultures (Zhao and Barnstable, 1996).

To determine the genes that may be involved in the bFGF-induced RPE transdifferentiation, we tested a number of paired-box, homeobox, POU-domain, and protooncogenes using RT-PCR. Homeobox gene *Chx10* and POU-domain gene *Brn-3.2* were activated in E13 RPE treated with bFGF for 3 days (Fig. 7). A study suggested that *Chx10* may play an important role in determination of retinal fate at the optic vesicle stage (Liu *et al.*, 1994). *Chx10* could be one of the upstream factors that initiated the transdifferentiation process. It has been shown that *Brn-3.2* is expressed in postmitotic retinal ganglion cells (Xiang *et al.*, 1993, 1995; Turner *et al.*, 1994). This indicates that ganglion cells were already generated in the transdifferentiating RPE within 3 days in culture.

We were unable to demonstrate an effect of aFGF even at concentration of 100 ng/ml. This result differs from that seen in chick cultures in which aFGF was more potent than bFGF in promoting cell growth and transdifferentiation of RPE (Guillemot and Cepko,1992). On the other hand, *in vivo* experiments on retinal regeneration from RPE in chick found that bFGF was more potent than aFGF (Park and Hollenberg, 1991). These differences may reflect problems in using mammalian growth factors in avian species and also that the growth factor active *in vivo* may be an embryonic member of the FGF gene family rather than either aFGF or bFGF.

Most actions of FGFs are believed to be mediated by a family of transmembrane tyrosine kinase receptors (Dionne *et al.*, 1991; Miller and Rizzino, 1994). Two of the FGF receptors (types 1 and 2) are expressed in

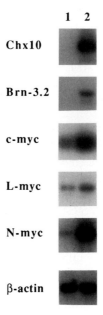

FIG. 7 Gene expression in E13 RPE cultured for 3 days in serum-free medium UC (lane 1) or UC supplemented with bFGF (lane 2) as detectd by RT-PCR amplification. Neuron-specific genes *Chx10* and *Brn-3.2* were activated in the transdifferentiating RPE. Expression of protooncogene N-*myc* was significantly enhanced in RPE treated with bFGF, whereas levels of c-*myc* and L-*myc* expression increased to a less extent. Primers specific to β-actin were used as the PCR control.

chick RPE at least from E4 to Postnatal Day 1.5 (Tcheng *et al.*, 1994). This suggests that the developmental decline in responsiveness to bFGF we and others have observed in RPE cells is not due to loss of FGF receptors but rather to a change in intracellular signaling. Our results show that bFGF leads to a decrease in tyrosinase mRNA and increases in expression of N-CAM, neurofilament and retinal cell markers, Thy-1, HPC-1, and rhodopsin. Thus, at some point, the actions of bFGF result in changes in gene expression. Our experiments demonstrate that bFGF needs to be present for a certain period of time but that after this time its effects become irreversible and it is no longer required. This suggests that bFGF does not act directly to regulate transcription of tyrosinase, N-CAM, neurofilaments, Thy-1, HPC-1, or rhodopsin genes but rather acts on factors that are capable of determining the RPE or retinal differentiation pathways. Such factors might include those genes thought to define the early lineages of retina and RPE such as *Chx10, Rx,* or *Otx-2* (Liu *et al.*, 1994; Mathers and Jamrich, 1995; Simeone *et al.*, 1993).

## D. Oncogenes and Change of the RPE Phenotype

Many oncogenes, including the *myc* gene family, are known to play important roles in regulation of cell proliferation and differentiation (Cooper, 1990; Marcu *et al.*, 1992; Bernard *et al.*, 1992; Wakamatsu *et al.*, 1993). Expression levels of N-*myc* in E13 rat RPE were significantly elevated as a result of treatment with bFGF, whereas expression of c-*myc* and L-*myc* was also increased but to a lesser extent (Fig. 7). N-*myc* is normally expressed at a high level in the developing central nervous system and its amplification or overexpression is found in some neural tumors (DePinho *et al.*, 1986; Sawai *et al.*, 1991; Stanton *et al.*, 1992). It is, therefore, plausible that a high level of N-myc protein is necessary for the optic vesicle neuroepithelium to choose the retinal fate. Agata *et al.* (1993) have shown that expression of c-*myc* gene was enhanced during the dedifferentiation process of chick RPE and they suggested that c-*myc* gene expression is intrinsically correlated with dedifferentiation in general.

A family of GTPases, Ras proteins, is involved in regulation of many aspects of cell growth, differentiation, and other actions in response to extracellular signals (McCormick, 1994; Khosravi-Far and Der, 1994). Wilcox and Vergnes (1989) found that metaplastic human RPE cells exhibited early and abnormally high levels of N-*ras* expression compared to normal proliferating RPE cells. The increase in N-*ras* expression preceded the vitreous-induced phenotypic change from an epithelial to a fibroblastic morphology. By transfecting primary cell cultures with viral DNA E1A and protooncogenes H-*ras* and c-*myc*, Dutt *et al.* (1993) established cell lines from adult human RPE. The cell lines derived from H-*ras* transfection contained cells with a neuronal phenotype. This feature was not observed in cell lines established with other two oncogenes. The transdifferentiated cells expressed neuronal markers such as neurofilaments and neuron-specific enolases. This study suggests that although mature mammalian RPE cells have lost their ability to respond to exogenous factors such as bFGF and transdifferentiate to neuronal cells, it is still possible to change their phenotype by altering the intracellular signaling pathway.

It is interesting to note that avian retinal cells can be transformed into pigmented cells after infection with MC29 virus (Martin *et al.*, 1992). This avian myelocytomatosis virus strain transduces v-*myc* oncogene into the cell. It was shown that transformation of retinal cells into pigmented cells was correlated with the decline of mRNA level of *Pax-QNR*, the quail homolog of the *Pax-6* gene. Although the molecular mechanism of cell phenotype change caused by the viral oncogene is unknown, this study combined with the one mentioned previously indicate that neither phenotype of the mature RPE nor of retinal cells is permanently fixed and one can be changed to the other by altering the internal state of the cell.

## V. Concluding Remarks

The RPE is a unique cell type in at least two respects. First, it is derived from the neuroepithelium that also gives rise to the neural retina. It provides an interesting system for developmental biologists to study phenotype specification of neuronal and nonneuronal cells. Second, the RPE is different from other epithelia in its major functions and structure. This is determined by its roles in supporting the survival and facilitating the metabolism of photoreceptor cells. As discussed earlier, a number of homeobox and paired-box genes are shown to be involved in the determination of neural retinal and RPE fates during the early eye development, but the exact roles of these genes are still unknown. Recent *in vitro* and *in vivo* studies suggest that bFGF may be the critical factor that specifies the anterior of the optic vesicle as the neural retina. Our work shows that bFGF activated expression of some of the homeobox genes. However, whether bFGF induces formation of neural retina *in vivo* by turning on these homeobox and paired-box genes remains to be investigated. The different fates of anterior and posterior of the optic vesicle may result from exposure of the anterior to a much higher level of FGF (De Iongh and McAvoy, 1993). Alternatively, the presumptive RPE may be prevented from becoming neural tissue by an inhibitory signal. It was shown that bone morphogenesis protein 4 (Bmp-4), a relative of the growth factor activin, is an epidermis inducer but a neural fate inhibitor during *Xenopus* development (Wilson and Hemmati-Brivanlou, 1995; Sasai *et al.*, 1995). This raises the possibility that the RPE fate is determined by a factor such as Bmp-4.

Study of RPE development, structure, and functions not only contributes to our current understanding of cellular physiology in general, but it will also have profound impacts on finding cures for ocular diseases such as photoreceptor degeneration caused by pathological conditions of the RPE.

### Acknowledgments

Work from the authors' laboratories described in this chapter has been supported by NIH Grants EY 00785 (CJB), NS 20483 (CJB), and EY 08694 (LJR), as well as by grants from the Foundation Fighting Blindness, the Connecticut Lions Eye Research Foundation, and the Kemper Fund. SZ was the recipient of a Dr. Charles A. Perera Fellowship of the Fight for Sight research division of Prevent Blindness America and CJB is a Jules and Doris Stein Research to Prevent Blindness Professor.

### References

Agata, K., Kobayashi, H., Itoh, Y., Mochii, M., Sawada, K., and Eguchi, G. (1993). Genetic characterization of the multipotent dedifferentiated state of pigmented eepithelial cells in vitro. *Development* **118,** 1025–1030.

Anderson, D., Stern, W., Fisher, S., Erickson, P., and Borgula, G. (1983). Retinal detachment in the cat: The pigment epithelial–photoreceptor interface. *Invest. Opthalmol. Vis. Sci.* **24,** 906–926.

Balda, M. S., and Anderson, J. M. (1993). Two classes of tight junctions are revealed by ZO-1 isoforms. *Am J. Physiol. 264 (Cell Physiol.* **33**), C918–C924.

Barnstable, C. J. (1987). A molecular view of mammalian retinal development. *Mol. Neurobiol.* **1,** 9–46.

Barnstable, C. J. (1991). Molecular aspects of development of mammalian optic cup and formation of retinal cell types. *Prog. Retina Res.* **10,** 45–67.

Barnstable, C. J., and Drager, U. C. (1984). Thy-1 antigen: A ganglion cell specific marker in rodent retina. *Neuroscience* **11,** 847–855.

Barnstable, C. J., Hofstein, R., and Akagawa, K. (1985). A marker of early amacrine cell development in rat retina. *Dev. Brain Res.* **20,** 286–290.

Båvik, C.-O., Lévy, F., Hellman, U., Wernstedt, C., and Eriksson, U. (1993). The retinal pigment epithelial membrane receptor for plasma retinol-binding protein: Isolation and cDNA cloning of the 63-kDa protein. *J. Biol. Chem.* **268,** 20540–20546.

Beermann, F., Ruppert, S., Hummler, E., Bosch, F. X., Muller, G., Ruther, U., and Schütz, G. (1990). Rescue of the albino phenotype by introduction of a functional tyrosinase gene into mice. *EMBO J.* **9,** 2819–2826.

Beermann, F., Schmid, E., Ganss, R., Schütz, G., and Ruppert, S. (1992a). Molecular characterization of the mouse tyrosinase gene: Pigment cell-specific expression in transgenic mice. *Pigment Cell Res.* **5,** 295–299.

Beermann, F., Schmid, E., and Schütz, G. (1992b). Expression of the mouse tyrosinase gene during embryonic development: Recapitulation of the temporal regulation if transgenic mice. *Proc. Natl. Acad. Sci. USA* **89,** 2809–2813.

Bentley, N. J., Eisen, T., and Goding, C. R. (1994). Melanocyte-specific expression of the human tyrosinase promoter: Activation by the microphthalmia gene procuct and role of the initiator. *Mol. Cell. Biol.* **14,** 7996–8006.

Bernard, O., Drago, J., and Sheng, H. (1992). L-myc and N-myc influence lineage determination in the central nervous system. *Neuron* **9,** 1217–1224.

Bhatt, N. S., Newsome, D. A., French, T., Hessburg, T. P., Diamond, J. G., Miceli, M. V., Kratz, K. E., and Oliver, P. D. (1994). Experimental transplantation of human retinal pigment epithelial cells on collagen substrates. *Am. J. Ophthalmol.* **117,** 214–221.

Bok, D. (1982). Autoradiographic studies on the polarity of plasma membrane receptors in retinal pigment epithelial cells. *In* The Structure of the Eye. (J. Hollyfield, Ed.), pp. 247–256. Elsevier Biomedical, New York.

Bok, D., O'Day, W., and Rodriguez-Boulan, E. (1992). Polarized budding of vesicular stomatitis and influenza virus from cultured human and bovine retinal pigment epithelium. *Exp. Eye Res.* **55,** 853–860.

Bradl, M., Klein-Szanto, A., Porter, S., and Mintz, B. (1991). Malignant melanoma in transgenic mice. *Proc. Natl. Acad. Sci. USA* **88,** 164–168.

Burke, J. M., Skumatz, C. M. B., Irving, P. E., and McKay, B. S. (1996). Phenotypic heterogeneity of retinal pigment epithelial cells in vitro and in situ. *Exp. Eye Res.* **62,** 63–73.

Burmeister, M., Novak, J., Liang, M.-Y., Basu, S., Ploder, L., Hawes, N. L., Vidgen, D., Hoover, F., Goldman, D., Kalnins, V. I., Roderick, T. H., Taylor, B. A., Hankin, M. H., and McInnes, R. R. (1996). Ocular retardation mouse caused by *Chx10* homeobox null allele: Impaired retinal progenitor proliferation and bipolar cell differentiation. *Nature Genet.* **12,** 376–384.

Cavallaro, T., Martone, R. L., Dwork, A. J., Schon, E. A., and Herbert, J. (1990). The retinal pigment epithelium is the unique site of transthyretin synthesis in the rat eye. *Invest. Ophthalmol. Vis. Sci.* **31,** 497–501.

Chang, C. W., Roque, R. S., Defoe, D. M., and Caldwell, R. B. (1991). An improved method for isolation and culture of pigment epithelial cells from rat retina. *Curr. Eye Res.* **10,** 1081–1086.

Chu, P., and Grunwald, G. B. (1990a). Identification of an adhesion-associated protein of the retinal pigment epithelium. *Invest. Ophthalmol. Vis. Sci.* **31,** 847–855.

Chu, P., and Grunwald, G. B. (1990b). Generation and characterization of monoclonal antibodies specific for the retinal pigment epithelium. *Invest. Ophthalmol. Vis. Sci.* **31,** 856–862.

Chu, P., and Grunwald, G. B. (1991). Identification of the 2A10 antigen of retinal pigment epithelium as a $\beta 1$ subunit of integrin. *Invest. Ophthalmol. Vis. Sci.* **32,** 1757–1762.

Clark, V. M. (1986). The cell biology of the retinal pigment epithelium. *In* The Retina: A Model for Cell Biology Studies, Part II. (A. Ruben and D. Farber, Eds.), pp. 129–168. Academic Press, Orlando.

Cooper, G. M. (1990). Oncogenes. Jones and Bartlett, Boston.

Coulombre, J. L., and Coulombre, A. J. (1965). Regeneration of neural retina from the pigmented epithelium in the chick embryo. *Dev. Biol.* **12,** 79–92.

Davis, A. A., Bernstein, P. S., Bok, D., Turner, J., Nachtigal, M., and Hunt, R.C. (1995). A human retinal pigment epithelial cell line that retains epithelial characteristics after prolonged culture. *Invest. Opthalmol. Vis. Sci.* **36,** 955–964.

De Iongh, R., and McAvoy, J. W. (1993). Spatio-temporal distrbution of acidic and basic FGF indicates a role for FGF in rat lens morphogenesis. *Dev. Dynamics* **198,** 190–202.

DePinho, R. A, Legouy, E., Feldman, L. B., Kohl, N. K., Yancopoulos, G. D., and Alt, F. W. (1986). Structure and expression of the murine N-myc gene. *Proc. Natl. Acad. Sci. USA* **83,** 1827–1831.

Dermietzel, R., and Krause, D. (1991). Molecular anatomy of the blood–retinal barrier as defined by immunocytochemistry. *Int. Rev. Cytol.* **127,** 57–109.

Dionne, C. A., Jaye, M., and Schlessinger, J. (1991), Structural diversity and binding of FGF receptors. *Ann. N.Y. Acad. Sci.* **638,** 161–166.

Donoso, L. A., Braunagel, S. C., Newsome, D. A., and Organisciak, D. T. (1988). Retinal pigment epithelial cell plasma membrane: A monoclonal antibody study. *Hybridoma* **7,** 265–272.

Dunn, K. C., Aotaki-Keen, A. E., Putkey, F. R., and Hjelmeland, L. M. (1996). ARPE-19, a human retinal pigment epithelial cell line with differentiated properties. *Exp. Eye Res.* **62,** 155–169.

Dutt K., Scott, M., Sternberg, P. P., Linser, P. J., and Srinivasan, A. (1993). Transdifferentiation of adult human pigment epithelium into retinal cells by transfection with an activated H-ras proto-oncogen. *DNA Cell Biol.* **12,** 667–673.

Eguchi, G., (1987). Instability in cell commitment of vertebrate pigmented epithelial cells and their transdifferentiation into lens cells. *Curr. Topics. Dev. Biol.* **20,** 21–37.

Eguchi, G., and Kodama, R. (1993). Transdifferentiation. *Curr. Opin. Cell Biol.* **5,** 1023–1028.

Elner, S. G., Elner, V. M., Nielsen, J. C., Torczynski, E., Yu, R., and Franklin, W. A. (1992). CD68 antigen expression by human retinal pigment epithelial cells. *Exp. Eye Res.* **55,** 21–28.

Fadool, J. M., and Linser, P. J. (1993). 5A11 Antigen is a cell recognition molecule which is involved in neuronal–glial interactions in avian neural retina. *Dev. Dynamics* **196,** 252–262.

Fadool, J. M., and Linser, P. J. (1994). Spatial and temporal expression of the 5A11/HT7 antigen in the chick embryo: Association with morphogenetic events and tissue maturation. *Roux's Arch. Dev. Biol.* **203,** 328–339.

Fujisawa, H., Morioka, H., Watanabe, K., and Nakamura, H. (1976). A decay of gap junctions in association with cell differentiation of neural retina in chick embryonic development. *J. Cell Sci.* **22,** 585–596.

Furuse, M., Hirase, T., Itoh, M., Nagafuchi, A., Yonemura, S., Tsukita, S., and Tsukita, S. (1993). Occludin: A novel integral membrane protein localizing at tight junctions. *J. Cell Biol.* **123,** 1777–1788.

Ganss, R., Montoliu, L., Monaghan, A. P., and Schütz, G. (1994a). A cell-specific enhancer far upstream of the mouse tyrosinase gene confers high level and copy number-related expression in transgenic mice. *EMBO J.* **13,** 3083–3093.

Ganss, R., Schmidt, A., Schütz, G., and Beermann, F. (1994b). Analysis of the mouse tyrosinase promoter in vitro and in vivo. *Pigment Cell Res.* **7,** 275–278.

Ganss, R., Schütz, G., and Beermann, F. (1994c). The mouse tyrosinase gene: Promoter modulation by positive and negative regulatory elements. *J. Biol. Chem.* **269,** 29808–29816.

Green, M. C. (1989). Catalog of mutant genes and polymorphic loci. *In* Genetic Variants and Strains of the Laboratory Mouse. (M. F. Lyon and A. G. Searle, Eds.), pp. 12–403. Oxford Univ. Press, Oxford, UK.

Grierson, I., Hiscott, P., Hogg, P., Robey, H., Mazure, A., and Larkin, G. (1994). Development, repair and regeneration of the retinal pigment epithelium. *Eye* **8,** 255–262.

Grisanti, S., and Guidry, C. (1995). Transdifferentiation of retinal pigment epithelial cells from epithelial to mesenchymal phenotype. *Invest. Ophthalmol. Vis. Sci.* **36,** 391–405.

Guérin, C. J., Anderson, D. H., and Fisher, S. K. (1990). Changes in intermediate filament immunolabeling occur in response to retinal detachment and reattachment in primates. *Invest. Ophthalmol. Vis. Sci.* **31,** 1474–1482.

Guillemot, F., and Cepko, C. L. (1992). Retinal fate and ganglion cell differentiation are potentiated by acidic FGF in an in vivo assay of early retinal development. *Development* **114,** 743–754.

Gundersen, D., Orlowski, J., and Rodriguez-Boulan, E. (1991). Apical polarity of Na,K-ATPase in retinal pigment epithelium is linked to a reversal of the ankyrin-fodrin submembrane cytoskeleton. *J. Cell Biol.* **112,** 863–872.

Gundersen, D., Powell, S. K., and Rodriguez-Boulan, E. (1993). Apical polarization of N-CAM in retinal pigment epithelium is dependent on contact with the neural retina. *J. Cell Biol.* **121,** 335–343.

Hamburger, V., and Hamilton, H. L. (1951). A series of normal stages in the development of the chick embryo. *J. Morph.* **88,** 49–82.

Hamel, C. P., Tsilou, E., Harris, E., Pfeffer, B. A., Hooks, J. J., Detrick, B., and Redmond, T. M. (1993a). A developmentally regulated microsomal protein specific for the pigment epithelium of the vertebrate retina. *J. Neurosci. Res.* **34,** 414–425.

Hamel, C. P., Tsilou, E., Pfeffer, B. A., Hooks, J. J., Detrick, B., and Redmond, T. M. (1993b). Molecular cloning and expression of RPE65, a novel retinal pigment epithelium-specific microsomal protein that is post-transcriptonally regulated in vitro. *J. Biol. Chem.* **268,** 15751–15757.

Hamel, C. P., Jenkins, N. A., Gilbert, D. J., Copeland, N. G., and Redmond, T. M. (1994). The gene for the retinal pigment epithelium-specific protein RPE65 is localized to human 1q31 and mouse 3. *Genomics* **20,** 509–512.

Hammerton, R. W., Krzeminski, K. A., Mays, R. W., Ryan, T. A., Wollner, D. A., and Nelson, W. J. (1991). Mechanism for regulating cell surface distribution of $Na^+K^+$-ATPase in polarized epithelial cells. *Science* **254,** 847–850.

Hasson, T., Heintzelman, M. B., Santos-Sacchi, J., Corey, D. P., and Mooseker, M. S. (1995). Expression in cochlea and retina of myosin VIIa, the gene product defective in Usher syndrome type 1B. *Proc. Natl. Acad. Sci. USA* **92,** 9815–9819

Hearing, V. J., and King, R. A. (1993). Determinants of skin color: Melanocytes and melanization. *In* Pigmentation and Pigmentary Disorders (N. Levine, Ed.). CRC Press, Boca Raton, FL

Hearing, V. J., and Tsukamoto, K. (1991). Enzymatic control of pigmentation in mammals. *FASEB J.* **5,** 2902–2909.

Hemesath, T. J., Steingrimsson, E., McGill, G., Hansen, M. J., Vaught, J., Hodgkinson, C. A., Arnheiter, H., Copeland, N. G., Jenkins, N. A., and Fisher, D. E. (1994). Microphthal-

mia, a critical factor in melanocute development, defines a discrete transcription factor family. *Genes Dev.* **8,** 2770–2780.
Hero, I., Farjah, M., and Scholtz, C. L. (1991). The prenatal development of the optic fussure in colobomatous microphthalmia. *Invest. Ophthalmol. Vis. Sci.* **32,** 2622–2635.
Hertwig, P. (1942). Neue mutationen und kopplungsgruppen bei der hausmaus. *Z. Indukt. Abstammungs-Vererbungsl.* **80,** 220–246.
Hilfer, S. R., Brady, R. C., and Wang, J. J. W. (1981). Intracellular and extracellular changes during early ocular development in the chick embryo. *In* Ocular Size and Shape: Regulation during Development. (S. R. Hilfer and J. B. Sheffield, Eds.), pp. 47–78. Springer-Verlag, New York.
Hodgkinson, C. A., Moore, K. J., Nakayama, A., Steingrimsson, E., Copeland, N. G., Jenkins, N. A., and Arnheiter, H. (1993). Mutations at the mouse microphthalmia locus are associated with defects in a gene encoding a novel basic-helix-loop-helix-zipper protein. *Cell* **74,** 395–404.
Höfer, D., and Drenckhahn, D. (1993). Molecular heterogeneity of the actin filament cytoskeleton associated with microvilli of photoreceptors, Müller's glial cells and pigment epithelial cells of the retina. *Histochemistry* **99,** 29–35.
Hooks, J. J., Detrick, B., Percopo, C., Hamel, C., and Siraganian, R. P. (1989). Development of characterization of monoclonal antibodies directed against the retinal pigment epithelial cell. *Invest. Ophthalmol. Vis. Sci.* **30,** 2106–2113.
Hu, J. G., Gallemore, R. P., Bok, D., Lee, A. Y., Frambach, D. A. (1994). Localization of Na-K ATPase on cultured human retinal pigment epithelium. *Invest. Ophthalmol. Vis. Sci.* **35,** 3582–3888.
Hudspeth, A. J., and Yee, A. G. (1973). The intercellular junctional complexes of retinal pigment epithelia. *Invest. Ophthalmol.* **12,** 354–365.
Hughes, A. E., Newton, V. E., Liu, X. Z., and Read, A. P. (1994). A gene for Waardenburg syndrome type 2 maps close to the human homologue of the microthalmia gene at chromosome 3p12-p14.1. *Nature Genet.* **7,** 509–512.
Hughes, M. J., Lingrel, J. B., Krakowsky, J. M., and Anderson, K. P. (1993). A helix–loop–helix transcription factor-like gene is located at the mi locus. *J. Biol. Chem.* **268,** 20687–20690.
Hunt, R. C., and Davis, A. A. (1990). Altered expression of keratin and vimentin in human retinal pigment epithelial cells in vivo and in vitro. *J. Cell. Physiol.* **145,** 187–199.
Huotari, V., Sormunen, R., Lehto, V.P., and Eskelinen, S. (1995). The polarity of the membrane skeleton in retinal pigment epithelial cells of developing chicken embryos and in primary culture. *Differentiation* **58,** 205–215.
Ito, S. (1993). Biochemistry and physiology of melanin. *In* Pigmentation and Pigmentary Disorders. (N. Levine, Ed.), pp. 33–59. CRC Press, Boca Raton, FL.
Janssen, J. J., Janssen, B. P., and Vugt, A. H. (1994). Characterization of monoclonal antibodies recognizing retinal pigment epithelial antigens. *Invest. Ophthalmol. Vis. Sci.* **35,** 189–198.
Kapur, R. P., Bigler, S. A., Skelly, M., and Gown, A. M. (1992). Anti-melanoma monoclonal antibody HMB45 identifies an oncofetal glycoconjugate associated with immature melanosomes. *J. Histochem. Cytochem.* **40,** 207–212.
Khosravi-Far, R., and Der, C. J. (1994). The Ras signal transdcution pathway. *Cancer Metastasis Rev.* **13,** 67–89.
Kim, R. Y., and Wistow, G. J. (1992). The cDNA RPE1 and monoclonal antibody HMB-50 define gene products preferentially expressed in retinal pigment epithelium. *Exp. Eye Res.* **55,** 657–662.
Kitaoka, T., Sharif, M., Hanley, M. R., and Hjelmeland, L. M. (1994). Expression of the MAS proto-oncogene in the retinal pigment epithelium of the rhesus macaque. *Curr. Eye Res.* **13,** 345–351.
Klein, L. R., MacLeish, P. R., and Wiesel, T. N. (1990). Immunolabelling by a newt retinal pigment epithelium antibody during retinal development and regeneration. *J. Comp. Neurol.* **293,** 331–339.

Klein-Szanto, A., Bradl, M., Porter, S., and Mintz, B. (1991). Melanosis and associated tumors in transgenic mice. *Proc. Natl. Acad. Sci. USA* **88,** 169–173.

Klüppel, M., Beermann, F., Ruppert, S., Schmid, E., Hummler, E., and Schütz, G. (1991). The mouse tyrosinase promoter is sufficient for expression in melanocytes and in the pigmented epithelium of the reina. *Proc. Natl. Acad. Sci. USA* **88,** 3777–3781.

Kobayashi, H., Ueda, M., and Honda, Y. (1991). Molecules specific to pigment epithelial cells: Expression durng in situ development and in vitro lens transdifferentiation of chick emryo pigment epithelium. *Ophthal. Res.* **23,** 309–319.

Kobayashi, T., Urabe, K., Winder, A., Tsukamoto, K., Brewington, T., Imokawa, G., Potterf, B., and Hearing, V. J. (1994a). DHICA oxidase activity of TRP1 and interactions with other melanogenic enzymes. *Pigment Cell Res.* **7,** 227–234.

Kobayashi, T., Urabe, K., Winder, A., Jimenez-Cervantes, C., Imokawa, G., Brewington, T., Solano, F., Garcia-Borron, J. C., and Hearing, V. J. (1994b). Tyrosinase related protein (TRP1) functions as a DHICA oxidase in melanin biosynthesis. *EMBO J.* **13,** 5818–5825.

Korte, G. E., and Wanderman, M. C. (1993). Distribution of Na$^+$K$^+$-ATPase in regenerating retinal pigment epithelium in the rabbit. A study by electron microscopic cytochemistry. *Exp. Eye Res.* **56,** 219–229.

Korte, G. E., Perlman, J. I., and Pollack, A. (1994). Regeneration of mammalian retinal pigment epithelium. *Int. Rev. Cytol.* **152,** 223–263.

Krakowsky, J. M., Boissy, R. E., Neuman, J. C., and Lingrel, J. B. (1993). A DNA insertinal muation results in microphthalmia in transgenic mice. *Transgenic Res.* **2,** 14–20.

Kwon, B. S. (1993). Pigmentation genes: The tyrosinase gene family and the pml 17 gene family. *J. Invest. Dermatol.* **100**(Suppl. 2), 134S–140S.

LaVail, M. M., Li, L., Turner, J. E., and Yasumura, D. (1992). Retinal pigment epithelial cell transplantation in RCS rats: Normal metabolism in rescued photoreceptors. *Exp. Eye Res.* **55,** 555–562.

Li, H-S., Yang, J. M., Jacobson, R. D., Pasko, D., and Sundin, O. (1994), Pax-6 is first expressed in a region of ectoderm anterior to the early neural plate: Implications for stepwise determunation of the lens. *Dev. Biol.* **162,** 181–194.

Lin, W. L., and Essner, E. (1994). Differential localization of type IV collagen in Bruch's membrane of the eye: Comparison of two polyclonal antibodies. *Cell. Mol. Biol.* **40,** 851–858.

Liu, I., Chen, J-D., Ploder, L., Vidgen, D., van der Kooy, D., Kalnins, V. I., and McInnes, R. R. (1994), Developmental expression of a novel murine homeobox gene (*Chx10*): Evidence for roles in determination of the neroretina and inner nuclear layer. *Neuron* **13,** 377–393.

Lopashov, G. V., and Sologub, A. A. (1972). Artificial metaplasia of pigmented epithelium into retina in tadpoles and adult frogs. *J. Embryol. Exp. Morphol.* **28,** 521–546.

Lopez, R., Gouras, P., Brittis, M., and Kjeldbye, H. (1987). Transplantation of cultured rabbit retinal epithelium to rabbit retina using a closed-eye method. *Invest. Ophthalmol. Vis. Sci.* **28,** 1131–1137.

Machemer, R. (1988). Proliferative vitreoretinopathy (PVR): A personal account of its pathogenesis and treatment. *Invest. Ophthalmol. Vis. Sci.* **29,** 1771–1783.

Machemer, R., and Laqua, H. (1975). Pigment epithelial proliferaiton in retinal detachment (massive periretinal proliferation). *Am. J. Ophthalmol.* **80,** 1–23.

Machemer, R., van Horn, D., and Aaberg, T. M. (1978). Pigment epithelial proliferation in human retinal detachment and massive periretinal proliferation. *Am. J. Ophthalmol.* **85,** 181–191.

Mandelcorn, M. S., Machemer, R., Fineberg, E., and Hersch, S. B. (1975). Proliferation and meaplasia of intrvitreal retinal pgmnet epithelial cell autotransplants. *Am. J. Ophthalmol.* **80,** 227–237.

Marcu, K. B., Bossone, S. A., and Patel, A. J. (1992). *myc* function and regulation. *Annu. Rev. Biochem.* **61,** 809–860.

Martin, P., Carriere, C., Dozier, C., Quatannens, B., Mirabel, M., Vandenbunder, B., Stehelin, D., and Saule, S. (1992). Characterization of a paired box- and homeobox-containing quail gene (*Pax-QNR*) expressed in the neuroretina. *Oncogene* **7,** 1721–1728.

Mathers, P. H., and Jamrich, M. (1995). Induction of a novel homeodomain gene, Rx, in the anterior neural plate and its role in retinal proliferation. *Invest. Ophthalmol. Vis. Sci.* **36,** S213.

Mazure, A., and Grierson, I. (1992). In vivo studies of the contractility of cell types involved in proliferative vitreoretinopathy. *Invest. Ophthalmol. Vis. Sci.* **33,** 3407–3416.

McCormick, F. (1994). Activators and effectors of ras p21 proteins. *Curr. Opin. Genet. Dev.* **4,** 71–76.

McKay, B. S., and Burke, J. M. (1994). Separation of phenotypically distinct subpopulations of cultured human retinal pigment epithelial cells. *Exp. Cell Res.* **213,** 85–92.

McKay, B. S., Suson, J. D., and Burke, J. M. (1992). Subpopulations of human RPE in vitro. *Invest. Opthalmol. Vis. Sci.* **33,** 1204.

Miller, K., and Rizzino, A. (1994). Developmental regulation and signal transductuion pathways of fibroblast growth factors and their receptors. *In* Growth Factors and Signal Transduction in Development (M. Nilsen-Hamilton, Ed.) Wiley-Liss, New York.

Mircheff, A. K., Miller, S. S., Farber, D. B., Bradley, M. E., O'Day, W. T., and Bok, D. (1990). Isolation and provisional identification of plasma membrane populations from cultured human retinal pigment epithelium. *Invest. Ophthalmol. Vis. Sci.* **31,** 863–878.

Mizuno, R., Cavallaro, T., and Herbert, J. (1992). Temporal expression of the transthyretin gene in the developing rat eye. *Invest. Ophthalmol. Vis. Sci.* **33,** 341–349.

Monaghan, A. P., Davidson, D. R., Sime, C., Graham, E., Baldock, R., Bhattacharya, S. S, and Hill, R. E. (1991). The Msh-like homeobox genes define domains in the developing vertebrate eye. *Development* **112,** 1053–1061.

Morse, D. E., and McCann, P. S. (1984). Neuroectoderm of the early embryonic rat eye. *Invest. Ophthalmol. Vis. Sci.* **25,** 899–907.

Nabi, I. R., Mathews, A. P., Cohen-Gould, L., Gundersen, D., and Rodriguez-Boulan, E. (1993). Immortalization of polarized rat retinal pigment epithelium. *J. Cell Sci.* **104,** 37–49.

Neill, J. M., and Barnstable, C. J. (1990). Expression of the cell surface antigens RET-PE2 and N-CAM by rat retinal pigment epithelial cells during development and in tissue culture. *Exp. Eye Res.* **51,** 573–583.

Neill, J. M., Thornquist, S. C., Raymond, M. C., Thompson, J. T., and Barnstable, C. J. (1993), RET-PE10: A 61 kD polypeptide epitope expressed late during vertebrate RPE maturation. *Invest. Ophthalmol. Vis. Sci.* **34,** 453–462.

Nornes, H. O., Dressler, G. R., Knapik, E. W., Deutsch, U., and Gruss, P. (1990). Spatially and temporally restricted expression of *Pax2* during murine neurogenesis. *Development* **109,** 797–809.

Okami, T., Yamamoto, A., Omori, K., Takada, T., Uyama, M., and Tashiro, Y. (1990). Immunocytochemical localization of $Na+,K+$-ATPase in rat retinal pigment epithelial cells. *J. Histochem. Cytochem.* **38,** 1267–1275.

Opas, M., and Dziak, E. (1994). bFGF-induced transdifferentiation of RPE to neuronal progenitors is regulated by the mechanical properties of the substratum. *Dev. Biol.* **161,** 440–454.

Packer, S. O. (1967). The eye and skeletal defects of two mutants alleles at the microphthalmia locus of Mus musculus. *J. Exp. Zool.* **165,** 21–45.

Park, C. M., and Hollenberg, M. J. (1989). Basic fibroblast growth factor induces retinal regeneration in vivo. *Dev. Biol.* **134,** 201–205.

Park, C. M., and Hollenberg, M. J. (1991). Induction of retinal regeneration in vivo by growth factors. *Dev. Biol.* **148,** 322–333.

Park, C. M., and Hollenberg, M. J. (1993). Growth factor-induced retinal regeneration in vivo. *Int. Rev. Cytol.* **146,** 49–74.

Philp, N. J., and Nachmias, V. T. (1987). Polarized distribution of integrin and fibronectin in retinal pigment epithelium. *Invest. Ophthalmol. Vis. Sci.* **28,** 1275–1280.

Philp, N. J., Yoon, M. Y., and Hock, R. S. (1990). Identification and localication of talin in chick retinal pigment epithelial cells. *Exp. Eye Res.* **51,** 191–198.

Philp, N., Chu, P., Pan, T.-C., Zhang, R. Z., Chu, M.-L., Stark, K., Boettiger, D., Yoon, H., and Kieber-Emmons, T. (1995). Developmental expression and molecular cloning of REMP, a novel retinal epithelial membrane protein. *Exp. Cell Res.* **219,** 64–73.

Pittack, C., Jones, M., and Reh, T. A. (1991). Basic fibroblast growth factor induces retinal pigment epithelium to generate neural retina *in vitro*. *Development* **113,** 577–588.

Ponnazhagan, S., Hou, L., and Kwon, B. S. (1994). Structural organization of the human tyrosinase gene and sequence analysis and characterization of its promoter region. *J. Invest. Dermatol.* **102,** 744–748.

Prota, G. (1992). Melanins and Melanogenesis. Academic Press, San Diego.

Quiring, R., Walldorf, U., Kloter, U., and Gehring, W. J. (1994). Homology of the *eyeless* gene of *Drosophila* to the *Small eye* gene in mice and *Aniridia* in humans. *Science* **265,** 785–789.

Reh, T. A., and Nagy, T. (1987). A possible role for the vascular membrane in retinal regeneration in *Rana catesbienna* tadpoles. *Dev. Biol.* **122,** 471–482.

Rizzolo, L. J. (1990). The distribution of Na$^+$K$^+$-ATPase in the retinal pigmented epithelium from chicken embryo is polarized in vivo but not in primary cell culture. *Exp. Eye Res.* **5,** 435–446.

Rizzolo, L. J. (1991). Basement membrane stimulates the polarized distribution of integrins but not the Na,K-ATPase in the retinal pigment epithelium. *Cell Regul.* **2,** 939–949.

Rizzolo, L. J., and Heiges, M. (1991). The polarity of the retinal pigment epithelium is developmentally regulated. *Exp. Eye Res.* **53,** 549–553.

Rizzolo, L. J., and Joshi, H. C. (1993). Apical orientation of the microtubule organizing center and associated $\gamma$-tubulin during the polarization of the pigment epithelium in vivo. *Dev, Biol.* **157,** 147–156.

Rizzolo, L. J., and Li, Z.-Q. (1993). Diffusible, retinal factors stimulate the barrier properties of junctional complexes in the retinal pigment epithelium. *J. Cell Sci.* **106,** 859–867.

Rizzolo, L. J., and Zhou, S. (1995). The distribution of Na$^+$,K$^+$-ATPase and 5A11 antigen in apical microvilli of the retinal pigment epithelium is unrelated to $\alpha$-spectrin. *J. Cell Sci.* **108,** 3623–3633.

Rizzolo, L. J., Zhou, S., and Li, Z. Q. (1994). The neural retina manitains integrins in the apical membrane of the RPE early in development. *Invest. Ophthalmol. Vis. Sci.* **35,** 2567–2576.

Rodriguez-Boulan, E., and Nelson, W. J. (1989). Morphogenesis of the polarized epithelial cell phenotype. *Science* **245,**: 718–725.

Rubin, L. L., Hall, D. E., Porter, S., Barbu, K., Cannon, C., Horner, H. C., Janatpour, M., Liaw, C. W., Manning, K., Morales, J., Tanner, L. I., Tomaselli, K. J.,. and Bard, F. (1991). A cell culture model of the blood–brain barrier. *J. Cell Biol.* **115,** 1725–1735.

Ryeom, S. W., Sparrow, J. R., and Silverstein, R. L. (1996). CD36 participates in the phagocytosis of rod outer segments by retinal pigment epithelium. *J. Cell Sci.* **109,** 387–395.

Sagara, H., and Hirosawa, K. (1991). Monoclonal antibodies which recognize endoplasmic reticulum in the retinal pigment epithelium. *Exp. Eye Res.* **53,** 765–771.

Sandig, M., and Kalnins, V. I. (1988). Subunits in zonulae adhaerentes and striations in the associated circumferential microfilament bundles in chicken retinal pigment epithelial cells in situ. *Exp. Cell Res.* **175,** 1–14.

Sandig, M., and Kalnins, V. I. (1990). Morphological changes in the zonula adhaerens during embryonic development of chick retinal pigment epithelial cells. *Cell Tissue Res.* **259,** 455–461.

Sarna, T. (1992). Properties and functions of the ocular melanin: a photobiophysical view. *J. Photochem. Photobiol.-B (Biol.)* **12,** 215–258.

Sasai, Y., Lu, B., Steinbeisser, H., and De Roberts, E. M. (1995). Regulation of neural induction by Chd and Bmp-4 antagonistic patterning signals in Xenopus. *Nature* **376**, 333–336.

Sawai, S., Shimono, A., Hanaoka, K., and Kondoh, H. (1991). Embryonic lethality resulting from disruption of both N-myc alleles in mouse zygotes. *New Biol.* **3**, 861–869.

Scott, M. P. (1992). Vertebrate homeobox gene nomenclature. *Cell* **71**, 551–553.

Seaton, A. D., Sheedlo, H. J., and Turner, J. E. (1994). A primary role for RPE transplants in the inhibition and regression of neovascularization in the RCS rat. *Invest. Ophthalmol. Vis. Sci.* **35**, 162–169.

Seulberger, H., Lottspeich, F., and Risau, W. (1990). The inducible blood–brain barrier specific molecule HT7 is a novel immunoglobulin-like cell surface glycoprotein. *EMBO J.* **9**, 2151–2158.

Seulberger, H., Unger, C. M., and Risau, W. (1992). HT7, neurothelin, basigin, gp-42 and OX-47—Many names for one developmentally regulated immuno-globulin-like surface glycoprotein on blood–brain barrier endothelium, epithelial tissue barriers and neurons. *Neurosci. Lett.* **140**, 93–97.

Sheedlo, H. J., Li, L., Gaur, V. P., Young, R. W., Seaton, A. D., Stovall, S. V., Jaynes, C. D., and Turner, J. E. (1992). Photoreceptor rescue in the dystrophic retina by transplantation of retinal pigment epithelium. *Int. Rev. Cytol.* **138**, 1–49.

Sheedlo, H. J., Li, L., Barnstable C. J., and Turner J. E. (1993). Synaptic and photoreceptor components in retinal pigment epithelial cell transplanted retinas of Royal College of Surgeons dystrophic rats. *J. Neurosci. Res.* **36**, 423–431.

Sheng, Y., Gouras, P., Cao, H., Berglin, L., Kjeldbye, H., Lopez, R., and Rosskothen, H. (1995). Patch transplants of human fetal retinal pigment epithelium in rabbit and monkey retina. *Invest. Ophthalmol. Vis. Sci.* **36**, 381–390.

Silvers, W. K. (1979). The Coat Colors of Mice. A Model for Mammalian Gene Action and Interaction. Springer-Verlag, New York.

Simeone, A., Acampora, D., Mallamaci, A., Stornaiuolo, A., Rosaria D'Apice, M., Nigro, V., and Boncinelli, E. (1993), A vertebrate gene related to *orthodenticle* contains a homeodomain of the bicoid class and demarcates anterior neuroectoderm in the gastrulating mouse embryo. *EMBO J.* **12**, 2735–2747.

Sologub, A. A. (1975). The development of differentiation of pigment epithelium in teleosts and its stimulation to metaplasia. *Ontogenez* **6**, 39–46.

Sparrow, J. R., Hicks, D., and Barnstable, C. J. (1990). Cell commitment and differentiation in explants of embryonic rat retina. Comparison to the developmental potential of dissociated retina. *Dev. Brain Res.* **51**, 69–84.

Sparrow, J. R., Ryeom, S. W., Abumrad, N. A., Ibrahimi, A., and Silverstein, R. L. (1996). CD36 expression is altered in retinal pigment epithelial cells of the RCS rat. *Exp. Eye Res.* In press.

Stanton, B. R., Perkins, A. S., Tessarollo, L., Sassoon, D. A., and Parada, L. F. (1992). Loss of N-myc function results in embryonic lethality and failure of the epithelial component of the embryo to develop. *Genes Dev.* **6**, 2235–2246.

Steinberg, R. H., and Wood, I. (1979). The relationship of the retinal pigment epithelium to photoreceptor outer segments in human retina. *In* The Retinal Pigment Epithelium (K. M. Zinn and M. F. Marmor, Eds.), pp. 32–44. Harvard Univ. Press, Cambridge, MA.

Steingrimsson, E., Moore, K. J., Lamoreux, M. L., Ferre-D'Amare, A. R., Burlwy, S. K., Sanders Zimring, D. C., Skow, L. C., Hodgkinson, C. A., Arnheiter, H., Copeland, N. G., and Jenkins, N. A. (1994). Molecular basis of mouse microphthalmia (mi) mutations helps explain their developmental and phenotypic consequences. *Nature Genet.* **8**, 256–263.

Stone, L. S. (1950). The role of retinal pigment cells in regenerating neuyral retinae of adult salamander eyes. *J. Exp. Zool.* **113**, 9–32.

Stone, L. S., and Steinitz, H. (1957). Regeneration of neural retina and lens from retina pigment cell grafts in adult newts. *J. Exp. Zool.* **135,** 301–318.
Stroeva, O. G. (1960). Experimental analysis of the eye morphogenesis in mammals. *J. Embryol. Exp. Morphol.* **8,** 349–368.
Stroeva, O. G., and Mitashov, V. I. (1983). Retinal pigment epithelium: proliferation and differentiation during development and regeneration. *Int. Rev. Cytol.* **83,** 221–293.
Sundin, O., Yang, J-M., and Kang. R. (1995). Homeobox gene expression during establishment of the eye primordia. *Invest. Ophthalmol. Vis. Sci.* **36,** S213.
Suzuki, Y., Ishiguro, S., and Tamai, M. (1993). Identification and immunohistochemistry of retinol dehydrogenase from bovine retinal pigment epithelium. *Biochim. Biophys. Acta* **1163,** 201–208.
Tachibana, M., Hara, Y., Vyas, D., Hodgkinson, C., Fex, J., Grundfast, K., and Arnheiter, H. (1992). Cochlear disorder associated with melanocyte anomaly in mice with a transgenic insertional mutation. *Mol. Cell. Neurosci.* **3,** 433–445.
Tachibana, M., Perez-Jurado, L. A., Nakayama, A., Hodgkinson, C. A., Li, X., Schneider, M., Mili, T., Fex, J., Francke, U., and Arnheiter, H. (1994). Cloning of MITF, the human homolog of the mouse microphthalmia gene and assignment to chromosome 3p14.1-p12.3. *Hum. Mol. Genet.* **3,** 553–557.
Tanaka, S., Yamamoto, H., Takeuchi, S., and Takeuchi, T. (1990). Melanization in albino mice trandformed by introducing clined mouse tyrosinase gene. *Development* **108,** 223–227.
Tarnowski, B. I., Shepherd, V. L., and McLaughlin, B. J. (1988). Expression of mannose receptors for pinocytosis and phagocytosis on rat retinal pigment epithelium. *Invest. Ophthalmol. Vis. Sci.* **29,** 742–748.
Tassabehji, M., Newton, V. E., and Read, A. P. (1994). Waardenburg syndrome type 2 caused by mutations in the human microphthalmia (MITF) gene. *Nature Genet.* **8,** 251–255.
Tcheng, M., Fuhrmann, G., Hartmann, M-P., Courtois, Y., and Jeanny, J-C. (1994). Spatial and temporal expression patterns of FGF receptor genes type 1 and type 2 in the developing chick retina. *Exp. Eye Res.* **58,** 351–358.
Tomasz, M., Lipman, R., Chowdary, D., Pawlak, J., Verdine, G. L., and Nakanishi, K. (1987). Isolation and structure of a covalent cross-link adduct between mitomycin C and DNA. *Science* **235,** 1204–1208.
Tso, M. (1979). Developmental, reactive, and neoplastic proliferation of the retinal pigment epithelium. *in* The Retinal Pigment Epithelium (K. M. Zinn and M. F. Marmor Eds.), pp 267–276. Harvard Univ. Press, Cambridge, MA.
Tsukamoto, K., Jackson, I. J., Urabe, K., Montague, P. N., and Hearing, V. J. (1992). A second tyrosinase-related protein, TRP-2, is a melanogenic enzyme termed DOPAchrome tautomerase. *EMBO J.* **11,** 519–526.
Turksen, K., Opas, M., Aubin, J. E., and Kalnins, V. I. (1983). Microtubules, microfilaments and adhesion patterns in differentiating chick retinal pigmented epithelial (RPE) cells in in vitro. *Exp. Cell Res.* **147,** 379–391
Turksen, K., Aubin, J. E., Sodek, J., and Kalnins, V. I. (1985). Localization of laminin, type IV collagen, fibronectin, and heparan sulfate proteoglycan in chick retinal pigment epithelium basement membrane during embryonic development. *J. Histochem. Cytochem.* **33,** 665–671.
Turner, E. E., Jenne, K. J., and Rosenfeld, M. G. (1994). Brn-3.2: A Brn-3-related transcription factor with distinctive cental nercous sustem ecpression and regulation by retinoic acid. *Neuron* **12,** 205–218.
Vinores, S. A., Orman, W., Hooks, J. J., Detrick, B., and Campochiaro, P. A. (1993). Ultrastructural localization of RPE-associated epitopes recognized by monoclonal antibodies in human RPE and their induction in human fibroblasts by vitreous. *Graefes Arch. Clin. Exp. Ophthalmol.* **231,** 395–401.

Vinores, S. A., Derevjanik, N. L., Mahlow, J., Hackett, S. F., Haller, J. A., deJuan, E., Frankfurter, A., and Campochiaro, P.A. (1995). Class III β-tubulin in human retinal pigment epithelial cells in culture and in epiretinal membranes. *Exp. Eye Res.* **60**, 385–400.

Wakamatsu, Y., Watanabe, Y., Shimono, A., and Kondoh, H. (1993). Transition of localization of the N-Myc protein from nucleus to cytoplasm in differentiating neurons. *Neuron* **10**, 1–9.

Walther, C., and Gruss, P. (1991). Pax-6, a murine paired box gene, is expressed in the developing CNS. *Development* **113**, 1435–1449.

Wilcox, D. K., and Vergnes, J. P. (1989). Differential expression of N-ras in metaplastic retinal pigmented epithelium. *Exp. Eye Res.* **48**, 271–280.

Williams, C. D., and Rizzolo, L .J. (1995). Molecular remodeling of tight junctions during development of the blood–retinal barier. *Mol. Biol. Cell* **6**, 193a.

Wilson, P. A., and Hemmati-Brivanlou, A. (1995). Induction of epidermis and inhibition of neural fate by Bmp-4. *Nature* **376**, 331–333.

Wongpichedchai, S., Weiter, J. J., Weber, P., and Dorey, C. K. (1992). Comparison of external and internal approaches for transplantation of autologous retinal pigment epithelium. *Invest. Ophthalmol. Vis. Sci.* **33**, 3341–3352.

Xiang, M., Zhou, L., Macke, J. P., Peng, Y-W., Eddy, R. L., Shows, T. B., and Nathans, J. (1993). Brn-3b: A POU-domain gene expressed in a subset of retinal ganglion cells. *Neuron* **11**, 689–701.

Xiang, M., Zhou, L., Macke, J. P., Yoshioka, T., Hendry, S., Eddy, R. L., Shows, T. B., and Nathans, J. (1995). The Brn-3 family of POU-domain factors: Primary structure, binding specificity, and expression in subsets of retinal ganglion cells and somatosensory neurons. *J. Neurosci.* **15**, 4762–4785.

Yamaguchi, K., Yamaguchi, K, Gaur, V. P., and Turner, J .E. (1993). Retinal pigment epithelial cell transplantation into aging retina: a possible approach to delay age-related cell death. *Jpn J. Ophthalmol.* **37**, 16–27.

Yamamoto, S., Du, J., Gouras, P., and Kjeldbye, H. (1993). Retinal pigment epithelial transplants and retinal function in RCS rats. *Invest. Ophthalmol. Vis. Sci.* **34**, 3068–3075.

Yasumoto, K-I., Yokoyama, K., Shibata, K., Tomita, Y., and Shibahara, S. (1994). Microphthalmia-associated transcription factor as a regulator for melanocyte-specific transcription of the human tyrosinase gene. *Mol. Cell. Biol.* **14**, 8058–8070.

Yavuzer, U., Keenan, E., Lowings, P., Vachtenheim, J., Currie, G., and Goding, C. R. (1995). The microphthalmia gene product interacts with the retinoblastoma protein in vitro and is a target for deregualtion of melanocyte-specific transcription. *Oncogene* **10**, 123–134.

Yokoyama, T., Silversides, D. W., Waymire, K. G., Kwon, B. S., Takeuchi, T., and Overbeek, P. A. (1990). Conserved cysteine to serine mutation in tyrosinase is responsible for the classical albino mutation in laboratory mice. *Nucleic Acids Res.* **18**, 7293–7298.

Yu, H. S., Hernandez, V., Haywood, M., and Wong, C. G. (1993). Melatonin inhibits the proliferation of retinal pigment epithelial cells in vitro. *In Vitro Cell. Dev. Biol. Anim.* **29A**, 415–418.

Zhao, S., and Barnstable, C. J. (1996). Differential effects of bFGF on development of the rat retina. *Brain Res.* **723**, 169–176.

Zhao, S., Thornquist, S. C., and Barnstable, C. J. (1995). In vitro transdifferentiation of embryonic rat pigment epithelium to neural retina. *Brain Res.,* **677**, 300–310.

Zhou, Y., and Opas, M. (1994). Cell shape, intracellular pH, and fibroblast growth factor responsiveness during transdifferentiation of retinal pigment epithelium into neuroepithelium in vitro. *Biochem. Cell Biol.* **72**,257–265.

Zhou, Y., Moszczynska, A., and Opas, M. (1994). Generation and characterization of antibodies to adhesion-related molecules of retinal pigment epithelial cells. *Exp. Eye Res.* **58**, 585–593.

Zinn, K. M., and Benjamin-Henkind, J. V. (1979). Anatomy of the human retinal pigment epithelium. *In* The Retinal Pigment Epithelium (K. M. Zinn and M. F. Marmor, Eds.), pp. 3–31. Harvard Univ. Press, Cambridge, MA.

# The Role of Endothelins in the Paracrine Control of the Secretion and Growth of the Adrenal Cortex

Gastone G. Nussdorfer,* Gian Paolo Rossi,† and Anna S. Belloni*
*Department of Anatomy and †Department of Clinical and Experimental Medicine, University of Padua, I-35121 Padua, Italy

Endothelins (ETs) are a family of vasoactive peptides (ET-1, ET-2, and ET-3) mainly secreted by vascular endothelium and widely distributed in the various body systems, where they play major autocrine/paracrine regulatory functions, acting via two subtypes of receptors ($ET_A$ and $ET_B$). Adrenal cortex synthesizes and releases ETs and expresses both $ET_A$ and $ET_B$. Zona glomerulosa possesses both $ET_A$ and $ET_B$, whereas zona fasciculata/reticularis is almost exclusively provided with $ET_B$. ETs exert a strong mineralocorticoid and a less intense glucocorticoid secretagogue action, mainly via $ET_B$ receptors. ETs also appear to enhance the growth and steroidogenic capacity of zona glomerulosa and to stimulate its proliferative activity. This trophic action of ETs is likely to be mediated mainly by $ET_A$ receptors. The intraadrenal release of ETs undergoes a multiple regulation, with the rise in blood flow rate and the local release of nitric oxide being the main stimulatory factors. Data are also available that indicate that ETs may also have a role in the pathophysiology of primary aldosteronism caused by adrenal adenomas and carcinomas.

**KEY WORDS:** Endothelins, Endothelin receptors, Adrenal cortex, Steroidogenesis regulation, Adrenocortical growth regulation, Adrenocortical pathophysiology.

## I. Introduction

Endothelins (ETs) are a family of 21-amino acid peptides, secreted by vascular endothelium, that exert a potent and long-lasting vasoconstrictor and pressor activity. The ET family consists of three isopeptides, ET-1, ET-2, and ET-3, that were originally isolated from porcine aortic endothelium

(Yanagisawa et al., 1988; Inoue et al., 1989). Each isopeptide is encoded by a separate gene; gene products (prepro-ETs) are clived intracellularly to generate 38- or 39-amino acid biologically inactive precursors (pro-ETs or big ETs) that subsequently are transformed to the active isopeptides by the endothelin-converting enzymes (ECEs) (Rubanyi and Polokoff, 1994). Two subtypes of ET receptors, $ET_A$ and $ET_B$, have been identified, that are mainly expressed in vascular smooth muscle cells and endothelial cells, respectively (Sakurai et al., 1992; Haynes et al., 1993; Bax and Saxena, 1994). The isopeptide-selective $ET_A$ receptor (binding potency, ET-1 $\geq$ ET-2 $\gg$ ET-3) mediates vasoconstriction, whereas the nonisopeptide-selective $ET_B$ receptor (binding potency, ET-1 = ET-2 = ET-3) is mainly involved in endothelium-dependent vasodilation mediated by nitric oxide release (Haynes et al., 1993; Rubanyi and Polokoff, 1994).

In the past decade cDNAs encoding big ETs, ECE, and ET receptors were cloned (Yanagisawa et al., 1988; Bloch et al., 1989, 1991; Arai et al., 1990; Lin et al., 1991; Ogawa et al., 1991; Schmidt et al., 1994; Shimada et al., 1994; Tsukahara et al., 1994; Xu et al., 1994), and selective ET receptor ligands were developed. These ligands include the $ET_A$-selective antagonist BQ-123 (Ihara et al., 1992), the $ET_B$ agonists BQ-3020 and IRL-1620 (Ihara et al., 1992), and the specific and potent $ET_B$ antagonist BQ-788 (Ishikawa et al., 1994). Four highly homologous cordiotoxic molecules, sarafotoxins (S), were also isolated from the venom of *Atractaspis engaddensis* (Kloog and Sokolovsky, 1989), and two of them (S6B and S6C) were found to be specific but weak $ET_B$ agonists (Williams et al., 1991). With the use of molecular biology techniques and pharmacologic tools investigators were able to demonstrate that ETs are synthesized not only in the cardiovascular apparatus but also in almost all the main biological systems, where they play important physiological and pathophysiological roles. The almost exclusive autocrine/paracrine mechanism(s) of action of ETs can be deduced by the fact that their circulating concentrations (ranging from 1 to 8 p$M$) are in general well below those able to evoke appreciable biological effects (Kennedy et al., 1993; Rubanyi and Polokoff, 1994).

It has recently been demonstrated that the adrenal gland synthesizes ETs and expresses ET receptors, and a large mass of data has accumulated indicating that these regulatory peptides play a major role in its secretory activity and growth under both physiological and pathological conditions. Although some excellent review articles have appeared concerning the involvement of ETs in the functional regulation of the endocrine system (Kennedy et al., 1993; Masaki, 1993; Naruse et al., 1994a), none of them specifically dealt with the adrenal cortex. In the following sections, we survey findings indicating that locally synthesized ETs are involved in the paracrine control of the secretion and growth of the adrenal cortex. The possible mechanisms underlying these actions of ETs will be discussed as well as those controlling intraadrenal ET release. Finally, the possible involvement of ETs in the pathology of the adrenal cortex will be reviewed.

## II. Endothelin Biosynthesis

### A. Gene Expression Studies

Preliminary evidence of low abundance expression of ET-1 mRNA in porcine adrenal gland was given by Nunez *et al.* (1990) with reverse transcription-polymerase chain reaction (RT-PCR) and Southern hybridization with a 21-mer $^{35}$S-labeled oligonucleotide probe complementary to a sequence within the cDNA between the primers. In humans, the prepro-ET-1 mRNA was first demonstrated by a study on homogenates of adrenals from patients with aldosterone-producing adenoma causing Conn's syndrome who were treated by adrenalectomy (Imai *et al.*, 1992). By using Northern blotting of poly(A)$^+$RNA and hybridization with a specific cDNA probe for the ET-1 gene, Imai and associates were able to clearly detect the prepro-ET-1 mRNA in the normal adrenal cortices surrounding the tumors. These results were in agreement with those obtained by Rossi *et al.* (1994, 1995b), who investigated the prepro-ET-1 gene expression by RT-PCR on homogenates of a larger number of normal human adrenal cortices and aldosterone-producing adenomas (Fig. 1). It was suggested that the ET-1 gene might also be expressed in the zona glomerulosa (ZG) cells of the normal adrenal cortex, but the use of tissue homogenates obviously prevented the attainment of definitive conclusions concerning the localization of the prepro-ET-1 gene in the different zones of the adrenal cortex. To further address this issue, Belloni *et al.* (1996b) carried out RT-PCR studies with specific primers on rat adrenocortical tissue; they investigated both adrenocortical homogenates and isolated ZG and zona fasciculata/reticularis (ZF/R) cells and were able to clearly detect the prepro-ET-1 mRNA in all tissues examined. The finding that the gene of the human ECE-1 was also expressed in homogenates of both adrenocortical and adrenomedullary human tissue is consistent with the possibility of local ET-1 biosynthesis in the adrenal cortex (Rossi *et al.*, 1995a).

### B. Immunohistochemical Studies

The presence of immunoreactive ET-1 in the adrenal cortex has been investigated with the use of specific polyclonal and monoclonal antibodies against ET-1 risen in different species (Kondoh *et al.*, 1990; Traish *et al.*, 1992).

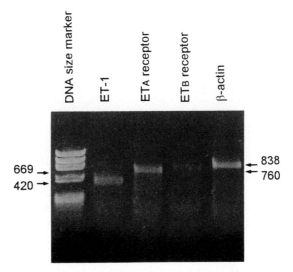

FIG. 1 An ethidium bromide-stained 1.5% agarose gel showing cDNA amplified with ET-1 and with $ET_A$ and $ET_B$ receptor-specific primers from the homogenate of a normal human adrenal cortex. The size marker lane was loaded with 200 ng of Phi × 174 *Hae*III. The resulting amplified fragments were of the expected size, which was 442, 669, and 760 bp for ET-1, $ET_A$, and $ET_B$, respectively. Amplification of a 838-bp fragment of the $\beta$-actin cDNA, as a positive control, is also shown.

Li *et al.* (1994), using a polyclonal rabbit anti-human ET-1 antiserum and deparaffinized 5-μm-thick sections, demonstrated the presence of ET-1-positive cells in normal human adrenals and especially in the ZF in which about 50% of parenchymal cells are immunostained. Positive cells were very few in the ZG and zona reticularis (ZR) and absent in adrenal medulla; vessels were not stained. Immunostaining may appear in the form of vacuoles, grains, or membranes. Further electron microscopic studies showed that vacuoles and grains correspond to ET-1 located around lipid droplets and in endoplasmic reticulum mitochondria, respectively; probably in membranes ET-1 is bound to its receptors (Li *et al.*, 1995).

## III. Endothelin Receptor Subtypes and Their Localization

### A. Gene Expression Studies

Expression of both the $ET_A$ and $ET_B$ receptor subtype genes was first reported in the rat adrenals by Hori *et al.* (1992). With *in situ* hybridization they found that the hybridization signal for $ET_B$ mRNA was diffusely distributed in both adrenal cortex and medulla, whereas that for the $ET_A$ mRNA was strong at the boundary of the adrenal cortex and medulla, and particularly strong over blood vessel cells at the corticomedullary junction. In humans, the expression of the two endothelin receptor subtype mRNAs was detected by Northern blotting analysis by Imai *et al.* (1992) in homogenates of three normal adrenal cortices surrounding aldosterone-producing adenomas. However, results of their [$^{125}$I]ET-1 competitive displacement binding were consistent with a single class of binding sites. The presence of both receptor subtypes in the human adrenal cortex was conclusively demonstrated by Rossi *et al.* (1994). The development of a sensitive RT-PCR technique allowed those authors to investigate the expression of both receptor subtype genes in a larger number of histologically normal adrenal cortices of patients adrenalectomized for kidney cancer (Fig. 1). They were able to detect the mRNA of both genes, a finding that was corroborated by the results of both [$^{125}$I]ET-1 displacement binding and autoradiographic studies.

To localize the $ET_A$ and $ET_B$ genes in the different zones of the rat adrenal cortex, Belloni *et al.* (1996b) investigated both adrenocortical homogenates and isolated ZG and ZF/R adrenocortical cells by RT-PCR. Although they found the mRNA of both receptors in tissue homogenates, they could detect only the $ET_B$ mRNA in isolated cells (Fig. 2). This finding was consistent with their observation that the ET-1 steroid secretagogue effect was inhibited by $ET_B$, but not $ET_A$ antagonists (see Section IV,E). Of interest, the expression of both receptor genes was reported by Fallo *et al.* (1995) in a human adrenocortical carcinoma cell line (see Section VII,D).

## B. Saturation and Displacement Binding Studies

The preparation of [$^{125}$I]ET-1 has allowed investigation of saturation and competitive displacement binding by the different endothelin isopeptides. These studies demonstrated specific [$^{125}$I]ET-1 binding to membranes of the normal adrenal cortex in rats (Koseki *et al.*, 1989b), bovines (Cozza *et al.*, 1989), and humans (Nunez *et al.*, 1990). In calf ZG cells in culture, [$^{125}$I]ET-1 binding was time dependent, saturable, and reached an apparent equilibrium after 60 min at room temperature (Cozza *et al.*, 1989). Scatchard analysis revealed the presence of a single class of high-affinity binding sites with an apparent dissociation constant ($K_d$) of 100 p$M$ and a maximal binding capacity ($B_{max}$) of 52,424 ± 7320 receptors/cell. Because angiotensin-II (ANG-II), adrenocorticotropic hormone (ACTH), vasopressin, apamin, and natriuretic peptides did not displace [$^{125}$I]ET-1 binding, the latter was considered to be

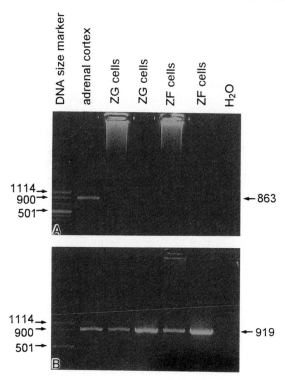

FIG. 2 Ethidium bromide-stained 1.5% agarose gels showing cDNA amplified with rat $ET_A$ (A) and $ET_B$-specific primers (B) from homogenates of rat adrenal cortices and from dispersed and purified ZG or ZF cells. The size marker lane was loaded with 200 ng of Phi × 174 HaeIII. The resulting amplified fragments were of the expected size, which was 780 and 919 bp for $ET_A$ and $ET_B$, respectively. No amplification product was detectable in lanes loaded with water and the PCR mixture but not template cDNA.

specific. Of interest, the isopeptide ET-3 was only about 40% as potent as ET-1. Therefore, Cozza et al. (1989) concluded that calf ZG cells were provided with a single class of specific high-affinity binding sites whose density was about five times greater than that of vascular smooth muscle cells. A high density of [$^{125}$I]ET-1 binding sites, which was surpassed only by that found in the lung, was also reported by Koseki et al. (1989a,b) with both in vitro and in vivo studies. However, the adrenal medulla was more heavily labeled than the cortex in that study. With the use of S6B, Gomez-Sanchez et al. (1990) were able to show that, although ET-1 and S6B were equipotent in their ability to stimulate aldosterone secretion in calf ZG cells (see Section IV,A), S6B displaced [$^{125}$I]ET-1 with only 3% of the potency of ET-1. This discrepancy suggested the possibility of a second adrenal receptor for this peptide, a hypothesis that was further supported by the observation that S6B was more effective than ET-1 in downregulating the higher affinity, lower capacity binding site.

In human membranes from three adenomas, however, [$^{125}$I]ET-1 saturation isotherms attained an equilibrium after 30 min at 30°C, and competitive displacement by ET-2 and ET-3 was consistent with the presence of a single class of high-affinity binding sites (Nunez et al., 1990). The development of antagonists specific for the $ET_A$ and $ET_B$ receptor subtypes has recently provided a powerful tool for investigating the density and kinetic properties of the endothelin receptors, both in isolated cells in culture and in membrane preparations of different adrenal tissues. With the use of [$^{125}$I]ET-1 binding and displacement studies with cold ET-1, ET-3, the weak $ET_B$ agonist S6C, and the $ET_A$ antagonist BQ-123 the issue of the presence of specific ET-1 receptors was studied by Rossi et al. (1994). They were able to conclusively establish the presence of both $ET_A$ and $ET_B$ receptor subtypes in homogenates of normal human adrenal cortices and to characterize their density and binding properties.

## C. Immunohistochemical Studies

The lack of widely available antibodies specific for each endothelin receptor subtype has prevented extensive histochemical studies thus far. To our knowledge, only one study describing the immunochemical characterization and localization of the bovine $ET_B$ receptor subtype with the use of a highly specific rabbit antiserum has been reported (Hagiwara et al., 1993). This antiserum did not show any cross-reactivity with the $ET_A$ receptors and proved to be suitable for localization of the $ET_B$ receptor subtype in bovine tissue. In the bovine adrenal cortex staining was prominent on the endothelial cells of inner adrenocortical zones and not evident on the steroidogenic cells. Because $ET_B$ mRNA was found to be diffusely distributed throughout the rat adrenal cortex by Hori et al. (1992), these findings may suggest that the density of $ET_B$ on the adrenal ZG is not high enough to be visualized with the antiserum and/or that wide differences of density from one species to another exist.

## D. Autoradiographic Studies

The autoradiographic demonstration of ET binding sites has been performed in the human, pig, and rat adrenals. Davenport et al. (1989), using in vitro labeling and quantitative densitometry, showed that [$^{125}$I]ET-1 binding is about two-fold higher in the ZG than in the ZF/ZR of human and rat adrenal cortex; however, a less intense difference in the density of the binding sites between outer and inner zones was found to occur in the pig adrenal cortex. Kohzuki *et al.* (1989, 1991) confirmed these findings in the rat and also calculated by computerized densitometry an affinity constant ($K_a$) for ET-1 binding in the ZG and adrenal medulla of $7.1 \times 10^{-9}$ and $9.5 \times 10^{-9}$ $M$, respectively. Although Koseki *et al.* (1989a,b) reported in rats heavier labeling of the medulla than of the cortex after *in vivo* intravenous injection of [$^{125}$I]ET-1, Neuser *et al.* (1989, 1991), using whole body autoradiography, showed equal [$^{125}$I]ET-1 enrichments in the ZG and adrenal medulla. Neuser *et al.* (1991) also demonstrated a diffused binding in the entire adrenal gland after the injection of [$^{125}$I]big-ET-1.

The kind and distribution of the ET receptor subtypes in the adrenal cortex have been investigated by the use of selective ligands. Kohzuki *et al.* (1991) showed that [$^{125}$I]S6B binding in rat adrenal is similar, though less intense, to that of [$^{125}$I]ET-1. The study of the displacement of [$^{125}$I]ET-1 binding by selective ligands of $ET_A$ and $ET_B$ gave similar results in the human (Belloni *et al.*, 1994; Rossi *et al.*, 1994) and rat adrenals (Belloni *et al.*, 1995, 1996b). Total [$^{125}$I]ET-1 ($10^{-9}$ $M$) binding is intense in the ZG, whereas in the ZF it appears weak and mainly confined between parenchymal cords; the muscular wall of the extracapsular arterioles is heavily labeled. Cold ET-1 ($10^{-7}$ $M$) virtually displaced all the binding. BQ-123 ($10^{-7}$ $M$) eliminated labeling in the vessel wall and markedly decreased it in the ZG without apparently affecting that in the ZF; S6C or BQ-788 ($10^{-7}$ $M$) decreased labeling in the ZG and completely displaced it in the ZF without affecting that in the *Tunica muscularis* of vessels (Figs. 3 and 4). After BQ-788 exposure, labeling appeared to be distributed in a spot-like manner in the subcapsular portion of the rat ZG (Fig. 4D). In light of their findings, these authors concluded that in both human and rat adrenals, ZG possesses both $ET_A$ and $ET_B$ receptors, whereas ZF is exclusively provided with $ET_B$.

As described previously, on the basis of immunohistochemical evidence, Hagiwara *et al.* (1993) attributed $ET_B$ receptors exclusively to the capillary endothelium. However, this does not seem to be the case in the human and rat adrenals, where $ET_B$ are too abundant to be attributed only to the thin endothelial lining. Whether $ET_A$'s in the ZG are located on the parenchymal cells or on the smooth muscle cells of subcapsular arterioles, which are present in the outermost portion of the ZG, is not known. However, the subcapsular spot-like distribution of $ET_A$ receptors, evidenced in the rat ZG after BQ-788-induced displacement of $ET_B$, suggests that the second possibility is the more probable one at least in this species.

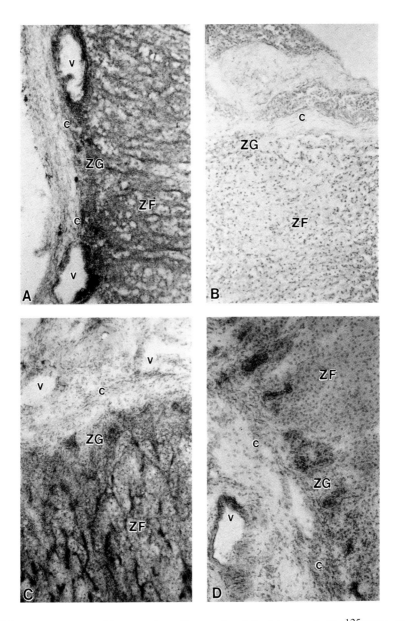

FIG. 3 Autoradiographs of frozen sections of human adrenal glands incubated with [$^{125}$I]ET-1. (A) Binding is intense in ZG and weak in ZF; extracapsular vessels are also heavily labeled. (B) [$^{125}$I]ET-1 binding is completely displaced by the addition of an excess of unlabeled ET-1. (C) BQ-123 displaces [$^{125}$I]ET-1 binding to the *Tunica muscularis* of vessels, attenuates labeling in ZG, and does not affect it in ZF. (D) BQ-788 eliminates [$^{125}$I]ET-1 binding in ZF and decreases it in ZG; labeling in extracapsular vessels is not modified. c, gland capsule; v, extracapsular vessels. Magnification, ×75.

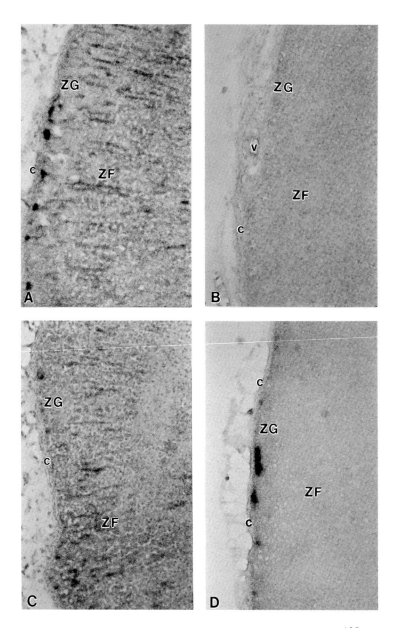

FIG. 4 Autoradiographs of frozen sections of rat adrenal glands incubated with [$^{125}$I]ET-1. (A) Binding is intense in the outer subcapsular portion of ZG and less intense in the ZF, where it appears to be located among the parenchymal cords. (B) [$^{125}$I]ET-1 binding is completely eliminated by addition of an excess of cold ET-1. (C) BQ-123 strongly decreases labeling in ZG without apparently affecting it in ZF. (D) DQ-788 eliminates binding in ZF and markedly attenuates it in ZG, where it remains localized just beneath the capsule. c, gland capsule; v, extracapsular vessels. Magnification, ×75.

## IV. Effects of Endothelins on the Secretory Activity of the Adrenal Cortex

### A. Zona Glomerulosa and Mineralocorticoid Secretion

The mineralocorticoid secretagogue effects of ETs has been investigated in humans, bovines, rabbits, and rats mainly by the use of *in vitro* preparations (dispersed ZG cells, capsule ZG strips, or adrenal slices).

Using a mixture of dispersed human ZG and ZF cells, Hinson *et al.* (1991b) observed a very strong stimulatory effect of both ET-1 and ET-3 on basal aldosterone secretion. The minimal effective concentration for ET-1 and ET-3 was $10^{-14}$ and $10^{-13}$ $M$, respectively, and the maximal effective one, which elicited about a five-fold rise, was $10^{-7}$ $M$ for both peptides. Similar but much less conspicuous effects were observed employing adrenal slices obtained from kidney donors or fragments of normal adrenal tissue adjacent to Conn's adenomas (Zeng *et al.*, 1992). Minimal and maximal effective concentrations of ET-1 in evoking basal aldosterone response were $10^{-11}$ $M$ (33% rise) and $10^{-8}$ $M$ (60% rise), respectively.

In a series of investigations employing cultured bovine ZG cells, Cozza's and Gomez-Sanchez's groups demonstrated a marked aldosterone secretagogue action of ETs that, however, is far less potent than that of angiotensin II (ANG-II). ET-1 concentration dependently increased basal aldosterone secretion, with its minimal and maximal effective concentrations $10^{-10}$ $M$ (about a two-fold increase) and $10^{-9}$ $M$ (about a four-fold increase) (Cozza *et al.*, 1989, 1992; Cozza and Gomez-Sanchez, 1990). ET-3 was less effective than ET-1, with an aldosterone response about 60–70% of that evoked by equimolar concentrations of ET-1 (Cozza *et al.*, 1989), whereas S6B was equipotent (Gomez-Sanchez *et al.*, 1990). Cozza *et al.* (1992) found that ET-1 maximally potentiates ANG-II ($10^{-9}$ $M$)-stimulated aldosterone secretion ($10^{-10}$ $M$ ET-1, 70%; $10^{-8}$ $M$ ET-1; 100% increase), acting on both early and late steps of steroidogenesis (*i.e.,* the conversion of cholesterol to pregnenolone and the conversion of corticosterone to aldosterone, respectively). ET-1 also potentiated, although to a lesser extent, ACTH-stimulated aldosterone response; the response to K$^+$ was unaffected. ET-2 and S6B were as potent as ET-1, whereas ET-3 and big-ET-1 were effective only at a $10^{-8}$ $M$ concentration; pro-ET(110–130) was ineffective. Quite different findings were reported by Rosolowsky and Campbell (1990), who did not observe any effect of ET-1 on either basal or ANG-II-stimulated aldosterone secretion of cultured bovine ZG cells over a range of concentrations from $10^{-11}$ to $10^{-6}$ $M$. However, ET-1 potentiated both ACTH (3 × $10^{-10}$ $M$)- and dibutyryl cyclic adenosine 3', 5'-monophosphate (dbcAMP)

($10^{-2}\,M$)-stimulated aldosterone output with a maximal effective concentration of ET-1 of $10^{-8}\,M$, which elicited 52 and 42% increases, respectively. The chronic (2-day) exposure of cultured cells to ACTH was found to enhance the acute aldosterone response to $5 \times 10^{-8}\,M$ ET-1, whereas chronic ET-1 exposure was without any effect. Rosolowsky and Campbell concluded that ACTH sensitizes bovine ZG cells to ET-1 via the activation of adenylate cyclase.

Morishita et al. (1989) reported that dispersed rabbit ZG cells respond to ET-1 by raising their basal production of aldosterone but not of corticosterone. The secretagogue action of the peptide was not intense (maximal effective concentration of ET-1 evoked a 26% rise), but ZG cells were very sensitive to ET-1 (EC50, $5 \times 10^{-14}\,M$).

The findings obtained using rat ZG preparations consistently indicated that ETs enhance basal aldosterone production, but marked differences were reported with regard to the level of the response and the sensitivity to the peptides. Minimal effective concentration of ET-1 was $10^{-10}\,M$ (Belloni et al., 1995), $10^{-11}\,M$ (Belloni et al., 1996b), $10^{-12}\,M$ (Woodcock et al., 1990b), $10^{-13}\,M$ (Hinson et al., 1991a,b), or $10^{-14}\,M$ (Mazzocchi et al., 1990a); Hinson and co-workers reported a minimal effective concentration for both ET-2 and ET-3 of $10^{-14}\,M$. Maximal effective concentration for ET-1 was $10^{-8}/10^{-7}\,M$, but the elicited rise in basal aldosterone output varied from 50% (Woodcock et al., 1990b) to 2- or 3-fold (Mazzocchi et al., 1990a; Belloni et al., 1995, 1996b) or even to 30-fold (Hinson et al., 1991a,b). The reasons for these differences are not clear, but they are likely to depend on the experimental conditions. With regard to the effects of ET-1 on agonist-stimulated responses, Mazzocchi et al. (1990a) observed an ET-1-induced potentiation of aldosterone response to $10^{-8}\,M$ ACTH (minimal and maximal effective concentrations were $10^{-12}$ and $10^{-8}\,M$ and evoked 40 and 50% increases, respectively) but not to $10^{-8}\,M$ ANG-II or 10 m$M$ K$^+$.

*In vivo* studies are scarce and concern humans, dogs, and rats. In human volunteers, ET-1 infusion (from 1 to 5 ng/kg min) was found not to influence plasma renin activity (PRA) and aldosterone concentration, although plasma levels of ET-1 achieved increases of about 30-fold (Vierhapper et al., 1990). These results were confirmed by the same investigators (Vierhapper et al., 1995) by infusing ET-1 at a rate of 2 pmol/kg/min for 15 min and 1 pmol/kg min for 105 min. They, however, observed that this procedure evokes a 75% potentiation of the plasma aldosterone response to the *bolus* injection of 0.25 mg ACTH. In contrast, the infusion with 30 ng/kg/min ET-1 was reported to induce in the dog sizable increases in both PRA and the plasma level of aldosterone (Goetz et al., 1988; Miller et al., 1989, 1993).

In the rat, Cao and Banks (1990) reported that the intravenous infusion of ET-1 (100 ng/kg/min) elicits a two- or three-fold increase of plasma aldosterone, which is unaffected by captopril pretreatment. These findings were at variance with those of Mazzocchi et al. (1990a), who showed that the *bolus* intravenous administration of 0.5 μg/kg of ET-1 does not change plasma aldosterone concentration within 60 min, although it does evoke a net rise in the blood pressure and a decrease in PRA. However, the same group of investigators (Mazzocchi et al., 1996c) reported that an intraperitoneal *bolus* injection of ET-1 dose dependently increases not only blood pressure (100 pmol/rat, 30% rise) but also basal aldosterone concentration (minimal and maximal effective doses were 1 pmol and 10 nmol and induced 56 and 90% rises). Mazzocchi and associates also demonstrated that ET-1 potentiates an ANG-II-evoked rise in the plasma level of aldosterone (minimal effective dose: 100 pmol, 40% rise; maximal effective dose: 10 nmol, 50% rise) but not in blood pressure. According to Mazzocchi et al. (1990b), the prolonged subcutaneous infusion of rats with ET-1 (0.2 μg/kg/h for 7 days) caused significant increases in blood pressure (15%) and plasma aldosterone concentration (40%), coupled with a net lowering in PRA (−34%). These discrepancies may be, at least in part, reconciled by the results of Pecci et al. (1993), who observed that ET-1 exerts biphasic effects on plasma aldosterone concentration and cytosolic aldosterone content of ZG cells: The effects appear to be inversely related to the dose (from 50 to 110 ng/kg/min) and directly related to the infusion rate (from 50 to 180 μl/min). Pecci and associates also showed that ET-1 stimulates both early and late steps of aldosterone synthesis, but a clear-cut effect on the conversion of cholesterol to pregnenolone can be observed only by administering the lower ET-1 dose. Enalapril-induced blockade of angiotensin-I converting enzyme does not influence the previously described effects of ET-1, thereby suggesting that they are independent, according to Cao and Banks (1990), of any change in the renin–angiotensin system (RAS) activity.

## B. Zona Fasciculata Reticularis and Glucocorticoid Secretion

Controversial results were obtained *in vitro*. According to Zeng *et al.* (1992), ET-1 did not affect basal cortisol secretion of human adrenal slices. At variance with their findings on ZG cells, Rosolowsky and Campbell (1990) did not find any effect of ET-1 on either basal or ACTH-stimulated cortisol production by cultured bovine ZF cells. In contrast, Hinson et al. (1991b), using a mixture of dispersed human ZG and ZF cells, observed a very strong glucocorticoid response to both ET-1 and ET-3. Minimal effective concentrations ($10^{-14}$ M for both peptides) elicited a 2.6-fold rise in cortisol yield, and maximal effective concentrations ($10^{-7}$ M) evoked 31- and 11-fold increases for ET-1 and ET-3, respectively.

Similarly, Hinson et al. (1991b) reported that both ET-1 and ET-3 exerted a very intense stimulatory action on dispersed rat ZF/ZR cells. Minimal effective concentration ($10^{-13}/10^{-14}$ $M$) evoked about a 2.5-fold rise in basal corticosterone yield, and the maximal effective concentration ($10^{-7}$ $M$) caused 15- to 18-fold increases. Hinson et al. (1991a) confirmed these findings and additionally showed that ET-2 exerts an intense glucocorticoid secretagogue action: Minimal and maximal effective concentrations ($10^{-14}$ and $10^{-7}$ $M$) elicited 25% and 7.5-fold rises in corticosterone output, respectively. Belloni et al. (1995, 1996b) described a significant stimulatory effect of ET-1 on corticosterone production of both rat adrenal slices and dispersed inner adrenocortical cells, but the sensitivity and the secretory response were far less potent than those reported by Hinson and associates. Minimal and maximal effective concentration ($10^{-10}$ and $10^{-9}/10^{-8}$ $M$) induced 60–70% and 2- to 2.5-fold increases, respectively.

At variance with the *in vitro* studies, the few *in vivo* investigations agree on the lack of effect of ET-1 on glucocorticoid secretion (as evaluated by their plasma concentration) in both humans (Vierhapper *et al.*, 1995) and rats (Mazzocchi *et al.*, 1990a,b).

## C. Interrenal Cells and Steroid Secretion

The effects of ET-1 on the interrenal gland were studied in the frog by Vaudry's goup. Delarue *et al.* (1990), using an *in vitro* perifusion system, demonstrated that ET-1 ranging from $10^{-11}$ to $10^{-8}$ $M$ concentration dependently stimulates basal output of both aldosterone and corticosterone. The responses of the two steroids were similar and achieved their maximum (60% increase) at an ET-1 concentration of $5 \times 10^{-9}$ $M$. Repeated 20-min pulses of $10^{-9}$ $M$ ET-1 at a frequency of one every 90 min resulted, after the second pulse, in a net lowering of the secretory response. The prolonged exposure (3 h) to ET-1 induced a rapid increase in aldosterone and corticosterone output, followed by a gradual decline. These authors concluded that ET-1 is a potent acute stimulator of steroidogenesis of frog interrenal cells and that prolonged or repeated administration of ET-1 induces a rapid desensitization phenomenon.

Further studies (Remy-Jouet *et al.*, 1994) have shown that cytochalasin-B ($5 \times 10^{-5}$ $M$), a microfilament disrupting agent, reversibly blocks the secretory response of frog interrenal cells to $5 \times 10^{-9}$ $M$ ET-1. In contrast, the antimicrotubular agent, vinblastine ($10^{-5}$ $M$), and the intermediate filament inhibitor, $\beta$-$\beta^1$-iminodipropionitrile ($10^{-3}$ $M$), were ineffective. Remy-Jouet and co-workers suggested that the integrity of the microfilament network is required for ET-1 to exert its corticosteroid secretagogue action. However, cytochalasin-B also lowered basal steroid output by frog interrenal preparations, a finding that casts doubts about the specificity of this drug effect on ET-1 secretagogue action. Accordingly, cytochalasin-B has been found to inhibit in frogs and humans the adrenocorticotropic action of various agonists including ACTH, ANG-II, serotonin, and acetylcholine (Netchitaïlo et al., 1985; Feuilloley et al., 1988, 1994). The involvement of the cytoskeleton in the basic mechanism underlying steroid synthesis in adrenocortical cells is well known (Feuilloley and Vaudry, 1996), and therefore it is conceivable that cytochalasin-B acts aspecifically by disrupting the steroidogenic machinery.

## D. Direct Mechanisms of Action

### 1. Stimulation of Phospholipase C and Increase in Cytosolic $Ca^{2+}$ Concentration

A good deal of evidence indicates that in many cell systems ETs activate phospholipase C (PLC), thereby releasing diacylglycerol and active inositol [1,4,5]-triphosphate, leading to activation of protein kinase C (PKC) and increasing cytosolic $Ca^{2+}$ concentration ($[Ca^{2+}]_i$) by enhancing both $Ca^{2+}$ release from intracellular stores and $Ca^{2+}$ influx (for review, see Simonson and Dunn, 1990; Rubanyi and Polokoff, 1994). Proof is available that ET-1, like other agonists, exerts its aldosterone secretagogue effect by activating PLC and $Ca^{2+}$ influx in ZG cells.

Woodcock et al. (1990a,b) showed that both ET-1 and ANG-II induce inositol phosphate accumulation in dispersed rat ZG cells as well as an increase in $[Ca^{2+}]_i$, ANG-II being about three-fold more potent at a $10^{-7}$ $M$ concentration than ET-1 with regard to intracellular $Ca^{2+}$. They also observed that the effects of the maximal effective concentrations ($10^{-7}$ $M$) of the two peptides are not additive, which suggests that they act on the same ZG cell population.

Using bovine ZG cell cultures, Rosolowsky and Campbell (1990) did not find any effect of ET-1 on $[Ca^{2+}]_i$. In contrast, nicardipine, a $Ca^{2+}$ channel blocker, was found to inhibit aldosterone response to ET-1 of both human adrenal slices (Zeng et al., 1992) and dispersed rabbit ZG cells (Morishita et al., 1989). Moreover, perfusion with $Ca^{2+}$-free medium blocks ET-1 secretagogue action on frog interrenal cells (Delarue et al., 1990).

Cozza and Gomez-Sanchez (1993) found that incubation of bovine ZG cell cultures in low $Ca^{2+}$ media or in the presence of the $Ca^{2+}$ channel antagonist verapamil suppresses basal aldosterone response to ET-1, but not ET-1 potentiation of ANG-II secretagogue effect. Conversely, PKC inhibition by staurosporin and PKC desensitization by ACTH or prolonged phorbol 12-myristate 13-acetate (PMA), an activator of PKC, did not affect basal aldosterone response to ET-1 but inhibited ET-1-induced potentiation of ANG-II aldosterone secretagogue effect. In light of these findings, they hypothesized that ET-1 activates two different pathways in cultured bovine ZG cells: one pathway that is $Ca^{2+}$ dependent (mainly related to $Ca^{2+}$ influx) and PKC independent that mediates the *per se* effect of ET-1 on aldosterone secretion; and the second pathway that is $Ca^{2+}$ independent and PKC dependent and involved in the ET-1-induced potentiation of ANG-II secretory effect. This hypothesis, however, does not appear to apply to rat ZG cells. Kapas and Hinson (1996) demonstrated that $10^{-7}$ M ET-1, like PMA and ANG-II (Kapas *et al.*, 1995), alters the subcellular distribution of PKC activity in ZG cells by inducing the translocation of the enzyme from cytosol to the membrane and nuclear fractions. Ro 31-8220 (150 n$M$), a specific inhibitor of PKC activity, prevented this effect of ET-1 and decreased by 75% the ET-1-induced basal aldosterone production of ZG cells, indicating that in rats the *per se* effect of ET-1 on aldosterone secretion is dependent on PKC activation.

### 2. Stimulation of Thyrosine Kinase Activity

Thyrosine kinase (TK) activation is currently recognized as an important final step of the intracellular signaling pathways, which start early events associated to mitogenesis. These pathways include those initiated by PLC and PKC activation or adenylate cyclase and protein kinase A (PKA) activation (Malarkey *et al.*, 1995). It has been demonstrated that ANG-II, at least in part via PKC activation, stimulates TK activity in bovine (Bodart *et al.*, 1995) and rat ZG cells (Kapas *et al.*, 1995) and that the TK pathway plays an essential role in aldosterone response to ANG-II. In fact, ANG-II-induced aldosterone production by ZG was significantly depressed or completely abolished by TK inhibitors such as tyrphostin-23 and genistein.

Kapas and Hinson (1996) showed that ET-1 and PMA concentration dependently stimulates TK activity in dispersed rat ZG cells, an effect completely abolished by tyrphostin-23 and partially reversed by Ro 31-8220. Tyrphostin-23 (5 n$M$) caused a moderate but significant (30%) lowering of the maximum aldosterone response to both ET-1 and PMA. ACTH and dbcAMP, which *per se* do not affect TK activity in rat ZG (Kapas *et al.*, 1995), were found to significantly inhibit the TK response to ET-1, ANG-II, or PMA. These observations allowed these authors to conclude that (i) the TK pathway plays a role in the mediation of the aldosterone secretagogue effect of both ETs and ANG-II; (ii) TK activation involves both a PKC-dependent and a PKC-independent pathway; and (iii) in the rat ZG cells there is cross-talk between different G-protein-coupled receptor-linked (PKC and PKA) pathways and TK-linked receptors, whose significance in the regulation of ZG function remains to be addressed.

## 3. Stimulation of Ouabain-Sensitive $Na^+/K^+$-ATPase

Pecci et al. (1994) demonstrated that intravenous infusion of ET-1 (80 ng/kg/min for 30 min), in addition to increasing blood pressure and adrenal aldosterone content, also causes a marked activation of microsomal $Na^+/K^+$-ATPase in rat ZG cells. Similar results, except the rise in the blood pressure, were obtained by infusing S6B, but the low $Na^+/K^+$-ATPase activity present in inner adrenocortical cells was unaffected. Interestingly, these investigators observed a small stimulatory effect of ET-1 on $Na^+/K^+$-ATPase activity in rat aorta, which is at variance with previous findings of an inhibitory effect of the peptide on the enzyme contained in the epithelial cells of renal tubules (Zeidel et al., 1989). Although this effect may play a role in the mechanism underlying the ET-1-induced potentiation of the aldosterone secretagogue action of ANG-II, its functional relevance remains to be investigated. ANG-II has no direct effect on adrenal $Na^+/K^+$-ATPase (Meuli and Müller, 1982), but it activates $Na^+-H^+$ exchange, resulting in a sizable increase in intracellular $Na^+$ concentration (Conlin et al., 1990, 1991). Stimulation of the $Na^+/K^+$-ATPase activity by ET-1 would decrease intracellular $Na^+$, thereby enhancing the sodium gradient across the plasma membrane and allowing faster intracellular alkalinization by ANG-II. In keeping with this contention, ouabain has been reported to abolish or decrease aldosterone response of ZG cells to ANG-II (Elliott et al., 1986). However, it must be recalled that ET-1 has been found to enhance $Na^+-H^+$ exchange, thus leading to intracellular alkalinization in vascular smooth muscle cells (Koh et al., 1990; Longchampt et al., 1991; Brock and Danthuluri, 1992).

## 4. Stimulation of Phospholipase $A_2$ and Arachidonic Acid Metabolism

As previously mentioned, ET-1 stimulates steroid secretion of perifused frog interrenal cells, an effect that causes rapid desensitization. Delarue et al. (1990) demonstrated that $10^{-9}$ M ET-1 strongly enhances the release of prostaglandin (PG) $E_2$ and 6-keto-PGF1α, the stable metabolite of $PGE_2$, by frog interrenals. The increase in $PGE_2$ biosynthesis occurred 10 min earlier than the ET-1-evoked peak of the corticosteroid response. Furthermore, indomethacin, a cyclooxygenase inhibitor, was found to suppress ET-1 secretagogue action on frog interrenal cells. These findings suggest that $PGE_2$ biosynthesis may contribute to such ET-1 action in these cells.

The possibility that stimulation of phospholipase $A_2$ may be involved in the mechanism underlying the secretagogue action of ETs on amphibian interrenal cells accords well with the results of many investigations indicating that ETs enhance PG synthesis and release in various tissues including smooth muscle cells (Reynolds et al., 1989), kidney (Stier et al., 1992), endometrium (Cameron et al., 1991), and intestine (Brown and Smith, 1991). Moreover, PGEs are well known to stimulate adrenal steroid secretion in several vertebrate species (Vinson et al., 1992).

### E. Possible Indirect Mechanisms of Action

Direct proof is not yet available that any indirect mechanism may mediate the *in vivo* effect of ETs on the adrenal cortex. Further studies are needed to confirm the possible mechanisms discussed below.

**1. Stimulation of Catecholamine Release by Adrenal Medulla**

Intramedullary catecholamines, epinephrine and norepinephrine, are deemed to exert a paracrine stimulatory control of the cortex function in many mammalian species. This effect is mediated by both $\alpha$- and $\beta$-adrenoceptors and concerns both mineralo- and glucocorticoid secretion (Ehrhart-Bornstein et al., 1995). Several investigations have also clearly shown that some regulatory molecules contained in adrenal glands exert an adrenocortical secretagogue action through an indirect mechanism, which involves the release of medullary catecholamines, because their stimulatory effect on *in vitro* adrenal preparations (adrenal slices of capsule ZG strips) is blocked by specific adrenoceptor antagonists. These regulatory peptides include (i) interleukin-1$\alpha$, which enhances corticosterone secretion in rats (Gwosdow et al., 1992; O'Connell et al., 1994); (ii) substance P (Mazzocchi et al., 1995), neuropeptide Y (Bernet et al., 1994; Mazzocchi et al., 1996a), and vasoactive intestinal peptide (Hinson et al., 1992; Mazzocchi et al., 1993; Bernet et al., 1994), which all increase aldosterone production in rats; and (iii) pituitary adenylate cyclase-activating polypeptide, which raises aldosterone yield in humans (Neri et al., 1996) and rats (Andreis et al., 1995).

Boarder and Marriott (1989, 1991) reported that ET-1 causes a marked enhancement of catecholamine efflux by cultured bovine adrenal chromaffin cells, with an $EC_{50}$ of about $10^{-9}$ $M$. S6B stimulated catecholamine release to a similar extent as ET-1, and the effect of both peptides was blocked by $10^{-6}$ $M$ nitrendipine. These findings were confirmed by Ohara-Imaizumi and Kumakura (1991), who also showed that the ET-1 effect is dependent on extracellular $Ca^{2+}$ and synergistic with acetylcholine-evoked catecholamine release. In anesthesized dogs, the intravenous ET-1 administration was found to raise splanchnic nerve-induced, but not basal, catecholamine output (Tacheuchi et al., 1992). Yamaguchi (1993,1995) observed that the 60-s infusion of ET-1 into the left adrenolumbary artery of anesthesized dogs dose dependently increases epinephrine and norepinephrine output in adrenal venous blood (0.1 and 1.0 μg ET-1 evoked rises of about 10- and 100-fold). Interestingly, the release of the catecholamines was found to be long lasting because it remained significantly elevated over a 30-min period. Yamaguchi also demonstrated that this ET-1-induced catecholamine output was abolished by nifedipine, but not by pentolinium or atropine, administered 10 min before ET-1 infusion, strongly suggesting that ET-1-evoked catecholamine release by bovine and dog adrenal medulla is largely mediated by the activation of dihydropyridine-sensitive L-type $Ca^{2+}$ channels. Studies are needed to ascertain whether this effect may contribute to the secretagogue action exerted by ETs on intact adrenal glands.

## 2. Modulation of the Release and Activity of Intramedullary Regulatory Peptides

Arginine-vasopressin (AVP) is a well-known potent stimulator of aldosterone secretion. Proof is available that adrenal medulla contains and releases sizable amounts of AVP, which may affect the cortex in a paracrine manner (Tóth and Hinson, 1995). Evidence indicates that ET-1 and ET-3, either systemically or centrally injected, stimulate AVP secretion in the hypothalamo–neurohypophyseal axis of dogs and rats (Goetz et al., 1988; Shichiri et al., 1989; Nakamoto et al., 1991; Ritz et al., 1992; Wall and Ferguson, 1992; Yamamoto et al., 1992; Yasin et al., 1994). The possibility that ETs stimulate the release by medullary chromaffin cells of AVP, which in turn enhances aldosterone secretion by ZG cells in a paracrine manner, merits exploration.

Natriuretic peptides are a family of hormones exerting a potent inhibitory action on aldosterone secretion of ZG cells, probably by activating guanylate cyclase. All three members of this family, atrial natriuretic peptide, brain natriuretic peptide, and C-type natriuretic peptide (ANP, BNP, and CNP), are contained in adrenal medulla and may exert a paracrine suppression of ZG function (Lee et al., 1993, 1994; Kawai et al., 1996). Many lines of evidence indicate that ETs stimulate ANP synthesis and release by rat atrial cardiocytes through the $ET_A$ receptor subtype (Fukuda et al., 1989; Gardner et al., 1991; Uusimaa et al., 1992; Irons et al., 1993; Thibault et al., 1994). However, low concentrations of ET-1, like those conceivably present in adrenals, were found to cause an opposite effect in the rats (Shirakami et al., 1993). Moreover, it has been shown that ETs inhibit ANP-induced stimulation of guanylate cyclase in both glia cells (Levin et al., 1992) and aortic smooth muscle cells of the rat (Jaiswal, 1992). It remains to be settled whether these last effects of ETs may counteract the aldosterone antisecretagogue action of intramedullary natriuretic peptides.

## 3. Modulation of Intraadrenal Renin Angiotensin System

There is now a general consensus that an intraadrenal RAS, mainly located in the capsule ZG, plays an important role in the paracrine control of aldosterone secretion (Mulrow, 1992). Many studies have shown that ETs inhibit renin release by kidney juxtaglomerular cells (Takagi *et al.*, 1989; Moe *et al.*, 1991; Yamada and Yoshida, 1991; Naess *et al.*, 1993), but there are indications that they activate extrarenal RAS, *e.g.*, vascular RAS in rat mesenteric arteries (Rakugi *et al.*, 1990) and human decidual cells (Chao *et al.*, 1993). Hence, the appealing possibility that ETs may exert their aldosterone secretagogue action on intact adrenal tissue also by modulating capsule ZG RAS awaits exploration.

## 4. Regulation of Adrenal Blood Flow

According to Vinson and Hinson (1992), a very tight direct correlation between blood flow rate and corticosteroid secretion by adrenal glands exists. The mechanism underlying this phenomenon may involve (i) washout of secretory products, (ii) increase in oxygen and substrate supply, and (iii) enhancement in the rate of presentation of agonists. Hence, any regulatory molecule that is able to modulate adrenal blood flow may indirectly influence the steroid secretion by the gland.

ETs are potent vasoconstrictor peptides, acting on the *T. muscularis* of arterioles (Rubanyi and Polokoff, 1994). Cameron *et al.* (1994) investigated the effect of exogenous ET-1 on adrenal blood flow using isolated *in situ* perfused rat adrenal glands. In this model, the left adrenal is perfused via a cannula introduced in the coeliac artery; the perfusion medium reaches an isolated segment of the aorta from which adrenal arteries arise, passes through the adrenal, and is then collected via a cannula inserted in the renal vein. They found that the *bolus* injection into the afferent cannula of 10 pmol ET-1 causes an abrupt fall (about 50%) in the flow rate of the perfusion medium that rapidly subsides in about 20 min. The rapid recovery of the flow rate could suggest the existence of a local intraadrenal mechanism that antagonizes the characteristically long-acting and difficult to wash out vasoconstrictor effect of ET-1. Hence, it is possible to hypothesize that exogenously administered ET-1 reaching the gland periphery evokes an initial vasoconstriction of the capsule ZG arterioles, which is soon counteracted by the induction of an intraglandular vasodilatatory mechanism. Intraadrenal released (endogenous) ETs may thus act differently from the exogenous ET-1 injected in the perfusion cannula because muscular arterioles are almost completely absent in the adrenocortical parenchyma (see Section III,C).

ET-1 and ET-3 are known to stimulate the release of the endothelial-derived relaxing factor, nitric oxide, by the endothelial lining of several vascular tissues (Botting and Vane, 1990; Moncada et al., 1991; Namiki et al., 1992; Warner et al., 1992; Zellers et al., 1994). In isolated rat aorta rings, S6C was found to cause relaxation, which can be prevented by both removal of the endothelium and exposure to $N^G$-nitro-L-arginine (L-NAME), a nitric oxide synthase (NOS) inhibitor (Rubanyi and Polokoff, 1994). Quite similar results were obtained in the pig pulmonary vessels (Zellers et al., 1994). It has also been suggested that nitric oxide, by activating guanylate cyclase, may have a major role in the termination of ET signaling (Goligorsky et al., 1994).

NOS is contained in rat adrenal cytosol (Palacios et al., 1989; Bredt et al., 1990), and nitric oxide may play a role in modulating adrenal blood flow and catecholamine secretion (Breslow et al., 1992; Torres et al., 1994) as well as in controlling steroidogenesis (Adams et al., 1991). Convincing proof of this latter action of nitric oxide was provided by Cameron and Hinson (1993), who, by the use of in situ perfused rat adrenals, found that the injection of L-arginine, the substrate of NOS, dose dependently raises both perfusion medium flow rate and steroid output, whereas even in the presence of L-arginine, L-NAME markedly decreases these two parameters. Hinson et al. (1996) confirmed the effect of L-arginine on the rate of perfusion medium flow and also showed that the injection of sodium nitroprusside, an endothelium-independent guanylate cyclase activator, markedly enhances this parameter. This last finding is in keeping with the view that nitric oxide raises adrenal blood flow by enhancing cGMP production.

Additional mechanisms whereby ETs may increase local blood flow could involve stimulation of the release by vascular endothelial cells of prostacyclin ($PGI_2$), a potent vasodilator prostanoid, as demonstrated in humans (Kato et al., 1992), bovines (Emori et al., 1991; Filep et al., 1991), pigs (Zellers et al., 1994), and rats (D'Orleans-Juste et al., 1992).

F. Receptor Subtypes Involved in the Secretagogue Action of Endothelins

Naruse et al. (1994b) advanced the hypothesis that $ET_A$ may be involved in the mediation of aldosterone secretagogue effect of ET-1 on cultured bovine ZG cells, inasmuch as BQ-123 slightly but significantly depresses it. However, the bulk of investigations strongly suggests that $ET_B$ is the main receptor subtype mediating both mineralocorticoid and glucocorticoid secretory response to ETs. ET-3, an isopeptide that preferentially binds to the $ET_B$, was found to stimulate, though less potently than ET-1, aldosterone secretion by cultured bovine ZG cells (Cozza et al., 1989, 1992). Furthermore, S6B, an $ET_B$ agonist, was also found to be equipotent with ET-1 in evoking the aldosterone response (Cozza et al., 1992). Hinson et *al.* (1991b) reported that ET-1 and ET-3 are about equipotent in stimulating basal aldosterone and cortisol or corticosterone output by dispersed human and rat adrenocortical cells. Consistent with these findings, Pecci *et al.* (1994) observed that S6B, like ET-1, raises *in vivo* aldosterone content and $Na^+/K^+$-ATPase activity in rat adrenal glands; these effects were not suppressed by the $ET_A$ antagonist BQ-123, which effectively blocked the ET-1-induced rise in blood pressure.

The recent availability of $ET_B$-selective antagonists, including BQ-788, allowed the direct demonstration that $ET_B$ is the receptor subtype mainly involved in the direct secretagogue effect of ET-1 in rats. Belloni *et al.* (1996b) showed that BQ-788, but not BQ-123, suppresses either aldosterone or corticosterone response to $10^{-9}$ M ET-1 of dispersed ZG and ZF/ZR cells (Fig. 5). Consistent with these results, Mazzocchi *et al.* (1996c) reported that BQ-788, but not BQ-123, abrogates the ET-1-induced potentiation of aldosterone response to ANG-II of rats *in vivo*. According to Belloni *et al.* (1996b), minimal and maximal effective concentrations of BQ-788 were $10^{-9}$ and $10^{-7}$ M, respectively; however, $10^{-7}$ M BQ-788, either alone or with equimolar concentrations of BQ-123, although completely abolishing the glucocorticoid response of ZF/ZR cells, only decreased by about 70% the aldosterone response to ET-1 of ZG cells. In light of these findings, the possibility cannot be ruled out that in the rat a third subtype of ET receptor (Karne *et al.*, 1993; Bax and Saxena, 1994) or an alternatively spliced receptor (Miyamoto *et al.*, 1994) may also be involved in the aldosterone secretory response of ZG cells to ETs.

The $ET_B$ subtype appears to also take part in the mediation of the other possible indirect mechanisms underlying the ET secretagogue effect on adrenals (see Section IV,E). S6B was found to be equipotent with ET-1 in eliciting catecholamine release by medullary chromaffin cells. ET-3, similar to ET-1, stimulated AVP release by posterior pituitary and inhibited ANP-evoked cGMP production as well as enhancing nitric oxide and prostanoid release by vascular endothelium. The direct demonstration of $ET_B$ receptor coupling to NOS has been provided by Tsukahara *et al.* (1994) using Chinese hamster ovary cells stably transfected with $ET_B$ receptor cDNA.

## V. Effects of Endothelins on the Growth of the Adrenal Cortex

### A. Cell Hypertrophy and Steroidogenic Capacity

The effects of the prolonged administration of ET-1 on the adrenal cortex have been investigated by Mazzocchi *et al.* (1990b). Male adult rats were

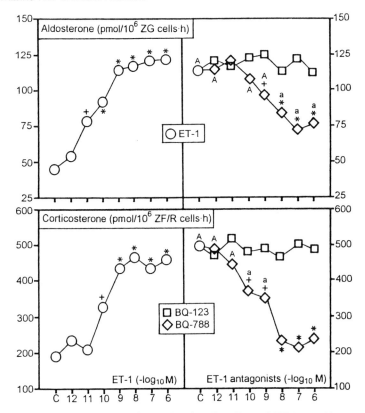

FIG. 5 A typical incubation experiment showing the effect of ET-1 on aldosterone and corticosterone secretion by dispersed rat ZG and ZF/R cells, respectively (left) and the effects of ET-1 receptor antagonists on aldosterone and corticosterone response to $10^{-9}$ $M$ ET-1 (right). C, control group. Data are means of three separate incubations, and SE are not indicated. $^+P<0.05$ and $^*P<0.01$ from C; $^aP<0.05$ and $^AP<0.01$ from baseline (C groups of left panels).

subcutaneously infused for a week with ET-1 (0.2 µg/kg h) using Alzet osmotic pumps. In addition to raising blood pressure and plasma aldosterone concentration (see Section IV,A), ET-1 infusion was found to cause a marked hypertrophy of the ZG (22% increase) and its parenchymal cells (35% increase), as evaluated by morphometry. Stereologic analysis of electron micrographs of ZG cells evidenced significant increases in the volumes of the mitochondrial compartment (49%) and smooth endoplasmic reticulum (SER) (39%), along with a sizable decrease in the volume of the lipid droplet compartment (−29%). Dispersed ZG cells obtained from ET-1-infused animals displayed significantly enhanced steroidogenic capacity: Basal production of aldosterone and corticosterone was about twofold, and maximally stimulated agonist (ANG-II, ACTH, or $K^+$) underwent a rise of about 40%. ET-1 infusion did not evoke any apparent effect on ZF cells.

Mazzocchi et al. (1990b) stated that their morphological and functional findings were in good agreement. In fact, the enzymes of aldosterone synthesis are located in mitochondria and SER, and changes in the surface area per cell of mitochondrial cristae and SER tubules are tightly coupled with corresponding changes in the activity per cell of some of these enzymes (e.g., 3β-hydroxysteroid dehydrogenase and 11/18-hydroxylase). In adrenocortical lipid droplets is stored cholesterol, which is the main precursor of steroid hormones: Hence, the decrease in the volume of the lipid droplet compartment may well be interpreted as the morphological counterpart of the increased utilization of cholesterol in aldosteronogenesis (Nussdorfer, 1986). Mazzocchi and associates concluded that, in the rat, ET-1 specifically stimulates the growth and expression of steroidogenic enzymatic machinery of ZG cells. The moderate but significant ET-1-induced increase in the average volume of the nucleus in hypertrophic ZG cells appeared to be in keeping with an enhanced transcription activity.

## B. Cell Proliferation

Using cultured bovine ZG cells, Cozza and Gomez-Sanchez (1990) observed that ET-1 concentration dependently decreased [$^3$H]thymidine incorporation: Minimal and maximal effective concentrations were $10^{-10}$ $M$ ($-23\%$) and $10^{-8}$ $M$ ($-57\%$), respectively. An analogous effect was induced by PMA, with minimal and maximal effective concentrations being $10^{-7}$ $M$ ($-47\%$) and $10^{-6}$ $M$ ($-65\%$). Additivity was found between the effects of the maximal effective concentrations of ET-1 and PMA.

Mazzocchi et al. (1992) showed that subcutaneous infusion for 24 h with ET-1 (30 pmol/min) induces a 9-fold increase in the mitotic index (percentage of metaphase-arrested cells) of rat ZG. Infusions with equimolar doses of ACTH, ANG-II, and AVP raised the ZG mitotic index 13-, 9-, and 10-fold, respectively. Combined infusion with ET-1 and ACTH increased the ZG mitotic index by about 20-fold, whereas no additivity was observed between the effects of ET-1 and ANG-II or AVP. Belloni et al. (1996a) reported that three intraperitoneal injections of 100 pmol ET-1 (at 8-h intervals) enhanced [$^3$H]thymidine incorporation into DNA (2.3-fold) and mitotic index (6-fold) in the rat ZG but not in the ZF/ZR.

On the basis of their findings, Mazzocchi et al. (1992) and Belloni et al. (1996a) suggested that ET-1 exerts a marked mitogenic effect on the rat adrenals in vivo that manifests itself exclusively in the ZG. This agrees with the "cell-migration theory" of adrenocortical cytogenesis, according to which ZG is the locus of cell renewal under both basal and stimulated conditions (Nussdorfer, 1986). Belloni et al. (1996a) tried to explain the discrepancy existing between their findings and those of Cozza and Gomez-Sanchez (1990) not only by assuming obvious interspecific variability of the response to ET-1 but also by taking into account that striking differences occur in the physiology of adrenocortical cells between *in vivo* conditions and *in vitro* culture. They recalled that ACTH, which exerts a potent mitogenic action on the adrenal cortex *in vivo*, induces the functional differentiation of adrenocortical cells cultured *in vitro* but simultaneously inhibits their proliferative activity (Nussdorfer, 1986).

## C. Mechanisms of Action

The additivity of the antimitogenic effects of ET-1 and PMA led Cozza and Gomez-Sanchez (1990) to rule out the possibility that ET-1 acts on cultured bovine ZG cells by activating PKC. In contrast, Mazzocchi et al. (1992) provided findings suggesting that in the rat ET-1 exerts its proliferogenic effect on ZG by a mechanism similar to that underlying the growth action of ANG-II.

The classical transduction mechanism of ANG-II growth signals involves the activation of phosphatidylinositol cascade and PKC, which in turn plays a key role in the activation of TK. The activation of these last proteins is known to be essential in the intracellular events leading to proliferation and growth (Malarkey et al., 1995). Activated PKC and TK have been found to induce expression of the protooncogene *fos/jun* gene family (Viard et al., 1992a,b; Kimura et al., 1993; Yu, 1993), and PKC has been reported to stimulate mitogen-activated protein kinase (MAPK) cascade (Rossomando et al., 1992). One mechanism that has been proposed to link hormone-receptor interaction with long-term trophic effect is the induction of transcription regulatory proteins, which modulate the activity of other regulatory genes including those of the *fos/jun* family; protooncogene proteins eventually mediate the growth effect of hormones (Schönthal, 1990). MAPK cascade is emerging as an important pathway of these sequential responses. MAPKs are activated by most growth factors (Pelech and Sanghera, 1992) and can phosphorylate many different substrates, including p90 ribosomal protein S6 kinase (Sturgil et al., 1988) and the protooncogene proteins *jun* and *fos* (Pulverer et al., 1991; Alvarez et al., 1992). The activation of MAPK cascade surely plays a pivotal role in $G_0-G_1$ and $G_2-M$ transition of the cell cycle (Wang et al., 1994).

ET-1 has been found to activate MAPK cascade in cultured ventricular cardiac myocytes (Bogoyevitch et al., 1993), vascular smooth muscle cells (Koide et al., 1992), and rat mesangial cells (Wang et al., 1992). Moreover, ET-1 rapidly increases the mRNA levels of the members of the *fos/jun* gene family in rat fibroblasts (Pribnow *et al.*, 1992) and vascular smooth muscle cells, in which the effect is prevented by inhibitors of PKC (Bobik *et al.*, 1990).

Whether these reviewed mechanisms of nuclear signal transduction mediate long-term effects of ETs on the ZG is not known at present, but it must be recalled that ET-1 was found to activate both PKC and TK in the rat adrenals (see Section IV,D).

### D. Receptor Subtypes Involved in the Growth Action of Endothelins

To the best of our knowledge, no information is currently available as to the receptor subtype involved in the ET-1-induced hypertrophy and enhanced steroidogenic capacity of rat ZG cells. A hypertrophic effect of ETs has been demonstrated in other cell types such as cultured rat cardiac myocytes that secrete ET-1 in the culture medium (Ito *et al.*, 1991). Tamamori *et al.* (1996) showed that ET-3 ($10^{-7}$ $M$) stimulates protein synthesis in cultured rat cardiomyocytes and induces their hypertrophy within 48 h. This effect was found to be blocked by both BQ-123 and BQ-788. ET-3 also induced a transient increase in ET-1 mRNA and enhanced ET-1 release by cardiomyocytes, two findings that led those investigators to hypothesize that locally generated ET-1 may contribute to the ET-3-induced hypertrophy of cardiomyocytes acting as an autocrine/paracrine factor via ETA receptors.

Belloni *et al.* (1996a) have provided evidence that $ET_A$ mediates the proliferogenic effect of ET-1 on rat ZG. In fact, an ET-1-induced rise in DNA synthesis and mitotic index was found to be completely abrogated by the simultaneous administration of BQ-123 but not BQ-788. This finding accords well with the presence of $ET_A$ receptors exclusively in the ZG (see Section III,C) as well as with the demonstration that $ET_A$ mediates the growth effect of ET-1 on the pulmonary vascular bed (Zamora *et al.*, 1993).

However, although rat ZG contains $ET_A$, its parenchymal cells appear to express only $ET_B$ mRNA (Belloni *et al.*, 1996b). It is conceivable that only a small population of ZG cells, possibly the "stem" cells located just beneath the capsule (Nussdorfer, 1986), is provided with $ET_A$ and that such a population is lost during the enzymatic and mechanical procedures employed to isolate ZG cells. Alternatively, it might be that ET-1, via $ET_A$ receptors, elicits in the rat ZG the release by extraparenchymal cells (*e.g.*, fibroblasts, pericytes, or endothelium) growth factors (GF), which in turn stimulate (via the previously discussed mechanisms) the proliferation of ZG parenchymal cells. This possibility is indirectly supported by the findings that ETs do not exert *per se* a mitogenic action on rat vascular smooth muscle cells but only potentiate the mitogenic action of GFs present in the calf serum of the culture medium (Weissberg et al., 1990; Janakidevi et al., 1992). With regard to the receptor subtype involved, Weissberg and associates (1990) observed that ET-1 and ET-2 are equipotent in stimulating vascular smooth muscle cell proliferation, whereas ET-3 is ineffective, a finding strongly suggesting the exclusive involvement of $ET_A$ receptors. At variance, according to Wang et al. (1994), ET-1, ET-3, and S6C were equipotent in stimulating MAPK cascade and cell proliferation in Chinese hamster ovary cells transfected with $ET_A$ and $ET_B$ cDNA. These authors concluded that both subtypes of ET receptors are involved in this system.

## VI. Regulation of Endothelin Release

### A. Increase in Adrenal Blood Flow

There is a general consensus that a moderate shear stress, like that generated by an increase in the blood flow rate, stimulates ET synthesis and release by vascular endothelium (Davies, 1995). This appears to also hold true for the adrenal glands. Using *in situ* perfused rat adrenals, Cameron *et al.* (1994) demonstrated a close direct correlation between the perfusion medium flow rate and ET-1 concentration in the venous effluent: By raising flow rate by mechanically increasing perfusion medium delivery from 0.5 to 2.0 ml/10 min, an increase in ET-1 release was observed from 15 to 68 fmol/10 min.

Obviously, all the molecules that can raise adrenal blood flow may also elicit intraglandular ET release. These substances include a number of intramedullary regulatory peptides (enkephalins, vasoactive intestinal peptide, neuropeptide Y, neurotensin, calcitonin gene-related peptide, and adrenomedullin) (Hinson *et al.*, 1994; Mazzocchi *et al.*, 1996b) as well as ACTH, whose adrenal hyperhaemizing effect is well known (Vinson *et al.*, 1992). The fact that shear-stress response elements, as well as potential binding sites for glucocorticoid receptors, have been identified on the regulatory region of the human ECE-1 gene can provide a rational basis for the coupling of shear stress with ET-1 synthesis (Valdenaire *et al.*, 1995).

According to Hinson *et al.* (1991b,c), the *bolus* injection into the perfusion medium of 300 fmol of ACTH increases medium flow rate by 60% and ET-1 release by 80%, a finding that has been confirmed by Cameron *et al.* (1994). The injection in the perfusion medium of compound 48/80, a mastcell degranulator, was found to mimic the ACTH effect, whereas the injection of disodium cromoglycate, a stabilizator of mastcell granules, 60 s before ACTH blocked the ACTH effect (Hinson *et al.*, 1991c). These authors hypothesized that ACTH-induced adrenal hyperhemia may be mediated by the release of histamine by intraglandular mastcells, and that the ensuing increase in the local concentration of ETs may concur with the secretagogue effect of ACTH. Vinson *et al.* (1992) hypothesized that ET release by capillary endothelium may well participate in the mechanism whereby the increase in blood flow enhances steroid secretion by adrenal gland (see Section IV,E). Although appealing, this hypothesis does not appear to fit with the fact that the secretion of glucocorticoid is much more dependent on blood flow rate than aldosterone. The possibility that shear stress may block the endothelial release of factors required for the maintenance of a normal basal aldosterone secretion cannot be disregarded. Rosolowsky and Campbell (1994) demonstrated the production by endothelium of a proteinaceous molecule, different from ANG-II, ETs, or bradykinin, that is able to stimulate basal aldosterone secretion by cultured bovine ZG cells but not cortisol production by ZF/ZR cells. Alternatively, shear stress could elicit the release of a hypothetical factor depressing aldosterone production (see below).

## B. Effect of Nitric Oxide

As described under Section IV,E, ETs stimulate the release by vascular endothelium of nitric oxide, which in addition to exerting a vasodilatatory action appears to have a role in terminating the response to ETs by altering ET receptor-mediated events. There are indications that nitric oxide decreases ET release inasmuch as inhibition of NOS results in an increase in ET-1 production (Boulanger and Luscher, 1990; Kuchan and Frangos, 1993). This finding could explain why elevated shear stresses, although markedly stimulating nitric oxide release, inhibit ET-1 production (Malek and Izumo, 1994).

Hinson *et al.* (1996), using *in situ* perfused rat adrenals, demonstrated that the addition of L-arginine (1 m$M$) to the perfusion medium, although increasing medium flow rate, inhibits ET-1 release by about 40%. Conversely, the addition of L-NAME (5 m$M$) raised ET-1 release in the perfusate by about 2.5-fold. This inhibitory effect of nitric oxide is unlikely to be due to an increase in cGMP production because the addition of 100 $\mu M$ sodium nitroprusside to the medium was found to increase both flow rate (75%) and ET-1 release (90%); moreover, L-arginine evoked a 50% lowering in the release of cGMP in the perfusate.

The mechanism(s) whereby nitric oxide inhibits ET-1 release in adrenals remains to be agreed on. However, the possibility of the existence in the rat adrenals of negative feedback mechanisms involved in the fine tuning of ET release, which could have relevance in the regulation of glandular blood flow and steroid secretion, must be explored.

## C. Effect of Natriuretic Peptides

The effects of ETs on natriuretic peptides have been discussed under Section IV,E. The bulk of evidence suggests that ETs enhance the release of ANP but inhibit ANP-induced production of cGMP in some cell system. Convincing proof has been provided that ANP inhibits basal ET secretion from cultured endothelial cells, acting via the C-receptor (Hu *et al.*, 1992), as well as stimulates ET production by mesangial (Kohno *et al.*, 1993) and vascular smooth muscle cells (Bokemeyer *et al.*, 1994).

## VII. Involvement of Endothelins in Pathophysiology

Based on the aforementioned observations that ET-1 is able to stimulate steroid biosynthesis from adrenocortical cells of different species (see Section IV) and to specifically enhance the cell growth and steroidogenic capacity of rat ZG (see Section V), the hypothesis of an involvement of ETs in the pathophysiology of the adrenal cortex has been put forward and extensively investigated.

### A. Aldosterone-Producing Tumors

The role of ETs in the pathophysiology of aldosterone-producing tumors has been investigated with different approaches. In an *ex vivo* study of adrenocortical tissue slices, Zeng *et al.* (1992) were able to show that ET-1 increases aldosterone secretion, although less potently that ANG-II, from normal adrenocortical tissue as well as from adrenal cortex adjacent to aldosterone-producing adenoma. In contrast, they were unable to detect any significant effect of the peptide on tumor slices and therefore suggested the possibility of downregulation of responsiveness to ET-1, possibly due to enhanced local synthesis of the peptide in the tumor. Evidence of local biosynthesis of ET-1 was provided by Imai *et al.* (1992), who found the expression of not only prepro-ET-l but also its two receptor subtype mRNAs by Northern blotting analysis in homogenates of adenomas obtained from patients with Conn's syndrome. The presence of specific endothelin binding sites was confirmed by experiments with [$^{125}$I]ET-1 saturation and competitive displacement binding by the different endothelin isopeptides, although evidence of a single class of receptor was attained (Imai et al., 1992). With regard to prepro-ET-1 mRNA, similar findings were consistently obtained by Rossi et al. (1994, 1995b) with RT-PCR, which because it is more sensitive allowed the investigation of a larger number of aldosterone-producing adenomas.

The possibility of local synthesis of ET-1 was also supported by the results of immunohistochemistry. Incubation of 5-$\mu$m-thick deparaffinized sections of eight aldosterone-producing adenomas, 87% of which were small (<2 cm diameter), with a polyclonal rabbit anti-human ET-1 antiserum and avidin–biotin complex staining with nickel enhancement allowed Li et al. (1994) to show that 50% or more of the tumor cells had ET-1 immunoreactivity and a moderate staining. This finding contrasted with the uniform lack of ET-1 immunoreactivity in cells of a large number of phaeochromocytomas. In adrenocortical adenomas, the immunoreactive products appeared similar to those found in normal and tumor-free cortex, but they were generally fewer in number per cell. The observation that aldosterone-producing adenomas express the human ECE-1 gene fully agrees with the concept of local ET-1 biosynthesis in these tumors (Rossi et al., 1995a). Conclusive evidence of the concurrence of both the $ET_A$ and $ET_B$ receptor subtypes in aldosterone-producing tumors was provided by Rossi et al. (1995b) with gene expression studies and [$^{125}$I]ET-1 competitive displacement binding by ET-3, BQ-123, and S6C and with autoradiography. However, marked differences in $ET_A$ receptor density between ZG-like and ZF-like tumors and even among histologically different areas of the same tumor were observed. Because no evidence of increased plasma levels of ET-1 in primary aldosteronism patients compared to normal subjects was found (Veglio et al., 1994), the hypothesis of an autocrine/paracrine downregulation of ET-1 receptors due to a local activation of ET-1 synthesis, which is unable to increase plasma ET-1 levels, was put forward (Rossi et al., 1995a).

B. Idiopathic Hyperaldosteronism

The involvement of ET-1 and its receptor subtype $ET_A$ and $ET_B$ in idiopathic hyperaldosteronism has not been investigated thus far, mainly due to the lack of adrenocortical tissue suitable for molecular and pharmacological in vitro studies. The plasma levels of ET-1 were not found to be increased, compared to normal subjects, in seven patients with primary aldosteronism and there was no evidence of tumor using computed tomography scan (Veglio et al., 1994). Nevertheless, due to the evidence supporting a role of ET-1 in the regulation of the function of the normal and tumorous adrenal gland, it is conceivable that the peptide may participate in the pathogenesis of this condition.

## C. Glucocorticoid- and Sexual Steroid-Secreting Tumors

The role of ETs has not been investigated extensively in patients with Cushing's syndrome mainly due to the fact that, although glucorticoids can influence ET-1 synthesis (Kanse et al., 1991; Calderon et al., 1994; Provencher et al., 1995), the opposite does not seem to be true (Kennedy et al., 1993). In an immunohistochemical study of six medium-sized (from 2 to 5 cm in diameter) adrenocortical adenomas obtained from patients with Cushing's syndrome, Li et al. (1994) found that the vast majority of the tumors show 50% or more of the cells with a moderate immunoreactive ET-1 staining. The immunoreactive products were similar to those found in aldosterone-producing and nonfunctioning adenomas.

## D. Adrenocortical Carcinomas

No immunoreactive ET-1 staining was found in two large (>5 cm diameter) virilizing carcinomas, but in a number of nonfunctioning carcinomas ET-1-stained cells ranged from none to less than 10% (Li et al., 1994). Accordingly, the hypothesis that ET-1 immunostaining may be of diagnostic value for identification of malignancy was put forward (Li et al., 1994) and remains to be exploited.

Both the prepro-ET-1 and the human ECE-1 mRNAs have been found by RT-PCR in a cell line of adrenocortical cancer (NCI-H295), which secretes adrenal steroids in a regulated fashion (Fallo et al., 1995). In the same cells, the presence of immunoreactive ET-1 was detected with immunohistochemistry with both a polyclonal and a monoclonal antibody. Because the expression of both the $ET_A$ and $ET_B$ receptor subtypes was detected at the mRNA as well as at the protein level, an autocrine/paracrine role of the peptide, which may be relevant for cell regulation and growth, was suggested.

## VIII. Concluding Remarks

The preceding sections of this chapter have shown that in the few years interlapsed from the discovery of ETs a large mass of investigations have accumulated, clearly suggesting that these peptides play an important role in the autocrine/paracrine regulation of adrenal–cortex functions under both physiological and pathological conditions.

This contention is largely based on studies carried out *in vitro*, and we stress that its confirmation would surely require additional investigations of the *in vivo* effects of ETs and their receptor agonists and antagonists. In fact, only the demonstration that short- or long-term administrations of ET antagonists, such as BQ-123, BQ-788, or the nonselective orally active bosentan, which is being made available for long-term treatment (Clozel *et al.*, 1993; Tamirisa *et al.*, 1995), elicit sizable adrenocortical effects could provide conclusive proofs that endogenous ETs are involved in the physiological or pathophysiological control of adrenocortical functions. Unfortunately, the wide distribution and the multiple activities of ETs in several organs and systems, whose functions are strictly connected with those of the adrenal gland (*e.g.*, hypothalamo–pituitary axis, kidney, and atrial miocardium), make it difficult to unequivocally interpret the results of *in vivo* studies. The use of the technique of *in situ* perfusion of the adrenal gland (see Section IV,E) might allow the study of the effects of ET and its receptor ligands on the intact and living adrenal gland in the absence of any necessary variable present in the classic *in vivo* studies and therefore could notably increase our knowledge of this field.

We shall now take the opportunity to mention some topics whose elucidation should be the task of future investigations.

ETs, ECEs, and ET receptors are surely expressed as mRNAs and proteins in the mammalian adrenals (see Sections II and III), but nothing is known about the regulation of their synthesis under physiological and pathophysiological conditions requiring, *e.g.*, enhanced or decreased steroidogenic activity of adrenocortical cells.

The possible indirect mechanisms underlying the adrenocortical secretagogue action of ETs (see Section IV) are still waiting for a definitive demonstration. These studies obviously require the structural integrity of the gland (*i.e.*, an intact morphologic interrelationship between cortex and medulla) and therefore could take advantage of the technique of *in situ* adrenal perfusion.

The *in vivo* mitogenic effect of ETs on the ZG (see Section V,B) awaits to be confirmed in different species and under different experimental conditions, and the mechanism(s) and the ET receptor subtype involved merit further investigative effort. Probably, as in the case of the study of the mitogenic action of ETs on vascular smooth muscle or mesangial cells, these issues could be addressed by using ZG cells cultured *in vitro*, although it is well known that this model does not strictly reproduce *in vivo* conditions of the cells (*e.g.*, in the absence of any agonist, cultured adrenocortical cells undergo a rapid dedifferentiation coupled with a striking increase in their proliferation rate). Alternatively, the *in situ* perfusion technique may be employed with the limitation that adrenal gland can survive in good conditions only for a relatively short period (no more than 3 h). The elucidation of these topics not only will increase our knowledge of the basic mechanism(s) regulating adrenocortical cytogenesis but will also increase our understanding of the possible involvement of ETs in the neoplastic growth of adrenal cortex (see Section VII).

The mechanism(s) modulating intraadrenal release of ETs (see Section VI) surely awaits further study, especially as far as the "ambiguous" role of nitric oxide is concerned. The possibility that intramedullary regulatory peptides may exert a paracrine control of steroid secretion and release by modifying local blood flow rate and ET release also merits investigation with the *in situ* perfusion technique.

Finally, we stress that the basic and clinical research dealing with ETs could be notably improved by the development of more and more potent and selective ET receptor ligands, including a currently unavailable $ET_A$ receptor agonist.

References

Adams, M. L., Nock, B., Trnong, R., and Cicero, T. J. (1991). Nitric oxide control of steroidogenesis: Endocrine effects of $N^G$-nitro-L-arginine and comparison to alcohol. *Life Sci.* **50**, PL35–PL40.

Alvarez, E., Northwood, I. C., Gonzalez, F. A., Lautor, D. A., Seth, A., Abate, C., Curran, T., and Davis, R. J. (1992). Pro-Leu-Ser/Thr-Pro is a consensus primary sequence for substrate protein phosphorylation. *J. Biol. Chem.* **266**, 15277–15285.

Andreis, P. G., Malendowicz, L. K., Belloni, A. S., and Nussdorfer, G. G. (1995). Effects of pituitary adenylate cyclase-activating peptide (PACAP) on the rat adrenal secretory activity: Preliminary *in vitro* studies. *Life Sci.* **56**, 135–142.

Arai, H., Hori, S., Aramori, I., Ohkubo, H., and Nakanishi, S. (1990). Cloning and expression of a cDNA encoding an endothelin receptor. *Nature (London)* **348**, 730–732.

Bax, W. A., and Saxena, P. R. (1994). The current endothelin receptor classification: Time for reconsideration? *Trends Pharmacol. Sci.* **15**, 379–386.

Belloni, A. S., Rossi, G. P., Zanin, L., Prayer-Galetti, T., Pessina, A. C., and Nussdorfer, G. G. (1994). *In vitro* autoradiographic demonstration of endothelin-1 binding sites in the human adrenal cortex. *Biomed. Res.* **15**, 95–99.

Belloni, A. S., Andreis, P. G., Neri, G., and Nussdorfer, G. G. (1995). Endothelin-1 (ET-1) and cyclosporine-A (CSA) stimulate steroid secretion of rat adrenal cortex: Evidence that both ET-1 and CSA secretagogue effects are mediated by the B subtype of ET-1 receptor. *Biomed. Res.* **16**, 287–294.

Belloni, A. S., Malendowicz, L. K., Gottardo, G., and Nussdorfer, G. G. (1996a). Endothelin-1 stimulates the proliferation of rat adrenal zona glomerulosa cells, acting via the ETA receptor subtype. *Med. Sci. Res.* **24**, 393–394.

Belloni, A. S., Rossi, G. P., Andreis, P. G., Neri, G., Albertin, G., Pessina, A. C., and Nussdorfer, G. G. (1996b). Endothelin adrenocortical secretagogue effect is mediated by the B receptor in rats. *Hypertension* **27**, 1153–1159.

Bernet, F., Bernard, J., Laborie, C., Montel, V., Maubert, E., and Dupony, J. P. (1994). Neuropeptide Y (NPY)- and vasoactive intestinal peptide (VIP)-induced aldosterone secretion by rat capsule/glomerulosa zone could be mediated by catecholamines via β1 adrenergic receptors. *Neurosci. Lett.* **166**, 109–112.

Bloch, K. D., Eddy, R. L., Shows, T. B., and Quertermous, T. (1989). cDNA cloning and chromosomal assignment of the gene encoding endothelin 3. *J. Biol. Chem.* **264**, 18156–18161.

Bloch, K. D., Hong, C. C., Eddy, R. L., Shows, T. B., and Quertermous, T. (1991). cDNA cloning and chromosomal assignment of the endothelin 2 gene: Vasoactive intestinal contractor peptide is rat endothelin 2. *Genomics* **10,** 236–242.

Boarder, M. R., and Marriott, D. B. (1989). Characterization of endothelin. 1. Stimulation of catecholamine release from adrenal chromaffin cells. *J. Cardiovasc. Pharmacol.* **13,** S223–S224.

Boarder, M. R., and Marriott, D. B. (1991). Endothelin-1 stimulation of noradrenaline and adrenaline release from adrenal chromaffin cells. *Biochem. Pharmacol.* **41,** 521–526.

Bobik, A., Grooms, A., Millar, J. A., Mitchell, A., and Grinpukel, S. (1990). Growth factor activity of endothelin on vascular smooth muscle. *Am. J. Physiol.* **258,** C408–C415.

Bodart, V., Ong, H., and De Léan, A. (1995). A role for protein kinase in the steroidogenic pathway of angiotensin II in bovine zona glomerulosa cells. *J. Steroid Biochem. Mol. Biol.* **54,** 55–62.

Bogoyevitch, M. A., Glennon, P. E., and Sugden, P. H. (1993). Endothelin-1, phorbol esters and phenylephrine stimulate MAP kinase activities in ventricular cardiomyocytes. *FEBS Lett.* **317,** 271–275.

Bokemeyer, D., Friedrichs, U., Backer, A., Drechsler, S., Kramer H. J., and Meyer-Lehnert, H. (1994). Atrial natriuretic peptide inhibits cyclosporine A-induced endothelin production and calcium accumulation in rat vascular smooth muscle cells. *Clin. Sci.* **87,** 383–387.

Botting, R. M., and Vane, J. R. (1990). Endothelins: Potent releasers of prostacyclin and EDRF. *Polish J. Pharmacol. Pharm.* **42,** 203–218.

Boulanger, C., and Luscher, T. F. (1990). Release of endothelin from the porcine aorta. Inhibition by endothelium-derived nitric oxide. *J. Clin. Invest.* **85,** 587–590.

Bredt, D. S., Hwang, P. M., and Snyder S. H. (1990). Localization of nitric oxide synthase indicating a neural role for nitric oxide. *Nature (London)* **347,** 768–770.

Breslow, M. J., Tobin, J. R., Bredt, D. S., Ferris, C. D., Snyder, S. H., and Traystman, R. J. (1992). Role of nitric oxide in adrenal medullary vasodilation during catecholamine secretion. *Eur. J. Pharmacol.* **210,** 105–106.

Brock, T. A., and Danthuluri, N. R. (1992). Cellular actions of endothelin in vascular smooth muscle. *In* "Endothelin" (G. M. Rubanyi, Ed.), pp. 103–124. Oxford Univ. Press, New York.

Brown, M. A., and Smith, P. L. (1991). Endothelin: A potent stimulator of intestinal ion secretion *in vitro*. *Regul. Pept.* **36,** 1–19.

Calderon, E., Gomez-Sanchez, C. E., Cozza, E. N., Zhou, M., Coffey, R. G., Lockey, R. F., Prockop, L. D., and Szentivanyi, A. (1994). Modulation of endothelin-1 production by a pulmonary epithelial cell line. I. Regulation by glucocorticoids. *Biochem. Pharmacol.* **48,** 2065–2071.

Cameron, I. T., Davenport, A. P., Brown, M. J., and Smith, S. K. (1991). Endothelin-1 stimulates prostaglandin F2 alpha release from human endometrium. *Prostaglandins Leukotrienes Essent. Fatty Acids* **42,** 155–157.

Cameron, L., Kapas, S., and Hinson, J. P. (1994). Endothelin-1 release from the isolated perfused rat adrenal gland is elevated acutely in response to increasing flow rates and ACTH(1-24). *Biochem. Biophys. Res. Commun.* **202,** 873–879.

Cameron, L. A., and Hinson, J. P. (1993). The role of nitric oxide derived from L-arginine in the control of steroidogenesis, and perfusion medium flow rate in the isolated perfused rat adrenal gland. *J. Endocrinol.* **139,** 415–423.

Cao, L., and Banks, R. O. (1990). Cardiorenal actions of endothelin. Part I: Effects of converting enzyme inhibition. *Life Sci.* **46,** 577–583.

Chao, H. S., Poisner, A., Poisner, R., and Handwerger, S. (1993). Endothelins stimulate the synthesis and release of prorenin from human decidual cells. *J. Clin. Endocrinol. Metab.* **76,** 615–619.

Clozel, M., Breu, V., Burri, K., Cassal, J. M., Fischli, W., Gray, G. A., Hirth, G., Loffler, B. M., Muller, M., Neidhart, W., and Ramuz, H. (1993). Pathophysiological role of endothelin revealed by the first orally active endothelin receptor antagonist. *Nature (London)* **365**, 759–761.

Conlin, P. R., Kim, S. Y., Williams, G. H., and Canessa, M. L. (1990). $Na^+$-$H^+$ exchanger kinetics in adrenal glomerulosa cells and its activation by angiotensin II. *Endocrinology* **127**, 236–244.

Conlin, P. R., Williams, G. H., and Canessa, M. L. (1991). Angiotensin II-induced activation of $Na^+$-$K^+$ exchange in adrenal glomerulosa cells is mediated by protein kinase C. *Endocrinology* **129**, 1861–1868.

Cozza, E. N., and Gomez-Sanchez, C. E. (1990). Effects of endothelin-1 on its receptor concentration and thymidine incorporation in calf adrenal zona glomerulosa cells: A comparative study with phorbol esters. *Endocrinology* **127**, 549–554.

Cozza, E. N., and Gomez-Sanchez, C. E. (1993). Mechanism of the ET-1 potentiation of angiotensin II stimulation of aldosterone production in calf adrenal glomerulosa cell cultures. *Am. J. Physiol.* **265**, E179–E183.

Cozza, E. N., Gomez-Sanchez, C. E., Foecking, M. F., and Chiou, S. (1989). Endothelin binding to cultured calf adrenal zona glomerulosa cells and stimulation of aldosterone secretion. *J. Clin. Invest.* **84**, 1032–1035.

Cozza, E. N., Chiou, S., and Gomez-Sanchez, C. E. (1992). Endothelin-1 potentiation of angiotensin II stimulation of aldosterone production. *Am. J. Physiol.* **262**, R85–R89.

Davenport, A. P., Nunez, D. J., Hall, T. A., Kaumann, A. J., and Brown, M. J. (1989). Autoradiographical localization of binding sites for [$^{125}$I]endothelin-1 in humans, pigs and rats: Functional relevance in humans. *J. Cardiovasc. Pharmacol.* **13**(Suppl. 5), S177–S180.

Davies, P. F. (1995). Flow-mediated endothelial mechanotransduction. *Physiol. Rev.* **75**, 519–560.

Delarue, C., Delton, I., Fiorini, F., Homo-Delarche, F., Fasolo, A., Braquet, P., and Vaudry, H. (1990). Endothelin stimulates steroid secretion by frog adrenal gland *in vitro*: Evidence for the involvement of prostaglandins and extracellular calcium in the mechanism of action of endothelin. *Endocrinology* **127**, 2001–2008.

D'Orleans-Juste, P., Telemaque, S., Claing, A., Ihara, M., and Yano, M. (1992). Human big endothelin 1 and endothelin 1 release prostacyclin via the activation of ET1 receptors in the rat perfused lung. *Br. J. Pharmacol.* **105**, 773–775.

Ehrhart-Bornstein, M., Bornstein, S. R., Gonzales-Hernandez, J., Holst, J. J., Waterman, M. R., and Scherbaum, W. A. (1995). Sympathoadrenal regulation of adrenocortical steroidogenesis. *Endocrine Res.* **21**, 13–24.

Elliot, M. E., Hadjokas, N. E., and Goodfriend, T. L. (1986). Effects of ouabain and potassium on protein synthesis and angiotensin-stimulated aldosterone synthesis in bovine adrenal glomerulosa cells. *Endocrinology* **118**, 1469–1475.

Emori, T., Hirata, Y., and Marumo, F. (1991). Endothelin 3 stimulates prostacyclin production in cultured bovine endothelial cells. *J. Cardiovasc. Pharmacol.* **17**(Suppl. 7), S140–S142.

Fallo, F., Pagotto, U., Albertin, G., Belloni, A. S., Pilon, C., Pistorello, M., Biasolo, M. A., Nussdorfer, G. G., and Rossi, G. P. (1995). Concomitant expression of endothelin-1 and its receptor A and B in the human NCI-H295 adrenocortical carcinoma cells. *In* "Proceedings of the Annual Meeting of the Endocrinology Society, 77th", p. 261. The Endocrine Society, Washington, DC.

Feuilloley, M., and Vaudry, H. (1996). Role of the cytoskeleton in adrenocortical cells. *Endocrine Rev.* **17**, 269–288.

Feuilloley, M., Netchitaïlo, P., Delarue, C., Leboulenger, F., Benyamina, M., Pelletier, G., and Vaudry, H. (1988). Involvement of the cytoskeleton in the steroidogenic response of frog adrenal glands to angiotensin II, acetylcholine and serotonin. *J. Endocrinol.* **118**, 365–374.

Feuilloley, M., Contesse, V., Lefebvre, H., Delarue, C., and Vaudry H. (1994). Effects of selective disruption of cytoskeletal elements on steroid secretion by human adrenocortical slices. *Am. J. Physiol.* **266**, E202–E210.

Filep, J. G., Battistini, B., Cote, Y. P., Beaudoin, A. R., and Sirois, P. (1991). Endothelin 1 induces protacyclin release from bovine aortic endothelial cells. *Biochem. Biophys. Res. Commun.* **177,** 171–176.

Fukuda, Y., Hirata, Y., Taketani, S., Kojima, T., Oikawa, S., Nakazato, H., and Kobayashi, Y. (1989). Endothelin stimulates accumulation of cellular atrial natriuretic peptide and its messenger RNA in rat cardiocytes. *Biochem. Biophys. Res. Commun.* **164,** 1431–1436.

Gardner, D. G., Newman E. D., Nakamura, K. K., and Nguyen, K. P. T. (1991). Endothelin increases the synthesis and secretion of atrial natriuretic peptide in neonatal rat cardiocytes. *Am. J. Physiol.* **261,** E177–E182.

Goetz, K. L., Wang, B. C., Madwed J. B., Zhu, J. L., and Leadley, R. J. (1988). Cardiovascular, renal, and endocrine responses to intravenous endothelin in conscious dogs. *Am. J. Physiol.* **255,** R1064–R1068.

Goligorsky, M. S., Tsukahara, H., Magazine, H., Andersen, T. T., Malik, A. B., and Bahou, W. F. (1994). Termination of endothelin signaling: Role of nitric oxide. *J. Cell. Physiol.* **158,** 485–494.

Gomez-Sanchez, C. E., Cozza, E. N., Foecking, M. F., Chiou, S., and Ferris, M. W. (1990). Endothelin receptor subtypes and stimulation of aldosterone secretion. *Hypertension* **15,** 744–747.

Gwosdow, A. R., O'Connell, N. A., Spencer, J. A., Kumar, M. S. A., Agarwal, R. K., Bode, H. H., and Abou-Samra, A. B. (1992). Interleukin-1-induced corticosterone release occurs by an adrenergic mechanism from the rat adrenal gland. *Am. J. Physiol.* **123,** E461–E466.

Hagiwara, H., Nagasawa, T., Yamamoto, T., Lodhi, K. M., Ito, T., Takemura, N., and Hirose, S. (1993). Immunochemical characterisation and localization of endothelin ETB receptor. *Am. J. Physiol.* **264,** R777–R783.

Haynes, W. G., Davenport, A. P., and Webb, D. J. (1993). Endothelin: Progress in pharmacology and physiology. *Trends Pharmacol. Sci.* **14,** 225–228.

Hinson, J. P., Kapas, S., Teja, R., and Vinson, G. P. (1991a). Effect of the endothelins on aldosterone secretion by rat zona glomerulosa cells *in vitro. J. Steroid Biochem. Mol. Biol.* **40,** 437–439.

Hinson, J. P., Vinson, G. P., Kapas, S., and Teja R. (1991b). The role of endothelin in the control of adrenocortical function: Stimulation of endothelin release by ACTH and the effects of endothelin-1 and endothelin-3 on steroidogenesis in rat and human adrenocortical cells. *J. Endocrinol.* **128,** 275–280.

Hinson, J. P., Vinson, G. P., Kapas, S., and Teja, R. (1991c). The relationship between adrenal vascular events and steroid secretion: The role of mastcells and endothelin. *J. Steroid Biochem. Mol. Biol.* **40,** 381–389.

Hinson, J. P., Kapas, S., Orford, C. D., and Vinson, G. P. (1992). Vasoactive intestinal peptide stimulation of aldosterone secretion by the rat adrenal cortex may be mediated by the local release of catecholamines. *J. Endocrinol.* **133,** 253–258.

Hinson, J. P., Purbrick, A., Cameron, L. A., and Kapas, S. (1994). The role of neuropeptides in the regulation of adrenal vascular tone: Effects of vasoactive intestinal polypeptide, substance P, neuropeptide Y, neurotensin, and met- and leu-enkephalin on perfusion medium flow rate in the intact perfused rat adrenal. *Regul. Pept.* **51,** 55–61.

Hinson, J. P., Kapas, S., and Cameron, L. A. (1996). Differential effects of endogenous and exogenous nitric oxide on the release of endothelin-1 from the intact perfused rat adrenal gland *in situ. FEBS Lett.* **379,** 7–10.

Hori, S., Komatsu, Y., Shigemoto, R., Mizuno, N., and Nakanishi, S. (1992). Distinct tissue distribution and cellular localization of two messenger ribonucleic acids encoding different subtypes of rat endothelin receptors. *Endocrinology* **130,** 1885–1895.

Hu, R. M., Levin, E. R., Pedram, A., and Frank H. J. L. (1992). Atrial natriuretic peptide inhibits the production and secretion of endothelin from cultured endothelial cells. Mediation through the C-receptor. *J. Biol. Chem.* **267,** 17384–17389.

Ihara, M., Noguchi, K., Saeki, T., Fukuroda, T., Tsuchida, S., Kimura, S., Fukami, T., Ishikawa, K., Nishikibe, M., and Yano, M. (1992). Biological profiles of highly potent novel endothelin antagonists selective for the ETA receptor. *Life Sci.* **50,** 247–256.

Imai, T., Hirata, Y., Eguchi, S., Kanno, K., Ohta, K., Emori, T., Sakamoto, A., Yanagisawa, M., Masaki, T., and Marumo, F. (1992). Concomitant expression of receptor subtype and isopeptide of endothelin by human adrenal gland. *Biochem. Biophys. Res. Commun.* **182,** 1115–1121.

Inoue, A., Yanasigawa, M., Kimura, S., Kasuya, Y., Miyauchi, T., Goto, K., and Masaki, T. (1989). The human endothelin family: Three structurally and pharmacologically distinct isopeptides predicted by three separate genes. *Proc. Natl. Acad. Sci. USA* **86,** 286–287.

Irons, C. E., Murray, S. F., and Glembotsky, C. C. (1993). Identification of the receptor subtype responsible for endothelin-mediated protein kinase-C activation and atrial natriuretic factor secretion from atrial myocytes. *J. Biol. Chem.* **268,** 23417–23421.

Ishikawa, K., Ihara, M., Noguchi, K., Mase, T., Mino, N., Saeki, T., Fukuroda, T., Fukani, T., Ozaki, S., Nagase, T., Nishikide, M., and Yano, M. (1994). Biochemical and pharmacological profile of a potent and selective endothelin B-receptor antagonist, BQ-788. *Proc. Natl. Acad. Sci. USA* **91,** 4892–4896.

Ito, H., Hirata, Y., Hiroe, M., Tsujino, M., Adachi, S., Takamoto, T., Nitta, M., Taniguchi, K., and Marumo, F. (1991). Endothelin-1 induces hypertrophy with enhanced expression of muscle specific genes in cultured neonatal rat cardiomyocytes. *Circ. Res.* **69,** 209–215.

Jaiswal, R. K. (1992). Endothelin inhibits the atrial natriuretic factor stimulated cGMP production by activating the protein kinase-C in rat aortic smooth muscle cells. *Biochem. Biophys. Res. Commun.* **182,** 395–402.

Janakidevi, K., Fisher, M. A., Del Vecchio, P. J., Tiruppathi, C., Figge, J., and Malik A. B. (1992). Endothelin-1 stimulates DNA synthesis and proliferation of pulmonary artery smooth muscle cells. *Am. J. Physiol.* **263,** C1295–C1301.

Kanse, S. M., Takahashi, K., Warren, J. B., Ghatei, M., and Bloom, S. R. (1991). Glucocorticoids induce endothelin release from vascular smooth muscle cells but not endothelial cells. *Eur. J. Pharmacol.* **199,** 99–101.

Kapas, S., and Hinson, J. P. (1996). Inhibition of endothelin- and phorbol ester-stimulated tyrosine kinase activity by corticotrophin in the rat adrenal zona glomerulosa. *Biochem. J.* **313,** 867–872.

Kapas, S., Purbrick, A., and Hinson, J. P. (1995). Role of tyrosine kinase and protein kinase C in the steroidogenic actions of angiotensin II, $\alpha$-melanocyte-stimulating hormone and corticotropin in the rat adrenal cortex. *Biochem. J.* **305,** 433–438.

Karne, S., Jayawickreme, C. K., and Lerner, M. R. (1993). Cloning and characterization of an endothelin-3 specific receptor (ETC receptor) from *Xenopus laevis* dermal melanophores. *J. Biol. Chem.* **268,** 19126–19133.

Kato, K., Sawada, S., Toyoda, T., Kobayashi, K., Shirai, K., Yamamoto, K., Tamagaki, T., Yamagami, M., Yoneda, M., Takada, O., Uno, M., Tsuji, H., and Nakagawa, M. (1992). Influence of endothelin on human platelet aggregation and prostacyclin generation from human vascular endothelial cells in culture. *Jpn. Circ. J.* **56,** 422–431.

Kawai, M., Naruse, M., Yoshimoto, T., Naruse, K., Shionoya, K., Tanaka, M., Morishita, Y., Matsuda, Y., Demura, R., and Demura, H. (1996). C-type natriuretic peptide as a possible local modulator of aldosterone secretion in bovine adrenal zona glomerulosa. *Endocrinology* **137,** 42–46.

Kennedy, R. L., Haynes, W. G., and Webb, D. J. (1993). Endothelins as regulators of growth and function in endocrine tissues. *Clin. Endocrinol.* **39,** 259–265.

Kimura, E., Sonobe, M. H., Armelin, M. C. S., and Armelin, H. A. (1993). Induction of *fos* and *jun* proteins by adrenocorticotropin and phorbol ester by not by 3′-5′-cyclic adenosine monophosphate derivatives. *Mol. Endocrinol.* **7,** 1463–1471.

Kloog, Y., and Sokolovsky, M. (1989). Similarities in mode and sites of action of sarafotoxins and endothelins. *Trends Pharmacol. Sci.* **10,** 212–214.
Koh, E., Morimoto, S., Kim, S., Nabata, T., Miyashita, Y., and Ogihara, T. (1990). Endothelin stimulates $Na^+/H^+$ exchange in vascular muscle cells. *Biochem. Int.* **20,** 375–380.
Kohno, M., Horio, T., Ikeda, M., Yokokawa, K., Fukui, T., Yasunari, K., Murakawa, K. I., Kurihara, N., and Takeda, T. (1993). Natriuretic peptides inhibit mesangial cell production of endothelin induced by arginine vasopressin. *Am. J. Physiol.* **264,** F678–F683.
Kohzuki, M., Johnston, C. I., Chai, S. Y., Casley, D. J., Rogerson, F., and Mendelsohn, F. A. O. (1989). Endothelin receptors in rat adrenal gland visualized by quantitative autoradiography. *Clin. Exp. Pharmacol. Physiol.* **16,** 239–242.
Kohzuki, M., Johnston, C. I., Abe, K., Chai, S. Y., Casley, D. J., Yasujima, M., Yoshinaga, K., and Mendelsohn, F. A. O. (1991). *In vitro* autoradiographic endothelin-1 binding sites and sarafotoxin S6B binding sites in rat tissues. *Clin. Exp. Pharmacol. Physiol.* **18,** 509–515.
Koide, M., Kawahara, Y., Tsuda, T., Ishida, Y., Shii, K., and Yokoyama, M. (1992). Endothelin-1 stimulates tyrosine phosphorylation and the activities of two mitogen activated protein kinases in cultured vascular smooth muscle cells. *J. Hypertension* **10,** 1173–1182.
Kondoh, M., Miyazaki, H., Watanabe, H., Shibata, T., Yanagisawa, M., Masaki, T., and Murakami, K. (1990). Isolation of anti-endothelin receptor monoclonal antibodies for use in receptor characterization. *Biochem. Biophys. Res. Commun.* **172,** 503–510.
Koseki, C., Imai, M., Hirata, Y., Yanagisawa, M., and Masaki, T. (1989a). Binding sites for endothelin-1 in rat tissues: An autoradiographic study. *J. Cardiovasc. Pharmacol.* **13**(Suppl. 5), S153–S154.
Koseki, C., Imai, M., Hirata, Y., Yanagisawa, M., and Masaki, T. (1989b). Autoradiographic distribution in rat tissues of binding sites for endothelin: A neuropeptide? *Am. J. Physiol.* **256,** R858–R866.
Kuchan, M. J., and Frangos, J. A. (1993). Shear stress regulates endothelin-1 release via protein kinase C and cGMP in cultured endothelial cells. *Am. J. Physiol.* **264,** H150–H156.
Lee, Y. J., Lin, S. R., Shin, S. J., and Tsai, J. H. (1993). Increased adrenal medullary atrial natriuretic polypeptide synthesis in patients with primary aldosteronism. *J. Clin. Endocrinol. Metab.* **76,** 1357–1362.
Lee, Y. J., Lin, S. R., Shin, S. J., Lai, Y. H., Lin, Y. T., and Tsai, J. H. (1994). Brain natriuretic peptide is synthesized in the human adrenal medulla and its messenger ribonucleic acid expression along with that of atrial natriuretic peptide are enhanced in patients with primary aldosteronism. *J. Clin. Endocrinol. Metab.* **79,** 1476–1482.
Levin, E. R., Frank, H. J., and Pedram, A. (1992). Endothelin receptors on cultured fetal rat diencephalic glia. *J. Neurochem.* **58,** 659–666.
Li, Q., Grimelius, L., Gröndal, S., Höög, A., and Johansson, H. (1994). Immunohistochemical localization of endothelin-1 in non-neoplastic and neoplastic adrenal gland tissue. *Virchows Arch.* **425,** 259–264.
Li, Q., Grimelius, L., Zhang, X., Lukinius, A., Gröndal, S., Höög, A., Johansson, H. (1995). Ultrastructural localization of endothelin-1 in nonneoplastic, hyperplastic, and neoplastic adrenal gland. *Ultrastruct. Pathol.* **19,** 489–494.
Lin, H. Y., Kaji, E. H., Winkel, G. K., Ives, H. E., and Lodish, H. F. (1991). Cloning and functional expression of a vascular smooth muscle endothelin 1 receptor. *Proc. Natl. Acad. Sci. USA* **88,** 3185–3189.
Longchampt, M. O., Pinelis, S., Goulin, J., Chabrier, P. E., and Braquet, P. (1991). Proliferation and $Na^+/H^+$ exchange activation by endothelin in vascular smooth muscle cells. *Am. J. Hypertension* **4,** 776–779.
Malarkey, K., Belham, C. M., Paul, A., Graham, A., McLees, A., Scott, P. H., and Plevin, R. (1995). The regulation of tyrosine kinase signaling pathways by growth factor and G-protein-coupled receptors. *Biochem. J.* **309,** 361–375.

Malek, A. M., and Izumo, S. (1994). Molecular aspects of signal transduction of shear stress in the endothelial cell. *J. Hypertension* **12**, 989–999.

Masaki, T. (1993). Endothelins: Homeostatic and compensatory actions in the circulatory and endocrine systems. *Endocrine Rev.* **14**, 256–268.

Mazzocchi, G., Malendowicz, L. K., and Nussdorfer, G. G. (1990a). Endothelin-1 acutely stimulates the secretory activity of rat zona glomerulosa cells. *Peptides* **11**, 763–765.

Mazzocchi, G., Rebuffat, P., Meneghelli, V., Malendowicz, L. K., Kasprzak, A., and Nussdorfer, G. G. (1990b). Effects of prolonged infusion with endothelin-1 on the function and morphology of rat adrenal cortex. *Peptides* **11**, 767–772.

Mazzocchi, G., Malendowicz, L. K., Meneghelli, V., and Nussdorfer, G. G. (1992). Endothelin-1 stimulates mitotic activity in the zona glomerulosa of the rat adrenal cortex. *Cytobios* **69**, 91–96.

Mazzocchi, G., Malendowicz, L. K., Meneghelli, V., Gottardo, G., and Nussdorfer, G. G. (1993). Vasoactive intestinal polypeptide (VIP) stimulates hormonal secretion of the rat adrenal cortex *in vitro*: Evidence that adrenal chromaffin cells are involved in the mediation of the mineralocorticoid, but not glucocorticoid secretagogue action of VIP. *Biomed. Res.* **14**, 435–440.

Mazzocchi, G., Malendowicz, L. K., Belloni, A. S., and Nussdorfer, G. G. (1995). Adrenal medulla is involved in the aldosterone secretagogue effect of substance P. *Peptides* **16**, 351–355.

Mazzocchi, G., Malendowicz, L. K., Macchi, C., Gottardo, G., and Nussdorfer, G. G. (1996a). Further investigations on the effects of neuropeptide-Y (NPY) on the secretion and growth of rat adrenal zona glomerulosa. *Neuropeptides* **30**, 19–27.

Mazzocchi, G., Musajo, F. G., Neri, G., Gottardo, G., and Nussdorfer, G. G. (1996b). Adrenomedullin stimulates steroid secretion by the isolated perfused rat adrenal gland *in situ*: Comparison with calcitonin gene-related peptide effects. *Peptides,* in press.

Mazzocchi, G., Rebuffat, P., Gottardo, G., Meneghelli, V., and Nussdorfer, G. G. (1996c). Evidence that both ETA and ETB receptor subtypes are involved in the *in vivo* aldosterone secretagogue effect of endothelin-1 in rats. *Res. Exp. Med.* **196**, 145–152.

Meuli, C., and Müller, J. (1982). Characterization of rat capsular adrenal (zona glomerulosa) ($Na^+$, $K^+$)-ATPase activity. *J. Steroid Biochem.* **16**, 129–132.

Miller, W. L., Redfield, M. M., and Burnett, J. C. (1989). Integrated cardiac, renal, and endocrine actions of endothelin. *J. Clin. Invest.* **83**, 317–320.

Miller, W. L., Cavero, P. G., Aarhus, L. L., Heublein, D. M., and Burnett, J. C. (1993). Endothelin-mediated cardiorenal hemodynamic and neuroendocrine effects are attenuated by nitroglycerin *in vivo*. *Am. J. Hypertension* **6**, 156–163.

Miyamoto, Y., Yoshimasa, T., Arai, H., Takaya, K., Ogawa, Y., Tanaka I., Ito, H., and Nakao, K. (1994). Two novel transcripts of the endothelin-A receptor generated by alternative splicing. *Hypertension* **24**, 375. [Abstract]

Moe, O., Tejedor, A., Campbell, W. B., Alpern, R. J., and Henrich, W. L. (1991). Effects of endothelin on *in vitro* renin secretion. *Am. J. Physiol.* **260**, E521–E525.

Moncada, S., Palmer, R. M. J., and Higgs, E. A. (1991). Nitric oxide: Physiology, pathophysiology and pharmacology. *Pharmacol. Rev.* **43**, 109–142.

Morishita, R., Higaki, J., and Ogihara, T. (1989). Endothelin stimulates aldosterone biosynthesis by dispersed rabbit adreno-capsular cells. *Biochem. Biophys. Res. Commun.* **160**, 628–632.

Mulrow, P. J. (1992). Adrenal renin. Regulation and function. *Front. Neuroendocrinol.* **13**, 47–60.

Naess, P. A., Christensen, G., and Kiil, F. (1993). Inhibitory effect of endothelin on renin release in dog kidneys. *Acta Physiol. Scand.* **148**, 131–136.

Nakamoto, H., Suzuki, H., Murakami, M., Kageyama, Y., Naitoh, M., Sakamaki, Y., Ohishi, A., and Saruta, T. (1991). Different effects of low and high doses of endothelin on haemodynamics and hormones in the normotensive conscious dog. *J. Hypertension* **9**, 337–344.

Namiki, A., Hirata, Y., Ishikawa, M., Moroi, M., Aikawa, J., and Machii, K. (1992). Endothelin-1 and endothelin-3 induced vasoditation via common generation of endothelium-derived nitric oxide. *Life Sci.* **50,** 677–682.

Naruse, M., Naruse, K., and Demura, H. (1994a). Recent advances in endothelin research on cardiovascular and endocrine systems. *Endocrine J.* **41,** 491–507.

Naruse, M., Naruse, K., Kawai, M., Yoshimoto, T., Tanabe, A., Tanaka, M., and Demura, H. (1994b). Endothelin as a local regulator of aldosterone secretion from bovine adrenocortical cells. *Am. J. Hypertension* **7,** 74A. [Abstract]

Neri, G., Andreis, P. G., Prayer-Galetti, T., Rossi, G. P., Malendowicz, L. K., and Nussdorfer, G. G. (1996). Pituitary adenylate cyclase-activating peptide (PACAP) enhances aldosterone secretion of human adrenal gland: Evidence for an indirect mechanism probably involving the local release of catecholamines. *J. Clin. Endocrinol. Metab.* **81,** 169–173.

Netchitaïlo, P., Perroteau, I., Feuilloley, M., Pelletier G., and Vaudry, H. (1985). In vitro effect of cytochalasin B on adrenal steroidogenesis in frog. *Mol. Cell. Endocrinol.* **43,** 205–213.

Neuser, D., Steinke, W., Theiss, G., and Stasch, J. P. (1989). Autoradiographic localization of [$^{125}$I]endothelin-1 and [$^{125}$I]atrial natriuretic peptide in rat tissue: A comparative study. *J. Cardiovasc. Pharmacol.* **13**(Suppl. 5), S67–S73.

Neuser, D., Steinke, W., Dellweg, H., Kazda, S., and Stasch, J. P. (1991). $^{125}$I-Endothelin and $^{125}$I-big endothelin-1 in rat tissue: Autoradiographic localization and receptor binding. *Histochemistry* **95,** 621–628.

Nunez, D. J., Brown, M. J., Davenport, A. P., Neylon, C. B., Schofield, J. P., and Wyse, R. K. (1990). Endothelin-1 mRNA is widely expressed in porcine and human tissues. *J. Clin. Invest.* **85,** 1537–1541.

Nussdorfer, G. G. (1986). Cytophysiology of the adrenal cortex. *Int. Rev. Cytol.* **98,** 1–405.

O'Connell, N. A., Kumar, A., Chatzipanteli, K., Mohan, A., Agarwal, R. K., Head, C., Bornstein, S. R., Abou-Samra, A. B., and Gwosdow, A. R. (1994). Interleukin regulates corticosterone secretion from the rat adrenal gland through a catecholamine-dependent and prostaglandin E2-independent mechanism. *Endocrinology* **135,** 460–467.

Ogawa, M., Nakao, K., Arai, H., Nagakawa, O., Hosoda K., Suga, S., Nakanishi, S., and Imura, H. (1991). Molecular cloning of a non-isopeptide selective human endothelin receptor. *Biochem. Biophys. Res. Commun.* **178,** 248–255.

Ohara-Imaizumi, M., and Kumakura, K. (1991). Dynamics of the secretory response evoked by endothelin-1 in adrenal chromaffin cells. *J. Cardiovasc. Pharmacol.* **17,** S156–S158.

Palacios, M., Knowles, R. G., Palmer, R. M. J., and Moncada, S. (1989). Nitric oxide from L-arginine stimulates the soluble guanylate cyclase in adrenal glands. *Biochem. Biophys. Res. Commun.* **165,** 802–809.

Pecci, A., Gomez-Sanchez, C. E., Bedners, M. E. O., Lantos, C. P., and Cozza, E. N. (1993). In vivo stimulation of aldosterone biosynthesis by endothelin: Loci of action and effects of doses and infusion rates. *J. Steroid Biochem. Mol. Biol.* **45,** 555–561.

Pecci, A., Cozza, E. N., Devlin, M., Gomez-Sanchez, C. E., and Gomez-Sanchez, E. P. (1994). Endothelin-1 stimulation of aldosterone and zona glomerulosa ouabain-sensitive sodium/potassium-ATPase. *J. Steroid Biochem. Mol. Biol.* **50,** 49–53.

Pelech, S. L., and Sanghera, J. S. (1992). Mitogen-activated protein kinases: Versatile transducers for cell signaling. *Trends Biochem. Sci.* **17,** 233–238.

Pribnow, D., Muldoon, L. L., Fajardo, M., Theodor, L., Chen, L. Y., and Magun, B. E. (1992). Endothelin induces transcription of *fos/jun* family genes: A prominent role for calcium ion. *Mol. Endocrinol.* **6,** 1003–1012.

Provencher, P. H., Saltis, J., and Funder, J. W. (1995). Glucocorticoids but not mineralocorticoids modulate endothelin-1 and angiotensin-II binding in SHR vascular smooth muscle cells. *J. Steroid Biochem. Mol. Biol.* **52,** 219–225.

Pulverer, B. J., Kyriakis, M. M., Avruch, J., Nikolakaki, E., and Woodgett, J. R. (1991). Phosphorylation of *c-jun* mediated by MAP kinases. *Nature (London)* **353,** 670–674.

Rakugi, H., Tabuchi, Y., Nakamaru, M., Nagamo, M., Higashimori, K., Mikani, H., and Ogihara, T. (1990). Endothelin activates the vascular renin-angiotensin system in rat mesenteric arteries. *Biochem. Int.* **21,** 867–872.

Remy-Jouet, I., Delarue, C., Feuilloley, M., and Vaudry, H. (1994). Involvement of the cytoskeleton in the mechanism of action of endothelin on frog adrenocortical cells. *J. Steroid Biochem. Mol. Biol.* **50,** 55–59.

Reynolds, E. E., Mok, L. L., and Kurokawa, S. (1989). Phorbol ester dissociates endothelin stimulated phosphoinositide hydrolysis and arachidonic acid release in vascular smooth muscle cells. *Biochem. Biophys. Res. Commum.* **160,** 868–873.

Ritz, M. F., Stuenkel, E. L., Dayanithi, G., Jones, R., and Nordmann, J. J. (1992). Endothelin regulation of neuropeptide release from nerve endings of the posterior pituitary. *Proc. Natl. Acad. Sci. USA* **89,** 8371–8375.

Rosolowsky, L. J., and Campbell, W. B. (1990). Endothelin enhances adrenocorticotropin-stimulated aldosterone release from cultured bovine adrenal cells. *Endocrinology* **126,** 1860–1866.

Rosolowsky, L. J., and Campbell, W. B. (1994). Endothelial cells stimulate aldosterone release from bovine adrenal zona glomerulosa cells. *Am. J. Physiol.* **266,** E107–E117.

Rossi, G. P., Albertin, G., Belloni, A. S., Zanin, L., Biasolo, M. A., Prayer-Galetti, T., Bader, M., Nussdorfer, G. G., Palù, G., and Pessina, A. C. (1994). Gene expression, localization and characterization of endothelin A and B receptors in the human adrenal cortex. *J. Clin. Invest.* **94,** 1226–1234.

Rossi, G. P., Albertin, G., Franchin, E., Sacchetto, A., Cesari, M., Palù, G., and Pessina, A. C. (1995a). Expression of the endothelin-converting enzyme gene in human tissues. *Biochem. Biophys. Res. Commun.* **211,** 249–253.

Rossi, G. P., Belloni, A. S., Albertin, G., Zanin, L., Biasolo, M. A., Nussdorfer, G. G., Palù, G., and Pessina, A. C. (1995b). Endothelin-1 and its receptors A and B in human aldosterone-producing adenomas. *Hypertension* **25,** 842–847.

Rossomando, A., Wu, J., Weber, M. J., and Sturgill, T. W. (1992). The phorbol ester-dependent activator of the mitogen-activated protein kinase p42MAPK is a kinase with specificity for the threonine and tyrosine regulatory sites. *Proc. Natl. Acad. Sci. USA* **89,** 5221–5225.

Rubanyi, G. M., and Polokoff, M. A. (1994). Endothelins: Molecular biology, biochemistry, pharmacology, physiology, and pathophysiology. *Pharmacol. Rev.* **46,** 325–415.

Sakurai, T., Yanagisawa, M., and Masaki, T. (1992). Molecular characterization of endothelin receptors. *Trends Pharmacol. Sci.* **13,** 103–108.

Schmidt, M., Kröger, B., Jacob, E., Seulberger, H., Subkowski, T., Otter, R., Meyer, T., Schmalzing, G., and Hillen, H. (1994). Molecular characterization of human and bovine endothelin converting enzyme (ECE-1). *FEBS Lett.* **356,** 238–243.

Schönthal, A. (1990). Nuclear protooncogene products: Fine-tuned components of signal transduction pathways. *Cell. Signal.* **2,** 215–225.

Shichiri, M., Hirata, Y., Kanno, K., Ohta, K., Emori, T., and Marumo, F. (1989). Effect of endothelin 1 on release of arginine vasopressin from perifused rat hypothalamus. *Biochem. Biophys. Res. Commun.* **163,** 1332–1337.

Shimada, K., Takahashi, M., and Tanzawa, K. (1994). Cloning and functional expression of endothelin-converting enzyme from rat endothelial cells. *J. Biol. Chem.* **269,** 18275–18278.

Shirakami, G., Nakao, K., Saito, Y., Magaribuchi, T., Mukoyama, M., Arai, H., Hosoda, K., Suga, S., Mori, K., and Imura, H. (1993). Low doses of endothelin-1 inhibit atrial natriuretic peptide secretion. *Endocrinology* **132,** 1905–1912.

Simonson, M. S., and Dunn, M. J. (1990). Endothelin. Pathways of transmembrane signaling. *Hypertension* **15,** I5–I12.

Stier, C. T., Jr., Quilley, C. P., and McGiff, J. C. (1992). Endothelin-3 effects on renal function and prostanoid release in the rat isolated kidney. *J. Pharmacol. Exp. Ther.* **262,** 252–256.

Sturgil, T. W., Ray, L. B., Erikson, E., and Maller, J. L. (1988). Insulin-stimulated MAP-2 kinase phosphorylates and activates ribosomal protein S6 kinase II. *Nature (London)* **334,** 715–718.

Tacheuchi, A., Kimura, T., and Satoh, S. (1992). Enhancement by endothelin-1 of the release of catecholamines from the canine adrenal gland in response to splanchnic nerve stimulation. *Clin. Exp. Pharmacol. Physiol.* **19,** 663–666.

Takagi, M., Tsukada, H., Matsuoka, H., and Yagi, S. (1989). Inhibitory effect of endothelin on renin release *in vitro. Am. J. Physiol.* **257,** E833–E838.

Tamamori, M., Ito, H., Adachi, S., Akimoto, H., Marumo, F., and Hiroe, M. (1996). Endothelin-3 induces hypertrophy of cardiomyocytes by endogenous endothelin-1-mediated mechanism. *J. Clin. Invest.* **97,** 366–372.

Tamirisa, P., Frishman, W. H., and Kumar, C. (1995). Endothelin and endothelin antagonism: Roles in cardiovascular health and disease. *Am. Heart J.* **130,** 601–610.

Thibault, G., Doubell, A. F., Garcia, R., Larivière, R., and Schiffrin E. L. (1994). Endothelin-stimulated secretion of natriuretic peptides by rat atrial myocytes is mediated by endothelin A receptors. *Circ. Res.* **74,** 460–470.

Torres, M., Ceballos, G., and Rubio, R. (1994). Possible role of nitric oxide in catecholamine secretion by chromaffin cells in the presence and absence of cultured endothelial cells. *J. Neurochem.* **63,** 988–996.

Tóth, I. E., and Hinson, J. P. (1995). Neuropeptides in the adrenal gland: Distribution, localization of receptors, and effects on steroid hormone synthesis. *Endocrine Res.* **21,** 39–51.

Traish, A. M., Moran, E., Daley, J. T., de Las Morenas, A., and Saenz de Tejada, I. (1992). Monoclonal antibodies to human endothelin-l: Characterization and utilization in radioimmunoassay and immunocytochemistry. *Hybridoma* **11,** 147–163.

Tsukahara, H., Ende, H., Magazine, H. I., Bahou, W. F., and Goligorsky, M. S. (1994). Molecular and functional characterization of the non-isopeptide-selective ETB receptor in endothelial cells. Receptor coupling to nitric oxide synthase. *J. Biol. Chem.* **269,** 21778–21785.

Uusimaa, P. A., Hassinen, I. E., Vuolteenaho, O., and Ruskoaho, H. (1992). Endothelin-induced atrial natriuretic peptide release from cultured neonatal cardiac myocytes. The role of extracellular calcium and protein kinase-C. *Endocrinology* **130,** 2455–2464.

Valdenaire, O., Rohrbacher, E., and Mattei, M. G. (1995). Organization of the gene encoding the human endothelin converting enzyme (ECE-1). *J. Biol. Chem.* **270,** 29794–29798.

Veglio, F., Melchio, R., Rabbia, F., and Chiandussi, L. (1994). Plasma immunoreactive endothelin-1 in primary hyperaldosteronism. *Am. J. Hypertension* **7,** 559–561.

Viard, I., Hall, S. H., Jaillard, C., Berthelon, M. C., and Saez, J. M. (1992a). Regulation of *c-fos, c-jun* and *jun-beta* messenger ribonucleic acids by angiotensin-II and corticotropin in ovine and bovine adrenocortical cells. *Endocrinology* **130,** 1193–1200.

Viard, I., Jaillard, C., Ouali, R., and Saez, J. M. (1992b). Angiotensin-II-induced expression of protooncogene (*c-fos, jun-B* and *c-jun*) messenger RNA in bovine adrenocortical fasciculata cells (BAC) is mediated by AT-1 receptors. *FEBS Lett.* **313,** 43–46.

Vierhapper, H., Wagner, O., Nowotny, P., and Waldhäusl, W. (1990). Effect of endothelin 1 in man. *Circulation* **81,** 1415–1418.

Vierhapper, H., Nowotny, P., and Waldhäusl, W. (1995). Effect of endothelin-1 in man: Impact on basal and adrenocorticotropin-stimulated concentrations of aldosterone. *J. Clin. Endocrinol. Metab.* **80,** 948–951.

Vinson, G. P., and Hinson, J. P. (1992). Blood flow and hormone secretion in the adrenal gland. *In* "The Adrenal Gland" (V. T. H. James, Ed.), pp. 71–86. Raven Press, New York.

Vinson, G. P., Whitehouse, B. J., and Hinson, P. J. (1992). "The Adrenal Cortex." Prentice Hall, Englewood Cliffs, NJ.
Wall, K. M., and Ferguson, A. V. (1992). Endothelin acts at the subfornical organ to influence the activity of putative vasopressin and oxytocin secreting neurons. *Brain Res.* **586**, 111–116.
Wang, Y. Z., Simonson, M. S., Pouyssegur, J., and Dunn, M. J. (1992). Endothelin rapidly stimulates mitogen-activated protein kinase activity in rat mesangial cells. *Biochem. J.* **287**, 589–594.
Wang, Y. Z., Rose, P. M., Webb, M. L., and Dunn, M. J. (1994). Endothelins stimulate mitogen-activated protein kinase cascade through either ETA and ETB. *Am. J. Physiol.* **267**, C1130–C1135.
Warner, T. D., Schmidt, H. H., and Murad, F. (1992). Interactions of endothelins and EDRF in bovine native endothelial cells: Selective effects of endothelin 3. *Am. J. Physiol.* **262**, H1600–H1605.
Weissberg, P. L., Witchell, C., Davenport, A. P., Hesketh, T. R., and Metcalfe, J. C. (1990). The endothelin peptides ET-1, ET-2, ET-3 and sarafotoxin S6B are comitogenic with platelet derived growth factor for vascular smooth muscle cells. *Atherosclerosis* **85**, 257–262.
Williams, D. L., Jones, K. L., Pettibone, D. J., Lis, E. V., and Clineschmidt, D. V. (1991). Sarafotoxin-S6C: An agonist which distinguishes between endothelin receptor subtypes. *Biochem. Biophys. Res. Commun.* **175**, 556–561.
Woodcock, E. A., Little, P. J., and Tanner, J. K. (1990a). Inositol phosphate release and steroidogenesis in rat adrenal glomerulosa cells. Comparison of the effects of endothelin, angiotensin II and vasopressin. *Biochem. J.* **271**, 791–796.
Woodcock, E. A., Tanner, J. K., Caroccia, L. M., and Little, P. J. (1990b). Mechanisms involved in the stimulation of aldosterone production by angiotensin II, vasopressin and endothelin. *Clin. Exp. Pharmacol. Physiol.* **17**, 263–267.
Xu, D., Emoto, N., Giaid, A., Slaughter, C., Kaw, S., De Wit, D., and Yanagisawa, M. (1994). ECE-1: A membrane-bound metalloprotease that catalyzes the proteolytic activation of big endothelin-1. *Cell* **78**, 473–485.
Yamada, K., and Yoshida, S. (1991). Role of endogenous endothelin in renal function during altered sodium balance. *J. Cardiovasc. Pharmacol.* **17**(Suppl. 7), S290–S292.
Yamaguchi, N. (1993). Inhibition by nifedipine of endothelin-induced adrenal catecholamine secretion in anesthetized dog. *Can. J. Physiol. Pharmacol.* **71**, 301–305.
Yamaguchi, N. (1995). Implication of L-type $Ca^{2+}$ channels in noncholinergic adrenal catecholamine secretion by endothelin-1 *in vivo*. *Am. J. Physiol.* **269**, R287–R293.
Yamamoto, S., Morimoto, I., Yamashita, H., and Eto, S. (1992). Inhibitory effects of endothelin 3 on vasopressin release from rat supraoptic nucleus *in vitro*. *Neurosci. Lett.* **141**, 147–150.
Yanagisawa, M., Kurihara, H., Kimura, S., Tomobe, Y., Kobayashi, M., Yazaki, Y., Goto, K., and Masaki, T. (1988). A novel potent vasoconstrictor peptide produced by vascular endothelial cells. *Nature (London)* **332**, 411–415.
Yasin, S., Costa, A., Navarra, P., Pozzoli, G., Kostoglou-Athanassiou, I., Forsling, M., and Grossman, A. (1994). Endothelin-1 stimulates the *in vitro* release of neurohypophyseal hormones but not corticotropin-releasing hormone via ETA receptors. *Neuroendocrinology* **60**, 553–558.
Yu, C. L. (1993). Attenuation of serum inducibility of immediate early genes by oncoproteins in tyrosine kinase signaling pathways. *Mol. Cell. Biol.* **13**, 2011–2019.
Zamora, M. A., Dempsey, E. C., Walchak, S. J., and Stelzuer, T. J. (1993). BQ-123, an ETA receptor antagonist, inhibits endothelin-1 mediated proliferation of human pulmonary artery smooth muscle cells. *Am. J. Respir. Cell Mol. Biol.* **9**, 429–433.
Zeidel, M. L., Brady, H. R., Kone, B. C., Gullans, S. R., and Brenner, B. M. (1989). Endothelin, a peptide inhibitor of $Na^+$-$K^+$-ATPase in intact renal tubular epithelial cells. *Am. J. Physiol.* **257**, C1101–C1107.

Zellers, T. M., McCormick, J., and Wu, Y. (1994). Interaction among ET-1, endothelium-derived nitric oxide, and prostacyclin in pulmonary arteries and veins. *Am. J. Physiol.* **267,** H139–H147.

Zeng, Z. P., Naruse, M., Guan, B. J., Naruse, K., Sun, M. L., Zang, M. F., Demura, H., and Shi, Y. F (1992). Endothelin stimulates aldosterone secretion *in vitro* from normal adrenocortical tissue, but not adenoma tissue, in primary aldosteronism. *J. Clin. Endocrinol. Metab.* **74,** 874–878.

# INDEX

## A

ABC transporters, *see* Proteins, ATP-binding cassette
Adrenal cortex, effect of endothelin
  on cell hypertrophy and steroidogenic capacity, 286–288
  on cell proliferation, 288–289
  mechanism, 289–290
  role of ET receptor subtypes, 290–291
Adrenal gland
  blood flow
    effect on endothelin release, 291–292
    regulation by endothelin, 284–285
  endothelin synthesis, 268–269
Adrenal medulla, release of catecholamine, stimulation by ET, 282–283
Aging
  normal, PHF-Tau proteins in, 197–198
  plant cell, effect of microgravity and clinostating, 53–56
Aldosterone, tumors producing, role of endothelin, 293–294
Algae
  cell walls, antibodies to, 105–107
  effects of microgravity, 25–30
ALS–PDC, *see* Amyotrophic lateral sclerosis–Parkinsonism dementia complex
Alzheimer's disease
  Tau protein aggregation, 191–195
  Tau protein phosphorylation, 188–191
Amyloid, as cofactor in Tau protein aggregation, 193–194
Amyotrophic lateral sclerosis–Parkinsonism dementia complex, Guamanian, PHF-Tau protein in, 196–197

*Anabaena azollae*, effect of microgravity, 30
Angiosperms, effects of microgravity
  biochemical content and enzyme activity, 38–39
  cells of generative organs, 41–42
  cell and tissue cultures, 42–44
  intracellular calcium balance, 39–40
  leaf photosynthetic cells, 36–38
  microviscosity and lipid peroxidation, 40–41
  organ culture, 42
  protoplast culture, 44–46
  respiration–heat discharge, 40
  root elongation and differentiation, 34–35
  root meristem, 32–34
  seed germination and seedling growth, 32
  shoot apical meristem, 35–36
  shoot primary structure, 36
  vegetative propagation, 41
Antibodies, *see* Monoclonal antibodies
Antigens
  associated with cell differentiation, 84–85
  membrane, expression during differentiation, 235–236
  plant cell wall and plasma membrane, 105
Apical cells, effect of microgravity
  on meristem shoot, 35–36
  on moss protonema, 23–25
Apical junction complexes, RPE, structure and composition, 238–240
Apolipoproteins, as cofactor in Tau protein aggregation, 191–193
*Arabidopsis*, root statocytes
  clinostating effects, 21–22
  microgravity effects, 22

309

*Arabidopsis thaliana*, effect of microgravity
    on generative organ cells, 41–42
    on photosynthetic cells, 37
Arabinogalactan-proteins
    antibodies to, development, 101–102
    developmentally regulated, detection, 103–104
    developmental regulation, 102–103
    role in plant cell development, 100–101
Arachidonic acid, metabolism, stimulation by endothelin, 281–282
ATPase
    activation in *Chlorella*, effect of clinostating, 28–29
    activity and stoichiometry, in drug pump model, 132–134
    ouabain-sensitive $Na^+/K^+$-ATPase, stimulation by ET, 281
ATP transport
    effect of MDR protein expression, 155–156
    hypothesis for ABC transporters, 156
Autoradiography, endothelin receptor subtypes, 273–274
Axon transport, by microtubules, 169

# B

Bases, weak, accumulation in cells, 149–153
Blood flow, adrenal gland
    effect on endothelin release, 291–292
    regulation by endothelin, 284–285
*Brassica napus*, protoplast culture, effect of microgravity, 44–45

# C

Calcium
    cytosolic concentration, effect of ET, 279–280
    intracellular, effect of microgravity, 39–40
Callose, antibodies to, generation, 93–94
Carcinoma, adrenocortical, role of endothelin, 295
Carrageenan, in algal cell walls, detection, 105–106
Catecholamine, release by adrenal medulla, stimulation by ET, 282–283

Cell differentiation
    plant cells
        angiosperm root, effect of microgravity, 34–35
        associated antigens, 84–85
        effect of microgravity and clinostating, 53–56
        space flight studies, 3
    retinal pigment epithelium, molecular markers, 235–237
Cell proliferation, effect of endothelin, 288–289
Cell separation, associated pectin epitopes, 88
Cell surface, plant, molecular probing with antibodies, 83–85
Cellulase, moss, effect of clinostating, 31
Cellulose, content in pea plant, effect of microgravity, 39
Cell walls
    algal, monoclonal antibodies to, 105–107
    plant
        antigens, 105
        components, taxonomic significance, 85
        interaction with microbes, 107–109
        polymers, probing with antibodies
            enzymes, 100
            extensin, 95–99
            glycine-rich proteins, 99–100
            lignin, 94–95
            matrix polysaccharides, 92–94
            pectin, 85–92
            proline-rich proteins, 99
        spatial architecture, 83–84
        structure and components, 79–80
Central nervous system, Tau protein structure in, 171–172
*Ceratodon purpureus*, apical cells, effect of microgravity, 24–25
CFTR, *see* Cystic fibrosis transmembrane conductance regulator
*Chara globularis*, rhizoids, effect of microgravity, 23
*Chlamydomonas reinhardtii*, effect of microgravity, 29–30
*Chlorella pyrenoidosa*, effect of microgravity
    on cell ultrastructure, 26–29
    on chloroplasts, 25–26
*Chlorella vulgaris*
    ATPase activation, effect of clinostating, 28–29

INDEX 311

effect of microgravity
  on cell ultrastructure, 26–29
  on chloroplasts, 25–26
Chlorophyll a, content in pea plant, effect of microgravity, 38–39
Chlorophyll b, content in pea plant, effect of microgravity, 38–39
Chloroplasts, effect of microgravity
  in angiosperm leaf, 37–38
  in *Chlorella*, 25–26
Chromatin, volume in plant cells, effect of microgravity, 51–52
Clinostating effects
  advantages, 9–10
  on angiosperm root elongation and differentiation, 35
  on ATPase activation in *Chlorella*, 28–29
  definition, 7
  limitations, 9
  on microviscosity and lipid peroxidation, 41
  on moss cellulosopectolytic enzymes, 31
  on plant cell aging, 53–56
  on plant cell differentiation, 53–56
  on plant cell and tissue cultures, 43–44
  on plant intracellular processes, 58
  on plant organelle structural–functional rearrangements, 46–53
  on root cap statocytes, 21–22
  types, 7–9
Cystic fibrosis transmembrane conductance regulator, and P-glycoprotein, and SUR, homology, 125–127
Cytoskeleton, cortical, RPE, structure and composition, 240–241

**D**

*Daucus carota*, effect of microgravity, 43–45
Degeneration
  corticobasal, with Tau 64 and 69, 200–201
  frontal lobe, with pathological Tau triplet, 199
Depolarization, MDR cells, 137–139
Dictyosome, volume in plant cell, effect of microgravity, 50
Differentiation, *see* Cell differentiation; Transdifferentiation

Displacement binding, endothelin receptor subtypes, 271–273
DNA, complementary, endothelins, cloning, 268
*Doritis pulcherrima*, enzyme activity, effect of microgravity, 39
Down's syndrome, PHF-Tau protein in, 195
*Drosophila melanogaster*, aging, effect of altered gravity, 55–56
Drug accumulation
  rate, effect of membrane potential, 144–149
  steady-state description, 141–144
Drug pumps
  ATPase activity and stoichiometry, 132–134
  coupling, 130
  function, 129–130
  kinetics, 134–135
  photolabeling, 136
  thermodynamics, 135
Drugs, cationic, effect of membrane potential
  on membrane flux, 139–141
  on steady-state drug accumulation, 141–144
Drug transport
  advantages of drug pump model, 132–136
  definitions, 128–129
  effect of membrane potential, 137–139, 153–155
    on cationic drug flux, 139–141
    on drug accumulation rate, 144–149
    on weak base accumulation, 149–153
  effect of $\Delta$pH, 153–155
  historical context, 130–132
  pump coupling, 130
  role of pumps, channels, and exchangers, 129–130

**E**

Electrical potential, membrane, *see* Membrane potential effects
Electron microscopy, in study of microgravity effects on statocytes, 11
Embryogenesis, vertebrate, RPE development during, 227–228
Endo-1,4-$\beta$-glucanase, *see* Cellulase

Endoplasmic reticulum, in plant statocytes, effect of microgravity, 50–51
Endothelin
  cDNA cloning, 268
  effects
    on adrenal blood flow, 284–285
    on adrenal cortex growth
      cell hypertrophy and steroidogenic capacity, 286–288
      cell proliferation, 288–289
      mechanism, 289–290
      role of ET receptor subtypes, 290–291
    on arachidonic acid metabolism, 281–282
    on catecholamine release by adrenal medulla, 282–283
    on cytosolic $Ca^{2+}$ concentration, 279–280
    on intraadrenal renin angiotensin system, 284
    on intramedullary regulatory peptides, 283
    on ouabain-sensitive $Na^+/K^+$-ATPase, 281
    on phospholipase $A_2$, 281–282
    on phospholipase C, 279–280
    on secretory activity
      interrenal cells, 278–279
      zona fasciculata reticularis cells, 277–278
      zona glomerulosa cells, 275–277
    on thyrosine kinase, 280
  family members, 267–268
  gene expression studies, 269
  immunohistochemical studies, 269–270
  release, affecting factors
    adrenal blood flow, 291–292
    natriuretic peptide, 293
    nitric oxide, 292–293
  roles
    in aldosterone-producing tumors, 293–294
    in idiopathic hyperaldosteronism, 294
    in tumors, 295
  secretagogue action, role of ET receptor subtypes, 285–286
  synthesis in adrenal gland, 268–269
Endothelin receptor subtypes
  autoradiographic studies, 273–274
  gene expression studies, 270–271
  immunohistochemical studies, 273
  roles
    in ET growth action, 290–291
    in ET secretagogue action, 285–286
    in idiopathic hyperaldosteronism, 294
  saturation and displacement binding studies, 271–273
Environment, *in vivo* effects on RPE differentiated states, 243–245
Enzymes, in plant cell wall, 100
*Epidendrum radicans*, effect of microgravity
  on leaf photosynthetic cells, 36–39
  on shoot primary structure, 36
Epithelium, *see* Retinal pigment epithelium
Esterification, pectin in plant cell walls, 87–92
ET, *see* Endothelin
*Euglena gracilis*, effect of microgravity, 30
Exo-1,4-β-glucanase, moss, effect of clinostating, 31
Extensin, structural role in plant cell wall, 95–99

## F

Ferns, effects of microgravity, 31
Flux, membrane, cationic drugs, 139–141
Frontal lobe, degeneration with pathological Tau triplet, 199
*Funaria hygrometrica*, effect of microgravity, 24, 30–31

## G

Genes
  endothelin, expression studies, 269
  endothelin receptor subtypes, expression studies, 270–271
  expression patterns in RPE
    *Chx-10* homeobox genes, 229–230
    *Mi* microphthalmia, 231
    Msh-like homeobox genes, 228–229
    *Otx-1* and *Otx-2* homeobox genes, 229
    *Pax-2* paired-box-containing gene, 229
    *pax-6* paired-box-containing gene, 229
    *Rx* homeodomain gene, 230
P-glycoprotein, 124–125
Tau proteins, organization, 170

tyrosinase, transcription, regulation, 233–235
Germination, angiosperm seed, effect of microgravity, 32
Gerstmann–Straussler–Scheinker disease, pathological Tau proteins in, 202–203
Glial cells, pathological Tau protein isoforms in, 207
Glucocorticoids, secretion, role of endothelin, 277–278, 295
Glycation, role in Tau protein aggregation, 194
Glycoproteins, algal cell wall, antibodies to, 106–107
Golgi apparatus, function in plant cell, effect of microgravity, 50
Growth
  adrenal cortex, effects of endothelin
    cell hypertrophy and steroidogenic capacity, 286–288
    cell proliferation, 288–289
    mechanism, 289–290
    role of ET receptor subtypes, 290–291
  angiosperm seedling, effect of microgravity, 32
Gymnosperms, effects of microgravity, 31–32

## H

*Haplopappus gracilis*, effect of microgravity, 42–44
Heat, discharge in angiosperms, effect of microgravity, 40
Hemicellulose, antibodies to, generation, 92–93
Hyperaldosteronism, idiopathic, role of endothelin, 294

## I

Immunohistochemistry
  endothelin, 269–270
  endothelin receptor subtypes, 273
Injury, neuronal, relationship to Tau protein, 203
Interrenal cells, steroid secretion, effect of endothelin, 278–279
Ion channels, in drug transport, 129–130

Ion exchangers, in drug transport, 129–130
Ion pumps, in drug transport, *see* Drug pumps

## K

Kinase, role in Tau proteins
  activity and specificity, 178–179
  Tau phosphorylation, 178
Kinetics, in drug pump model, 134–135

## L

Labeling, in drug pump model, 136
Leaf, photosynthetic cells, effect of microgravity, 36–38
Light microscopy, in study of microgravity effects on statocytes, 11
Lignin
  antibodies to, generation, 94–95
  content in mung bean, effect of microgravity, 39
Lipid peroxidation, in angiosperm seedlings, effect of microgravity, 40–41
Lipids, reserve in plant cells, effect of microgravity, 51

## M

Maize, root statocytes, effect of microgravity, 20–22
Markers, molecular, RPE during differentiation, 235–237
MDR, *see* Multidrug resistance
MDR1 protein, *see* P-glycoprotein
Melanin, synthesis in retinal pigment epithelium, 232–233
Membrane potential effects
  on cationic drug flux, 139–141
  on drug accumulation rate, 144–149
  on drug transport, 137–139, 153–155
  on steady-state drug accumulation, 141–144
  on weak base accumulation, 149–153
Membranes
  cytoplasmic, effect of microgravity on plant cells, 49
  primary sites, 58–62
  plasma, antigens, 105

Meristem cells, effect of microgravity
  on angiosperm root, 32–34
  shoot apical meristem, 35–36
Metabolism, arachidonic acid, stimulation
  by endothelin, 281–282
Microbes–plant cell wall interactions,
  107–109
Microgravity effects
  biological effects, range and
    mechanism, 2
  on cytoplasmic membrane, primary site,
    58–62
  on plant cells, 3–4
    adaptation strategy, 56–58
    aging, 53–56
    algae, 25–30
    angiosperms, see Angiosperms
    associated weightlessness, definition, 4
    differentiation, 53–56
    ferns, 31
    gymnosperms, 31–32
    moss, 30–31
    organelle structural–functional
      rearrangements, 46–53
    space vehicle cabin environment
      effects, 5
    specialized graviperceptive cells, 10–11,
      20–22
    techniques and instruments for
      measurement, 5–7
    tip-growing cells, 22–25
  simulated, see Clinostating
Microscopy, in study of microgravity effects
  on statocytes, 11
Microtubules
  assembly, role of Tau proteins
    associated domain, 176
    isoforms in neurons, 184–185
    other interactions, 181–182
    pathological Tau proteins, 204–206
    projection domain, 175–176
    proteins in nonneuronal cells, 185
    Tau phosphorylation
      associated kinases, 178–179
      associated phosphatases, 179–180
      regulations, 180–181
    Tau sorting in nerve cells, 182–184
    Tau–tubulin interactions, 176–177
  neuronal
    axon transport, 169
    neuron morphology, 168–169
  in nonneuronal cells, 169

Microvilli, retinal pigment epithelium,
  function, 238
Microviscosity, in angiosperm seedlings,
  effect of microgravity, 40–41
Mineralocorticoid, secretion, effect of
  endothelin, 275–277
Mitochondria, plant, effect of microgravity,
  46–48
Models
  direct and indirect, for MDR protein
    function, 127–128
  membrane potential effects
    on drug flux kinetics, 144–149
    on steady-state drug accumulation,
      141–144
Molecular markers, RPE during
  differentiation, 235–237
Monoclonal antibodies
  to algal cell wall components
    alginates and fucans, 106
    carrageenan, 105–106
    glycoproteins, 106–107
  to arabinogalactan-proteins,
    development, 101–102
  to melanoma, labeling of human RPE,
    236
  as molecular probes of plant cell surface,
    83–85
  probing of plant cell wall polymers
    enzymes, 100
    extensin, 95–99
    glycine-rich proteins, 99–100
    lignin, 94–95
    matrix polysaccharides, 92–94
    pectin, 85–92
    proline-rich proteins, 99
    recognition of plant plasma membranes,
      105
  in study of plant cell wall–microbe
    interactions, 107–109
Morphology, neuron, 168–169
Moss
  effects of microgravity, 30–31
  protonema apical cells, effect of
    microgravity, 23–25
Multidrug resistance, definition, 122–123
Multidrug resistance protein, see P-
  glycoprotein
Mung bean, effect of microgravity
  on lignin content, 39
  on respiration–heat discharge, 40

*Muscari racemosum*
  effect of altered gravity, 53–55
  effect of microgravity, 42
Muscle, pathological Tau proteins in, 207
Myotonic dystrophy, with Tau 55, 201–202

## N

Nerve cells, Tau protein sorting in, 182–184
Neurites, dystrophic, Tau proteins in, 187
Neurodegenerative disorders
  accumulation of Tau protein
    dystrophic neurites, 187
    neurofibrillary tangles, 185–187
    Pick bodies, 187–188
  without pathological Tau proteins, 203–204
Neurons
  injuries, relationship to Tau protein, 203
  morphology, 168–169
  pathological Tau protein variants
    in glial cells, 207
    in muscles, 207
    Tau distribution, 206
  Tau protein isoforms in, 184–185
*Nicotiana rustica,* protoplast culture, effect of microgravity, 44–45
*Nicotiana tabacum,* protoplast culture, effect of microgravity, 44–45
Niemann–Pick disease, type C, PHF-Tau protein in, 197
Nitric oxide, effect on endothelin release, 292–293
Nonneuronal cells
  microtubules in, 169
  Tau proteins in, 185

## O

O-glycosylation, Tau proteins, 174
Oncogenes, role in RPE phenotype change, 255
Organelles, plant cell, effect of microgravity and clinostating, 46–53
Organs
  generative, in plant cells, effect of microgravity, 41–42

potato, culture, effect of microgravity, 42
Ouabain, sensitive $Na^+/K^+$-ATPase, stimulation by ET, 281
Oxidation, role in Tau protein aggregation, 194–195

## P

Paired helical filaments–pathological Tau protein
  associated disorders
    Down's syndrome, 195
    Parkinsonism with dementia, 195–196
  phosphorylation sites, 189
Parkinson's disease, with dementia, PHF-Tau protein in, 195–196
Pea plant, effect of microgravity
  on biochemical content and enzyme activity, 38–39
  on microviscosity and lipid peroxidation, 41
  on root statocytes, 20–21
Pectin
  detection in various systems, 91
  epitopes
    distribution in plant cell wall, 87, 89
    modulation, 88
  role in pollen tube growth, 90–91
  role in primary cell walls, 87–88
  structure, 85–86
Pectinesterase, moss, effect of clinostating, 31
Peptides
  intramedullary regulatory, effect of ET, 283
  natriuretic, effect on endothelin release, 293
Peroxidase, in wheat seedling, effect of microgravity, 39
Peroxidation, lipid, effect of microgravity, 40–41
P-glycoprotein
  –ABC protein, forms, 157–158
  aspects of cellular expression
    alkalinization of cytoplasm, 149–153
    ATP transport, 155–156
    drug accumulation rate, 144–149
    membrane potential, 139–141, 153–155
  cell overexpressing, depolarization, 137–139

P-glycoprotein (*continued*)
and CFTR, and SUR, homology, 125–127
drug pump model
ATPase activity and stoichiometry, 132–134
kinetics, 134–135
photolabeling, 136
thermodynamics, 135
encoding genes, 124–125
function
associated hypotheses, 124–125
direct and indirect models, 127–128
historical context, 130–132
overproduction, 123–124
pH
ΔpH, effect on drug transport, 153–155
$pH_i$, effect on weak base accumulation, 149–153
PHF–Tau, *see* Paired helical filaments pathological Tau protein
Phosphatase, role in Tau proteins phosphorylation regulations, 180–181
regulations, 179–180
Tau dephosphorylation, 179
Phospholipase $A_2$, stimulation by endothelin, 281–282
Phospholipase C, stimulation by endothelin, 279–280
Phosphonium compounds, steady-state accumulation, 141–144
Phosphoproteins, Tau proteins as, 172–174
Phosphorylation
pathological Tau proteins, in Alzheimer's disease
abnormal phosphorylation, 188–189
sites in PHF-Tau and normal Tau, 189
Tau protein concepts, 189–191
site in Tau proteins, 172–174
state of Tau protein, relationship to neuronal injuries, 203
Tau proteins, regulations, 180–181
Photolabeling, in drug pump model, 136
Photoreceptors, retinal, RPE role in function, 226–227
Photosynthetic cells, angiosperm leaf, effect of microgravity, 36–38
Pick bodies, in neurodegenerative disorders, 187–188
Pick's disease, with Tau 55 and 64, 201
Pigmentation, role of tyrosinase gene, 232–235

*Pinus elliotti*, effect of microgravity, 32
*Pinus sylvestris*, effect of microgravity, 31–32
Plants
cell development, role of arabinogalactan-proteins, 100–101
cell differentiation, effect of microgravity and clinostating, 53–56
cell surface, molecular probing with antibodies, 83–85
cell wall polymers, probing with antibodies
enzymes, 100
extensin, 95–99
glycine-rich proteins, 99–100
lignin, 94–95
matrix polysaccharides, 92–94
pectin, 85–92
proline-rich proteins, 99
cell walls
–microbe interactions, 107–109
structure and components, 79–80
effect of microgravity
cell adaptation strategy, 56–58
space vehicle cabin environment, 5
techniques and instruments, 5–7
weightlessness, definition, 4
organelles, effect of microgravity and clinostating, 46–53
pea, *see* Pea plant
potato, organ culture, effect of microgravity, 42
structural–functional organization in altered gravity
algae, 25–30
angiosperms, *see* Angiosperms
ferns, 31
gymnosperms, 31–32
moss, 30–31
specialized graviperceptive cells, 10–11, 20–22
tip-growing cells, 22–25
in study of space flight biology, 2–4
wheat, effect of microgravity, 41
Plasma membranes, plant, antigens, 105
Plasticity, retinal pigment epithelium cells, 245–247
Plastids, plant, effect of microgravity, 48–49
Polarization
depolarization, MDR cells, 137–139
retinal pigment epithelium cells, 237–243
Pollen tubes, growth, role of pectin, 90–91

INDEX

Polygalacturonase, moss, effect of clinostating, 31
Polymers, plant cell wall, probing with antibodies
  enzymes, 100
  extensin, 95–99
  glycine-rich proteins, 99–100
  lignin, 94–95
  matrix polysaccharides, 92–94
  pectin, 85–92
  proline-rich proteins, 99
Potato plant, organ culture, effect of microgravity, 42
*Pottia intermedia,* protonema, apical cells, effect of microgravity, 24–25
Prokaryotes, growth, effect of altered gravity, 56
Propagation, vegetative, effect of microgravity, 41
Proteins
  amyloid precursor, in Tau protein aggregation, 193–194
  ATP-binding cassette
    ATP transport hypothesis, 156
    –MDR protein, forms, 157–158
  microsomal, RPE-specific, expression, 236
  microtubule-associated, Tau proteins as
    gene organization, 170
    splicing, 170–171
    structure
      MT assembly domain, 171–172
      projection domain, 171
  multidrug resistance, see P-glycoprotein
  structural role in plant cell wall
    glycine-rich proteins, 99–100
    proline-rich proteins, 99
  Tau, see Tau proteins
Protonema cells, moss, apical cells, effect of microgravity, 23–25
Protoplasts, effect of microgravity, 44–46

R

Renin angiotensin system, intraadrenal, effect of endothelin, 284
Respiration–heat discharge, in angiosperms, effect of microgravity, 40
Retinal pigment epithelium
  cells
    plasticity in culture, 245–247
    polarization, 237–243
  development during embryogenesis, 227–228
  differentiated state, effects of environmental changes, 243–245
  gene expression patterns, 228–232
  molecular marking during differentiation, 235–237
  phenotype change, role of oncogenes, 255
  role in retinal photoreceptor function, 226–227
  structure, 225–226
  tissue type specification, 228–232
  transdifferentiation
    amphibian RPE, 248–249
    chick RPE *in vitro,* 249
    rat RPE *in vitro,* 249–254
  transplantation, 247
  tyrosinase gene transcription regulation, 233–235
  tyrosinase and melanin synthesis, 232–233
Rhizoids, *Chara globularis,* effect of microgravity, 23
Roots, angiosperm, effect of microgravity
  on elongation and differentiation, 34–35
  on meristem cells, 32–34
RPE, *see* Retinal pigment epithelium

S

Saturation, endothelin receptor subtypes, 271–273
Secretion, effect of endothelin
  glucocorticoid from zona fasciculata reticularis, 277–278
  mineralocorticoid from zona glomerulosa, 275–277
  steroid from interrenal cells, 278–279
Seedlings, effect of microgravity
  angiosperms, 32, 40–41
  wheat peroxidase level, 39
Shoots, angiosperm, effect of microgravity
  on apical meristem cells, 35–36
  on primary structure, 36
*Solanum tuberosum,* protoplast culture, effect of microgravity, 44–45
Spaceflight
  biology
    experimental equipment complexity, 5–7
    fundamental tasks in, 2–4

Spaceflight (*continued*)
  microgravity in, effect on plants
    space vehicle cabin environment, 5
    techniques and instruments, 5–7
    weightlessness, definition, 4
    weightlessness as active factor in, 2
*Spirodela polyrrhiza*, vegetative propagation, effect of microgravity, 41
Splicing, Tau protein genes, 170–171
Statocytes
  plant, endoplasmic reticulum volume, effect of microgravity, 50–51
  root cap
    effects of microgravity, 11, 20–22
    structural characteristics, 10–11
Steele–Richardson–Olszewski syndrome, with Tau 64 and 69, 199–200
Steroids
  secretion, effect of endothelin, 278–279
  sexual, tumor secreting, role of endothelin, 295
Stoichiometry, ATPase, in drug pump model, 132–134
Structure–function relationship, plant, in altered gravity
  algae, 25–30
  angiosperms, *see* Angiosperms
  ferns, 31
  gymnosperms, 31–32
  moss, 30–31
  organelle rearrangements, 46–53
  specialized graviperceptive cells, 10–11, 20–22
  tip-growing cells, 22–25
Sugar, ethanol-soluble, in pea plant, effect of microgravity, 38
Sulfonyl urea receptor, and P-glycoprotein, and CFTR, homology, 125–127
SUR, *see* Sulfonyl urea receptor

# T

Tau proteins
  as microtubule-associated proteins
    gene organization, 170
    splicing, 170–171
    structure
      MT assembly domain, 171–172
      projection domain, 171
    O-glycosylation, 174
  pathological
    as biochemical marker for Alzheimer's disease
      Tau aggregation, 191–195
      Tau phosphorylation, 188–191
    in glial cells, 207
    isoform distribution in neuron population, 206
    in muscles, 207
    in neurodegenerative disorders
      dystrophic neurites, 187
      neurofibrillary tangles, 185–187
      Pick bodies, 187–188
    in neurofibrillary degeneration subtypes
      corticobasal degeneration, 200–201
      disorders with PHF-Tau, 195–199
      Gerstmann–Straussler–Scheinker disease, 202–203
      neuronal injuries, 203
      Tau 55, 201–202
      Tau 55 and 64, 201
      Tau 64 and 69, 199–200
      Tau triplet, 199
    role in microtubule assembly, 204–206
  as phosphoprotein, 172–174
  role in microtubule assembly
    isoforms in neurons, 184–185
    microtubule assembly domain, 176
    other interactions, 181–182
    projection domain, 175–176
    proteins in nonneuronal cells, 185
    Tau phosphorylation
      associated kinases, 178–179
      associated phosphatases, 179–180
      regulations, 180–181
    Tau sorting in nerve cells, 182–184
    Tau–tubulin interactions, 176–177
  role in pathological events, 207–208
Taxonomy, cell wall components, 85
Tetrameric protein transthyretin, RPE as source, 236–237
Thermodynamics, in drug pump model, 135
Thyrosine kinase, stimulation by endothelin, 280
Tissues
  plant, culture, effect of microgravity, 42–44
  specification of gene expression in RPE, 228–232

*Tradescantia paludosa*, generative organ cells, effect of microgravity, 42
Transcription, tyrosinase gene, regulation, 233–235
Transdifferentiation
  amphibian RPE, 248–249
  chick RPE *in vitro*, 249
  rat RPE *in vitro*, 249–254
Transplantation, retinal pigment epithelium, 247
Transport
  ATP
    effect of MDR protein expression, 155–156
    hypothesis for ABC transporters, 156
  axonal, by microtubules, 169
  drug, *see* Drug transport
*Tribolium confusum*, aging, effect of altered gravity, 55–56
Tubulin–Tau protein interactions, in microtubule assembly, 176–177
Tumors
  aldosterone-producing, role of endothelin, 293–294
  role of endothelin, 295
Tyrosinase
  gene transcription, regulation, 233–235
  synthesis in retinal pigment epithelium, 232–233

## U

Ubiquitination, role in Tau protein aggregation, 194

## V

Vacuolization, plant cell, effect of microgravity, 49–50
*Vicia faba*, photosynthetic cells, effect of microgravity, 38
Viscosity, in seedlings, effect of microgravity, 40–41

## W

Weightlessness
  as active factor in space flight, 2
  definition, 4
Wheat plant, effect of microgravity, 41

## Z

Zona fasciculata reticularis, glucocorticoid secretion, 277–278
Zona glomerulosa, mineralocorticoid secretion, effect of ET, 275–277